Advances in
CHROMATOGRAPHY

T0139123

Volume **53**

Advances in
CHROMATOGRAPHY

EDITORS
Eli Grushka • Nelu Grinberg

CRC Press
Taylor & Francis Group
Boca Raton London New York

CRC Press is an imprint of the
Taylor & Francis Group, an **informa** business

CRC Press
Taylor & Francis Group
6000 Broken Sound Parkway NW, Suite 300
Boca Raton, FL 33487-2742

First issued in paperback 2022

© 2017 by Taylor & Francis Group, LLC
CRC Press is an imprint of Taylor & Francis Group, an Informa business

No claim to original U.S. Government works

ISBN 13: 978-1-4987-2678-8 (hbk)
ISBN 13: 978-1-03-240237-6 (pbk)

DOI: 10.1201/9781315370385

Publisher's Note
The publisher has gone to great lengths to ensure the quality of this reprint but points out that some imperfections in the original copies may be apparent.

Visit the Taylor & Francis Web site at
http://www.taylorandfrancis.com

and the CRC Press Web site at
http://www.crcpress.com

Contents

Contributors

Vincenza Andrisano
Department of Life Quality Studies
   Rimini Campus
University of Bologna
Bologna, Italy

Leonid D. Asnin
Department of Chemistry and
   Biochemistry
Perm National Research Polytechnic
   University
Perm, Russia

Deirdre Cabooter
Pharmaceutical Analysis
Department of Pharmaceutical Sciences
Katholieke Universiteit Leuven
Leuven, Belgium

Alberto Cavazzini
Department of Chemistry and
   Pharmaceutical Sciences
University of Ferrara
Ferrara, Italy

Sander Deridder
Department of Chemical Engineering
Vrije Universiteit Brussel
Brussels, Belgium

Gert Desmet
Department of Chemical Engineering
Vrije Universiteit Brussel
Brussels, Belgium

Jessica Fiori
Department of Pharmacy and
   Biotechnology
University of Bologna
Bologna, Italy

Kathithileni Martha Kalili
Department of Chemistry and
   Biochemistry
University of Namibia
Windhoek, Namibia

Robert Kormány
EGIS Pharma
Budapest, Hungary

Teresa Kowalska
Institute of Chemistry
The University of Silesia
Katowice, Poland

Nicola Marchetti
Department of Chemistry and
   Pharmaceutical Sciences
University of Ferrara
Ferrara, Italy

Veronika R. Meyer
Empa, Swiss Federal Laboratories for
   Materials Science and Technology
Dübendorf, Switzerland

Imre Molnár
Institute of Applied Chromatography
Berlin, Germany

Marina Naldi
Department of Pharmacy and
   Biotechnology
University of Bologna
Bologna, Italy

Su Pan
Research and Development
Analytical and Bioanalytical
   Department
Bristol-Myers Squibb
New Brunswick, New Jersey

Hans-Jürgen Rieger
Institute of Applied Chromatography
Berlin, Germany

Mieczysław Sajewicz
Institute of Chemistry
The University of Silesia
Katowice, Poland

Joseph Sherma
Department of Chemistry
Lafayette College
Easton, Pennsylvania

Yueer Shi
Research and Development
Analytical and Bioanalytical
   Department
Bristol-Myers Squibb
New Brunswick, New Jersey

Angela De Simone
Department for Life Quality Studies
   Rimini Campus
University of Bologna
Bologna, Italy

André de Villiers
Department of Chemistry and Polymer
   Science
Stellenbosch University
Stellenbosch, South Africa

1 Solute–Stationary Phase Interaction in Chiral Chromatography

*Leonid D. Asnin, Alberto Cavazzini,
and Nicola Marchetti*

CONTENTS

1.1   INTRODUCTION

Since the beginning of chiral chromatography in 1930s, researchers have taken an interest in mechanisms responsible for the resolution of racemic mixtures on optically active adsorbents. It was understood that behind this phenomenon lies "the asymmetric character of the adsorbing surface, which causes it to react differently towards the enantiomorphous components of the racemic compound" [1]. From this, it immediately followed that the stereochemical configuration of enantiomers was the major factor affecting their interaction with the chiral surface. Hence, a study of the dependence of the adsorption affinity on the spatial configuration of solutes seemed to be a key step in the elucidation of the nature of enantioseparation [2]. This fundamental program of research took impulse after the famous publication by Dalgliesh [3], who attempted to explain the separation of enantiomers of amino acids on cellulose based on spatial considerations. At that time, there were no tools to investigate interactions between a solute and an adsorbent on the molecular level, so researchers used indirect integral characteristics, such as retention factor (k') and enantioselectivity (α), to elucidate mechanisms resulting in different migration velocities of optical antipodes in a chiral media. This approach is called *macroscopic* because it disregards the molecular structure of the system under investigation and operates with quantities averaged (in thermodynamic sense) over large ensembles of molecules and over a certain period of time. Tremendous improvements in molecular techniques made in the past three decades, in particular, in molecular modeling [4,5] as well as in spectrometric methods (NMR, FT-IR, VCD, etc.) and x-ray crystallography [6] allowed researchers to study solute–selector binding at the *microscopic* level.

    Both these approaches are important in chromatographic research. The application of the molecular techniques makes it possible to understand how a chiral selector discriminates between optical antipodes. These methods cannot, however, explain in full the phenomenon of retention on the chiral stationary phase (CSP). This is because the CSP is not a uniform array of identical chiral sites, each interacting in the same manner with a solute. A real CSP is a heterogeneous solid, including both enantioselective (chiral) sites and nonselective sites; each group of the sites may be, in its turn, heterogeneous as each individual site may slightly differ from other ones of the same type due to differences in the surrounding, minor conformational changes, location in the porous structure of a stationary phase, and so on. A mechanistic understanding of the interaction of a solute with a real stationary phase is a task of an utmost complexity, not yet resolved. Therefore, the macroscopic approach with its, some can say, rough but easy to handle and quite informative tools is applied to elucidate the solute behavior in the chromatographic column. Surprisingly, the macroscopic aspects of

TABLE 1.1
Literature on Chiral Recognition Mechanisms on Different CSPs

Class of CSP	Literature
General reviews (all classes)	[6,216–218]
Ligand exchange	[219]
Cyclodextrins	[198,220,221]
Crown ethers	[220]
Polysaccharides	[222,223]
Proteins	[224]
Molecularly imprinted polymers	[225,226]
Macrocyclic antibiotics	[220,227,228]
Pirkle type	[229]
Cinchona alkaloids	[230]

enantioselective adsorption have not been given much attention in the review literature unlike the problem of chiral recognition considered in a great number of publications, some important ones listed in Table 1.1 categorized by the type of chiral selector. The authors are aware of only two review papers [7,8] that describe macroscopic adsorption models and catalogue relevant publications. This disparity of attention encouraged us to write this chapter in order to revise existing concepts in this area and also to explain our own views on macroscopic adsorption mechanisms.

The organization of this review proceeds as follows. In Section 1.2, the structure of the CSP and its distinction from the achiral adsorbent is discussed. In Section 1.3, the thermodynamics of adsorption on chiral adsorbents is considered, with a special attention given to the problem of deriving information on solute–stationary phase interactions in chiral systems from thermodynamic data. In Sections 1.4 and 1.5, we discuss in detail microscopic retention mechanisms in chiral chromatography and how experimental conditions influence these mechanisms. A short excursus into the nature of enantioselective and nonselective interactions (Section 1.4.1) is made to facilitate the comprehension of the main theme. Finally, in Section 1.6, we describe the problem of adsorption of multichiral, that is, containing more than one asymmetric center, compounds.

1.2 CHIRAL STATIONARY PHASE

It is pertinent, before addressing issues of solute–stationary phase interaction, to define what the CSP is. Obviously, such an adsorbent must bear some elements of chirality. These may be intrinsic to the surface or to the whole bulk (if the latter is available to a solute) of a stationary phase or may be imparted to that by immobilization onto an achiral matrix of chiral entities, called chiral selectors. Quartz, in its *d*- or *l*-form [9,10], and crystalline powdered lactose [11], which were historically first materials used for enantioseparation, belong to the group of CSPs with a naturally chiral surface. These nonporous adsorbents have low surface area and are not used anymore because of their low efficacy. Porous or bulk permeable chiral materials, such as cellulose [12] and its derivatives, for example, cellulose triacetate [13,14], comprised the second generation of CSPs. Even if nonporous, these solids expose their entire volume to a solute owing to the permeation of solute molecules to the bulk solid phase. Chiral porous materials based on metal organic frameworks (MOFs) or periodic mesoporous silicate or organic polymers are recently developed adsorbents that belong to this group of CSPs [15–17]. In a chiral pore, a chiral environment is formed by asymmetrically arranged pore walls, and enantiorecognition depends entirely on accommodation of enantiomers within the pore space. Figure 1.1 illustrates a difference between a (surface) chiral site and a chiral pore. In general, adsorbents relying in separation on the diffusion of analytes into the bulk of a stationary phase are characterized by strong retention and slow intra-solid mass transfer leading to large analysis time and diffused peaks [18–20]. Therefore, such CSPs did not find a wide use in chromatography, although recent attempts to apply chiral MOFs in capillary gas chromatography [21] seem promising.

At present, the most popular sort of CSPs are those composed of an a chiral support with immobilized chiral selectors. The support is a solid, usually silica, occasionally zirconia [22] or porous polymers [23,24], with large pores (>10 nm) that allows fast intrapore mass transfer. The chiral component can be covalently attached to or coated onto the support. Coating, it can be supposed, results in a better shielding of the surface of a support as a coating substance should uniformly cover the whole surface of the support, not only the bounding sites as in the case of grafting. However, one cannot exclude the segregation of the coating phase causing exposure of parts of the bare support surface.

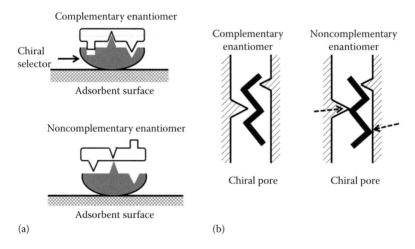

FIGURE 1.1 Schematic illustration of a surface chiral site (a) and a chiral pore (b). Dashed arrows in inset (b) indicate positions of steric hindrance for the inclusion of the noncomplementary enantiomer into a chiral pore.

1.2.1 Enantioselective and Nonselective Adsorption Sites

First, researchers considered the CSP as consisting of only enantioselective adsorption sites or of these sites distributed in an indifferent matrix that does not participate in adsorption. Eventually, it has been understood that chiral adsorbents have a more complex structure, comprising the enantioselective adsorption sites and the nonselective adsorption sites that exhibit the same affinity with respect to any stereoisomer of a chiral compound [25–27]. Any chiral structures on the surface or in the bulk of a stationary phase may play a role of an enantioselective site. Those can be chiral low molecular weight molecules, or macromolecules, or chiral pores. One molecule of a chiral polymer can bear a multitude of enantioselective sites. Moreover, chiral cavities created at the supramolecular level between adjacent polymer chains also constitute selective sites that contribute to enantioseparation [28].

The nonselective sites comprise the achiral elements of the stationary phase (silanol groups of the silica support, achiral fragments of chiral selectors, etc.) and those chiral entities that lost their enantioseparation ability. This later may be caused by random conformational changes, by blockage of a chiral cavity by mobile phase components, or by interfering effects of neighboring surface groups. Zhao and Cann [29] have shown through the molecular dynamic (MD) study of chiral recognition by the Whelk-O1 selector that at any instant of time roughly a third part of the analyte molecules coming into interaction with chiral selectors form an enantiodiscriminative docking complex while 60%–75% of the molecules interact with the selectors nondiscriminatively. It means that at any moment of time only a part of chiral selectors act as enantioselective sites. The mode of action for a particular selector is a random event, its probability depending on the nature of the solute–selector system and experimental conditions. The latter phenomenon has been well demonstrated by example of *Cinchona* alkaloids. It was established that these selectors exist in solvents in different so-called open and closed conformations, a conformer of the former type being preferentially enantioselective in certain processes [30,31]. The percentage of a particular conformer is a function of the protonation status, for example, pH, type of solvent, and so on [31,32]. Chromatographic evidences of chiral selectors acting as nonselective sites are discussed in the review [8].

The contributions of the enantioselective (es) and nonselective (ns) sites to the chromatographic retention factor k_i' of enantiomer i are usually assumed to be additive [8,33].

$$k_i' = k_{es,i}' + k_{ns,i}' \qquad (1.1)$$

supposing independent behavior of the enantioselective and nonselective sites. Table 1.2 illustrates a relative importance of these two contributions in different chiral adsorption systems. It follows from these data that nonselective adsorption is always present and frequently comparable by an order of magnitude to the effect of enantioselective adsorption.

TABLE 1.2

Relative Contributions of the Enantioselective Adsorption Sites to the Total Retention on Some Chiral Stationary Phases[a]

Solute	Stationary Phase[b]	α^c	$\alpha_{true}{}^c$	Relative Enantioselectivity Contribution, %		Reference
				1st Eluted	2nd Eluted	
Methyl mandelate	4-Methylcellulose tribenzoate	1.25	2.0	24.7	39.9	[231]
Ketoprofen	4-Methylcellulose tribenzoate	1.3	1.5	60.4	70.1	[232]
Propanolol	Cellobiohydrolase I, pH = 5	1.51	2.13	44.0	64.3	[61]
Propanolol	Cellobiohydrolase I, pH = 6	2.33	3.31	56.9	80.8	[61]
N-benzoyl-alanine	Bovine serum albumin	1.7	2.1	62.8	78.3	[83]
Mandelic acid	Bovine serum albumin	1.35	1.86	40.1	55.9	[233]
3-chloro-1-phenyl-propanol	Quinidine carbamate, $T = 22°C$	1.07	1.14	50.7	54.3	[85]
Naproxen	Whelk-O1, $T = 22°C$	1.83	1.95	87.5	93.1	[39]

[a] The contribution of the nonselective sites (ω_{ns}, %) is equal to $(100 - \omega_{es})$.
[b] The listed are chiral selectors; the support is silica in all cases.
[c] α, apparent (experimental) enantioselectivity; α_{true}, true enantioselectivity. See Section 1.3.1 for explanation.

1.2.2 HETEROGENEITY OF ADSORPTION SITES

Each group of the adsorption sites, that is, enantioselective and nonselective, is also not homogeneous even if a CSP bears a single type of chiral selectors. The affinity of the single-type active sites toward solute molecules varies due to surface defects, minor conformational changes, the influence of microenvironment, and so on. As a result, the population of the active sites is distributed over a range of local adsorption constants between the lowest (b_{min}) and the highest (b_{max}) values according to a distribution function $\varphi(b) = dq^*(b)/db$, where $q^*(b)$ is the cumulative concentration of the adsorption sites with an adsorption constant lower than a current value b. Figure 1.2 demonstrates a typical picture of an adsorption affinity distribution (AAD) for a single type selector CSP. It shows two modes, a low-energy and a high-energy mode, for both enantiomers. The high-energy mode associated with the enantioselective adsorption sites [34] has a small width, an

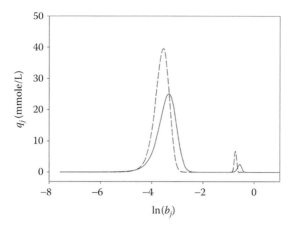

FIGURE 1.2 Adsorption affinity distributions of the R-(solid line) and S-2,2,2-trifluoro-1-(9-anthryl)-ethanol (dashed line) on a silica-bound quinidine carbamate CSP. q_j is the concentration of adsorption sites with adsorption constant in the range $\ln b_j \pm \Delta \ln b_j$; $\Delta \ln b_j = 0.0584$. (Reprinted with permission from Götmar, G. et al., *Journal of Chromatography A*, 1059, 43, 2004.)

order of 0.2 in terms of $\ln(b)$ or 0.5 kJ/mol in terms of adsorption energy $RT\ln(b)$. The width of the low-energy mode is an order of magnitude higher, equivalent to 5 kJ/mol in energy units. This mode was associated with the nonselective sites considering an essential overlaying of the respective AAD peaks for both enantiomers, the lack of full coincidence being attributed to experimental error [34]. A narrow width of the "enantioselective" mode is explained by the need for a particular conformation of a chiral selector that allows at least three simultaneous interactions with a solute molecule for enantiorecognition to take place, as it will be explained below. Any small deviation from this configuration would revoke the formation of a strong diastereomeric adsorbed complex and render this selector into a group of nonselective sites forming weaker adsorbed complexes lacking one or two interactions available for the undisturbed configuration. On the contrary, a wide range of selector conformations as well as various achiral elements of the stationary phase may play a role of a nonselective site that leads to a broad mode in an AAD.

Although bimodal adsorption systems are frequently found in liquid chromatography, CSPs with more diverse surface are not at all rare [8]. Kim and Guiochon have reported on 3–4 groups of adsorption sites (depending on mobile phase composition) on the surface of a molecular imprinted polymer (MIP) CSP [35–38]. The most abundant sites with the lowest affinity exhibited either nonselective or enantioselective properties with different solvents. The less numerous sites with intermediate and the highest affinity were selective. Asnin et al. have detected three different types of adsorption sites on the surface of the Whelk-O1 CSP by using naproxen as a probe [39,40]. Two of them were enantioselective, that is, chiral selectors in different conformations, and one site was nonselective, associated with the residual silanols of the bare silica support. Table 1.3 summarizes the concentrations q^* and respective adsorption constants b of all adsorption sites. Interestingly, the binding affinity of the nonselective sites exceeds that of the enantioselective entities, but the surface density of the former is by two orders of magnitude less than that of chiral selectors. One can suppose that the binding density is so high that only a very small fraction of bare silica is exposed to the solute. This is quite a common situation with many modern brash-type CSPs. Another important conclusion is that the concentration of adsorption sites depends on the mobile phase composition that demonstrates the importance of solvation effects for the availability of chiral selectors for binding. Indeed, if the solvation shell of a chiral selector is dense and strong, an energy gain of direct solute–selector interaction may be not high enough to compensate the disintegration of this shell. Then a chiral selector, although physically present on the surface will not be available to the given solute.

TABLE 1.3
Parameters of the Adsorption Sites of the Whelk-O1 CSP with Regard to Naproxen Enantiomers

	Methanol:Water Ratio in the Mobile Phase[b]		
Parameter[a]	80:20	85:15	90:10
$q^*_{es,l}$, mmol/L	800	728	580
$b_{es,l,S}$, L/mmol	0.0069	0.0051	0.0047
$b_{es,l,R}$, L/mmol	0.0089	0.0068	0.0071
$q^*_{es,h}$, mmol/L	47	42	12
$b_{es,h,S}{}^c$, L/mmol	$= b_{es,l,S}$	$= b_{es,l,S}$	$= b_{es,l,S}$
$b_{es,h,R}$, L/mmol	0.077	0.051	0.069
q^*_{ns}, mmol/L	5.9	9.1	8
b_{ns}, L/mmol	0.168	0.078	0.051

Source: L. Asnin and G. Guiochon, *Journal of Chromatography A*, 1217, 2871, 2010. With permission.

[a] $q^*_{es,l}$, surface concentration of the low-energy enantioselective adsorption sites; $b_{es,l,S}$ and $b_{es,l,R}$, adsorption equilibrium constants of the S- and R-naproxen, respectively, on the low-energy enantioselective sites; $q^*_{es,h}$, surface concentration of the high-energy enantioselective adsorption sites that were supposed to bind strongly only the R-enantiomer [40][c]; $b_{es,h,S}$ and $b_{es,h,R}$, adsorption equilibrium constants of the S- and R-naproxen, respectively, on the high-energy enantioselective sites; q^*_{ns}, surface concentration of the nonselective adsorption sites; b_{ns}, the adsorption equilibrium constant of the nonselective sites.

[b] Mobile phase: 0.01 M CH_3COOH in methanol–water mixture.

[c] This group of chiral selectors was supposed [40] to be in a conformation that allowed a strong binding of the R-enantiomer but interacted with the S-form as the low-energy conformation. Accordingly, $b_{es,h,S} = b_{es,l,S}$.

The above discussed heterogeneity of the enantioselective sites was caused by different configurations or distortion of configuration of a single chiral selector. A stationary phase may comprise different types of selectors. Spégel et al. [41] have prepared a multiple templated MIP in order to increase the application range of the resulting material. Protein molecules frequently have two or more chiral recognition centers that act as enantioselective sites of a protein-based CSP [42–45]. It is worth noting that polysaccharide CSPs are considered to bear multiple types of enantioselective sites formed at the molecular and supramolecular levels [6]. Respective partial AADs of all these site types were expected to overlap with each other to form a broad total AAD [34]. Experimental research, however, showed that AADs on cellulose derivatives immobilized on silica had a typical bimodal pattern [46,47], with a relatively narrow peak of the high-energy mode and a broad peak of the low-energy mode. Enmark et al. [48], who studied the adsorption of the enantiomers of methyl mandelate on the Kromasil Cellucoat CSP, reported a bimodal AAD for the stronger retained enantiomer and a broad monomodal AAD for the weaker retained enantiomer.

1.2.3 INTERFACE BETWEEN STATIONARY AND MOBILE PHASES

In liquid chromatography, the stationary phase contacts with the mobile phase, usually a mixture of two or more components, although single-component eluents are also occasionally used. It is well known that the structure and chemical composition (in case of mixed solvents) of a liquid at the liquid/solid interface differ from these in the bulk liquid. Since retention takes place at the interface, this circumstance will influence solute–stationary phase interactions. Surprisingly, this phenomenon has not been thoroughly studied. It is clear that the rule "similar-to-similar" applies to chiral adsorbents similar to any other solid. That is, the surface layer of the mobile phase in contact with nonpolar CSPs will be enriched in the less polar component of the mobile phase. On the contrary, the solution on the surface of polar CSPs will be enriched in the more polar component. Besides this general wisdom supported by MD simulations [49] there has been no much information available until recently.

One of the first studies on the adsorption of mobile phases on CSPs was made by Cavazzini and coworkers [50,51]. They investigated normal-phase systems. In the work [50], the adsorption of a MeOH/CH_2Cl_2 mixture on a polymeric DACH-ACR CSP and in the work [51] the adsorption of ethyl acetate/hexane and i-PrOH/hexane on a polysaccharide Chiralcel OD-I CSP were considered. In each case, the authors observed a positive excess adsorption of the polar constituent of a mobile phase over the entire composition range, from $x_1 = 0$ to 1, where x_1 is the molar fraction of the polar solvent. In the former case, the shape of the excess adsorption isotherm suggested the formation of a monomolecular surface layer saturated by methanol at low methanol concentration (~10 mol.%). In the case of a polysaccharide CSP in contact with hexane-based mobile phases, the surface layer can be deduced to be thicker and more diluted by the nonpolar solvent, although, as it was already mentioned, the polar modifier (ethyl acetate or i-propanol) is the preferentially adsorbed component.

The pattern of adsorption of hydro-organic mixtures on brash-type CSPs depends on the nature of chiral selector. On a Pirkle-type Whelk-O1 CSP, methanol was

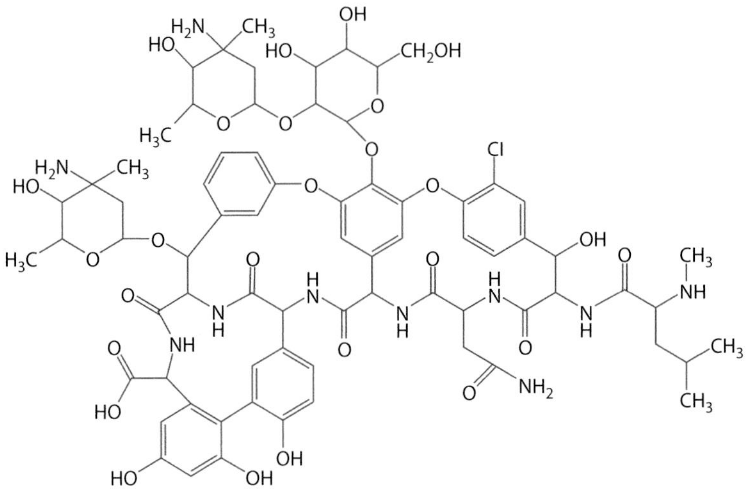

FIGURE 1.3   Structures of Whelk-O1 and eremomycin chiral selectors.

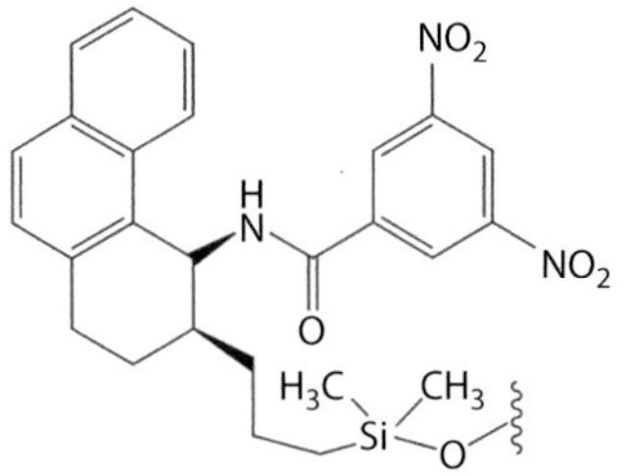

preferentially adsorbed from a water/MeOH mixture over the entire concentration range [39], whereas on a CSP with grafted antibiotic eremomycin (Nautilus-E) the surface solution was enriched in water over a wide concentration range, the upper limit of the range depending on the organic modifier. It was 75 mol.% for methanol and 90 mol.% for acetonitrile [52]. Obviously, such a difference in adsorption patterns is explained by the fact that Whelk-O1 contains a large hydrophobic phenanthren fragment, while the chiral selector of Nautilus-E has a hydrophilic exposed surface (Figure 1.3).

An organic component of a hydro-organic solvent affects the thickness of the adsorbed layer of the mobile phase on the surface of Nautilus-E. It is of the order of one molecular diameter for water/MeOH and contains 3–4 molecular layers when this CSP is in contact with water/MeCN [52]. Another interesting finding reported in the cited work concerns the structure of the interface layer. It was proven that the latter consists of two compartments (Figure 1.4). One compartment occupies the volume between bound chiral selectors. This part is available to solvent molecules but

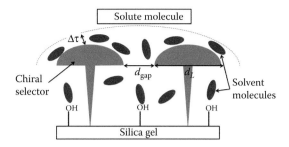

FIGURE 1.4 A model of the interface layer on the surface of the Nautilus-E (eremomycin) CSP. A dashed line outlines the volume that is inaccessible to large solute molecules but accessible to small solvent molecules. Note that the chiral selector consists of two parts: the chiral ligand proper and the tethering group anchoring the ligand to the surface. $\Delta\tau$ symbolizes the layer of the solvent molecules adsorbed on chiral selectors, an increment to the inaccessible volume. A distance between two neighboring chiral selectors d_{gap} is estimated to be 8 Å, a diameter of an eremomycin moiety d_L is ~24 Å. (Reprinted with minor modification with permission from Nikitina, Y. K. et al., *Journal of Chromatography. A*, 1363, 71, 2014.)

is unavailable to solute molecules if their size is larger than the gap between neighboring selectors. The second compartment includes the layer of the mobile phase adsorbed on the external surface of chiral selectors. This is the part of the adsorbed phase that contacts with organic solutes.

1.3 THERMODYNAMICS OF CHROMATOGRAPHIC ENANTIOSEPARATION

The phenomenon of chromatographic separation is accounted for by the reversible formation of transient complexes between solutes and adsorption sites of a stationary phase. If of two solutes one forms a more stable complex, it will be retained in a chromatographic column longer than the second component that forms a less stable complex. Thus, the retention factor k' directly relates to the strength of the bond between a solute and a stationary phase. The ratio of the k' values for a pair of solutes characterizes an ability of a given stationary phase to separate these two compounds. This characteristic designated α is called in chiral chromatography enantioselectivity as chiral chromatography concerns the separation of enantiomers.

As any equilibrium process, the transfer of a solute from the liquid (mobile) phase into the stationary phase (called thereinafter adsorption for the sake of brevity) can be characterized by respective thermodynamic quantities. The standard Gibbs free energy of adsorption ΔG^0 characterizes the stability of transient adsorption complexes, the standard adsorption enthalpy ΔH^0 measures the energy balance of interactions formed and destroyed as a result of adsorption, and the standard adsorption entropy ΔS^0 is a function of structure rearrangements accompanying the formation of an adsorption complex. These quantities are determined based on the retention factor and its temperature dependence. In order to avoid misinterpretation of experimental results, it is necessary to understand clearly relationship between elution and thermodynamic characteristics and physical meaning of the latter.

1.3.1 ELUTION CHARACTERISTICS AND CHROMATOGRAPHIC THERMODYNAMICS

The retention factor relates to the adsorption isotherm of a solute through the first derivative of the isotherm

$$k'(c) = F\left(\frac{dq}{dc}\right)$$
(1.2)

where:
q and c are solute concentrations in the liquid and stationary phases, respectively
F being the phase ratio or the ratio of the volumes of the stationary phase and the liquid phase in the column

Under infinite dilution conditions, where dq/dc becomes constant, $k'(c \to 0)$ designated thereafter k' can be expressed through the partition coefficient K:

$$k' = FK$$
(1.3)

The partition coefficient is the ratio of the solute concentration in the stationary phase (designated by subscript s) to that in the mobile phase (subscript m):

$$K = \frac{[E]_s}{[E]_m}$$
(1.4)

In other words, it is the equilibrium constant of the process of transfer of solute E from the mobile phase to the stationary phase:

$$(E)_m \leftrightarrow (E)_s$$
(1.5)

The standard Gibbs free energy ΔG^0 corresponding to adsorption equilibrium (Equation 1.5) can be expressed via the partition coefficient or the retention factor as

$$\Delta G^0 = -RT \ln K = -RT \ln(k'/F)$$
(1.6)

where:
R is the universal gas constant
T is the absolute temperature

The value ΔG^0 relates to the standard enthalpy and entropy of adsorption through the Gibbs–Helmholtz equation:

$$\Delta G^0 = \Delta H^0 - T\Delta S^0$$
(1.7)

Expressing the constant K in Equation 1.6 via k'/F and substituting the Gibbs free energy with Equation 1.7, we obtain the van't Hoff equation (1.8) which is used in chromatography to determine ΔH^0 and ΔS^0 from the temperature dependence of the retention factor.

$$\ln k' = -\frac{\Delta H^0}{RT} + \frac{\Delta S^0}{R} + \ln F \tag{1.8}$$

The enantioselectivity α relates to the difference in the Gibbs energies of adsorption for two separated enantiomers indicated below by subscripts 1 (first eluted) and 2 (second eluted):

$$\Delta\Delta G^0 = \Delta G_2^0 - \Delta G_1^0 = RT \ln\frac{k_2'}{k_1'} = RT \ln \alpha \tag{1.9}$$

The quantities ΔG^0, ΔH^0, and ΔS^0 derived by Equations 1.6 and 1.8 from experimentally measured retention factor are associated with equilibrium Equation 1.5. The application of this equation to adsorption in a chromatographic column supposes that a CSP is a homogeneous medium of infinite capacity, in which a solute is uniformly distributed after being transferred from the mobile phase. In fact, this does not hold true. As it was explained in the previous section, a real CSP contains two groups of the adsorption sites. Those that exhibit different affinities with respect to optical antipodes constitute a group of the enantioselective adsorption sites. There is another group of sites, called nonselective, that interact in the same way with any stereoisomer of a chiral compound. Thus, an equilibrium established in a column can be represented by a set of two equations describing a reversible binding of enantiomer E_i to an enantioselective site CS or to a nonselective site NS, each equilibrium characterized by a respective equilibrium constant, $b_{es,i}$ or $b_{ns,i}$ (index $i = 1, 2$ refers to a particular enantiomer):

$$E_i + CS \leftrightarrow E_iCS; b_{es,i} = [E_iCS]/([E_i]\cdot[CS]) \tag{1.10}$$

$$E_i + NS \leftrightarrow E_iNS; b_{ns,i} = [E_iNS]/([E_i]\cdot[NS]) \tag{1.11}$$

Each binding equilibrium is characterized by a set of thermodynamic quantities $\Delta G_{es,i}^0 = -RT \ln b_{es,i}$, $\Delta H_{es,i}^0$, $\Delta S_{es,i}^0$ (or $\Delta G_{ns,i}^0 = -RT \ln b_{ns,i}$, $\Delta H_{ns,i}^0$, $\Delta S_{ns,i}^0$) relating to each other through the Gibbs–Helmholtz equation. It is easy to show that the partition coefficient K is a linear combination of respective partial adsorption constants. Indeed,

$$K = \frac{[ECS]+[ENS]}{[E]} = \frac{[ECS]}{[E]} + \frac{[ENS]}{[E]} = b_{es}[CS] + b_{ns}[NS] \tag{1.12}$$

Under infinite dilution conditions when the equilibrium concentration of uncovered adsorption sites, either enantioselective or nonselective, does not deviate noticeably from the saturation capacities (q^*) ascribed to these groups of the sites, one can write

$$K = b_{es}q_{es}^* + b_{ns}q_{ns}^* \tag{1.13}$$

The above expression can be rewritten in terms of the retention factor

$$k' = Fb_{es}q_{es}^* + Fb_{ns}q_{ns}^* = k'_{es} + k'_{ns} \qquad (1.14)$$

This equation shows that retention in chiral chromatography includes two contributions, one for the enantioselective and another for the nonselective interactions. Each contribution depends not only on the strength of the binding between a solute and an adsorption site, but also on the concentration of respective sites. The experimental Gibbs free energy can then be expressed as follows:

$$\Delta G^0 = -RT \ln \left(k'_{es} + k'_{ns} \right) + RT \ln F \qquad (1.15)$$

illustrating that this is a lumped function not reduced to a linear combination of partial Gibbs energies ΔG_{es}^0 and ΔG_{ns}^0.

It follows from Equation 1.14 that enantioselectivity α does not reckon a difference in adsorption affinities of chiral selectors toward optical antipodes as it has frequently been asserted. Rather, it is a lumped function depending on both enantioselective and nonselective contributions to retention as given by [33,53]

$$\alpha = \frac{k'_2}{k'_1} = \frac{k'_{es,2} + k'_{ns}}{k'_{es,1} + k'_{ns}} \qquad (1.16)$$

While the enantioselective term k'_{es} differs for optical isomers ($k'_{es,1} \neq k'_{es,2}$), the nonselective term remains the same for either enantiomer, that is, $k'_{ns,1} = k'_{ns,2} = k'_{ns}$.

Guiochon, Fornstedt, and coauthors [33,54] proposed a quantity of true enantioselectivity

$$\alpha_{true} = \frac{k'_{es,2}}{k'_{es,1}} \qquad (1.17)$$

that accounts for the effect of only the enantioselective interactions. In order to emphasize difference between the experimental enantioselectivity α and the true enantioselectivity α_{true}, the former has been suggested to be called "apparent enantioselectivity" [54]. To find α_{true}, it is necessary to use the methods of nonlinear chromatography [55,56] that allow the determination of the enantioselective and nonselective contributions to retention. Table 1.2 illustrates a difference between α and α_{true} in several chiral adsorption systems.

1.3.2 ENTHALPY AND ENTROPY OF ADSORPTION ON CSPS

The standard enthalpy and entropy of adsorption found in a chromatographic experiment are lumped quantities including contributions from enantioselective interactions and from nonselective interactions. It can be shown that within the framework of the model considered the overall adsorption heat effect is a linear combination of the partial heat effects taken with respective weight coefficients $\delta_{es} = k'_{es}/k'$ and $\delta_{ns} = k'_{ns}/k'$:

$$\Delta H'^0 = \delta_{es} \Delta H_{es}^0 + \delta_{ns} \Delta H_{ns}^0 \qquad (1.18)$$

For the adsorption entropy, one can derive

$$\Delta S'^0 = R \ln \frac{\left(k'_{es} + k'_{ns}\right)}{b_{es}^{\delta_{es}} \cdot b_{ns}^{\delta_{ns}}} - R \ln F + \delta_{es}\Delta S_{es}^0 + \delta_{ns}\Delta S_{ns}^0 \tag{1.19}$$

$$\Delta S_{es(ns)}^0 = R \ln b_{es(ns)} + \Delta H_{es(ns)}^0 / T \tag{1.20}$$

The value of $\Delta H'^0$ ($\Delta S'^0$) is equal to ΔH^0 (ΔS^0) of Equation 1.8 if the number of the adsorption sites does not depend on temperature, which is not always observed in experiments. This subtle difference between the overall enthalpy and entropy in Equation 1.8 and those values in Equations 1.18 and 1.19 originates in the fact that ΔH^0 and ΔS^0 are found from a temperature dependence of the partition coefficient, whereas $\Delta H'^0$ and $\Delta S'^0$ are derived from those of the true equilibrium constants b_{es} and b_{ns}. This explains some apparent contradictions between overall and partial adsorption enthalpies compared in Table 1.4.

The separation of the contributions of enantioselective and nonselective interactions to a total thermodynamic effect is only possible if partial adsorption constants

TABLE 1.4
Site-Specific Thermodynamic Characteristics of Adsorption

CSP	Solute		$\Delta H_{ns(l)}^0$ [a] (kJ/mol)	ΔH_{es}^0 (kJ/mol)	ΔH^0 (kJ/mol)	$T\Delta S_{ns(l)}^0$ [a,b] (kJ/mol)	$T\Delta S_{es}^0$ [b] (kJ/mol)	References
CBH I[c], pH = 5.5	Propranolol	R	−4.6	−8.0	−10.0	+0.10	−3.2	[58]
		S	−4.6	+6.7	+3.0	+0.22	+14.5	
Amyloglucosidase	Propranolol	R	−7.8	−12.9	−9.5	−4.5	−10.6	[63]
		S	−7.8	−8.5	−8.3	−4.5	−5.3	
Chirobiotic T	Tryptophan	L	−9.0[d,e]	−34.3	na	−0.15[d,e]	−1.3	[66]
		D	−5.0[d,e]	−29.9	na	+0.36[d,e]	−0.69	
Chirobiotic TAG	Tryptophan	L	−12.2[d,e]	−29.2	na	−0.42[d,e]	−0.72	[66]
		D	−5.5[d,e]	−12.6	na	+0.51[d,e]	+1.3	
Chiris Chiral AX:QD1	TFAE[f]	R	−32.3	−43.9	−19.2	−40.5	−45.3	[34]
		S	−31.2	−46.4	−20.4	−39.0	−48.6	
Whelk-O1	Naproxen	S	−28	−19.8	−20.7	na	na	[39]
		R	−28	−23.0[g]	−26.6	na	na	
			−41					

Note: ns(*l*) = nonselective or low-energy sites; es = enantioselective sites.
[a] Unless otherwise mentioned the quantities $\Delta H_{ns(l)}^0$ and $\Delta S_{ns(l)}^0$ refer to the nonselective sites.
[b] $T = 298$ K.
[c] Cellobiohydrolase I.
[d] Computed based on the data given in the original reference.
[e] Low-energy enantioselective sites (no nonselective sites were found in this system).
[f] 2,2,2-trifluoro-1-(9-anthryl)-ethanol.
[g] There were supposed to be two types of the enantioselective sites specific to *R*-naproxen.

corresponding to these two types of interactions are measured and subjected to the van't Hoff analysis. Otherwise, the researcher is limited to the analysis of overall thermodynamic quantities, which may nevertheless provide useful information regarding the adsorption equilibrium in a column.

Typically, adsorption on a CSP is an exothermic process ($\Delta H^0 < 0$), although a few examples of an endothermic adsorption ($\Delta H^0 > 0$) have been reported [57–62]. Four of the mentioned publications [57–59,61] describe the adsorption of β-blockers, in particular propranolol, on protein cellobiohydrolase I (CBH I)-based CSPs. Fornstedt, Guiochon, and coworkers [58,61] proved that the binding of the stronger retained enantiomer to a chiral retention site is endothermic, while the nonselective interactions are exothermic for either enantiomer (Table 1.4). A predominance of the chiral retention mechanism over the nonchiral one, when happens [61], explains a global endothermic effect in accordance to Equation 1.18. It is interesting to note that the complexation of both enantiomers of propranolol with free CBH I in a solution is endothermic [57]. This example demonstrates a difference in behavior of immobilized and free enzymes. Table 1.4 summarizes information on site-specific thermodynamics available in the literature. It shows that endothermic enantiose-lective interactions are not common for protein-based CSPs. When another silica-immobilized enzyme (amyloglucosidase) was examined, both stereospecific and nonspecific binding appeared to be exothermic [63].

It should be understood that, unless one considers a gas phase process, ΔH^0 as well as ΔS^0 and ΔG^0 is composed of several contributions corresponding to desolva-tion of the solute in the liquid phase ("desolv"), desorption of the solvent from the surface of a stationary phase ("desorp"), formation of an transient complex on the surface (net adsorption = "netads"), resolvation of the transient complex ("resolv"), and dilution of the liquid phase by the solvent molecules desorbed from the chiral selector by the solute ("dil"):

$$\Delta H^0 = \Delta H^0_{desolv} + \Delta H^0_{desorp} + \Delta H^0_{netads} + \Delta H^0_{resolv} + \Delta H^0_{dil} \tag{1.21}$$

Only the third contribution corresponds to net adsorption, and this may be not domi-nating. Tentative estimation made in [64,65] shows that the enthalpies of solute sol-vation and net adsorption are comparable. Therefore, an endothermic effect may result not only from the formation of an unstable transient complex but also from endothermic contributions associated with desolvation/resolvation processes. As well a negative enthalpy change does not necessarily mean an enthalpy beneficial solute–selector association—although it would be a very likely explanation—but may be conditioned by exothermic solvation effects.

According to the Gibbs–Helmholtz equation, an enthalpy gain (or loss) due to the formation of a transient complex can be supplemented (or compensated) by the entropic term. The entropy change is a measure of structure rearrangement in a sys-tem. Entropy decreases when degree of order in a system increases. So, a nega-tive ΔS^0 is indicative of ordering processes accompanying solute–stationary phase binding (without specification to which sort of sites it is bound). As it was already noticed, those processes are not limited only to the formation of a transient com-plex. Therefore, although it is obvious that the entropy of a bound solute must be lower than the entropy of a free solute, the total entropy change can be positive due to the restructuring of the solvent surrounding it. Such a behavior is however rare [58–60,62,66] and for rare exceptions (see [60, table 1.5, 1st row] and [66, D-Trp on Chirobiotic TAG]) is limited to the case of endothermic adsorption, which in the virtue of Equation 1.7 is an entropically driven process. More frequently, an expected decrease in entropy is found ($\Delta S^0 < 0$) that according to Equation 1.7 partly cancels out the enthalpic gain from solute–stationary phase interactions. This effect is known as the enthalpy–entropy compensation and is a popular tool in the study of mechanisms of enantioseparation (see [6] and refs. therein).

An entropically driven adsorption is an interesting phenomenon consisting in adsorption occurring because of growth of disorder in a system. Much as contraintuitive it is, the entropically driven process does not require the formation of a stable surface complex. Quite contrary, a solute–surface site complex *per se* can be unstable but stabilized by surrounding disordering processes resulting in a negative Gibbs free energy due to a positive entropic term. Fornstedt et al. [58] revealed the machinery of the entropically driven adsorption using as a case study the binding of *S*-propranolol to the chiral site of CBH I enzyme. When inside of the chiral cavity of this site, water molecules, especially those in proximity to its nonpolar regions, are better organized than in the bulk solution, as well as the water molecules adjacent to a nonpolar fragment of a solute molecule. When the latter enters a chiral site, the ordered water molecules adjacent to the two surfaces are expelled, and a less ordered water structure is obtained. A corresponding positive entropic effect overcomes an increase in enthalpy due to the formation of a strained adsorption complex. *R*-propranolol forms with a chiral site a stable complex ($\Delta H^0 < 0$) probably because the configuration of the enantiomer does not allow an energetically strained convergence between the nonpolar regions of the solute and the selector. As a result, disintegration of the water structure nearby these regions does not happen.

1.3.3   Enthalpic and Entropic Factors in Enantioseparation

Commonly, the thermodynamic analysis of enantioseparation begins from the computation of differences in the thermodynamic quantities of adsorption for individual enantiomers 1 (first eluted) and 2 (second eluted):

$$\Delta\Delta X^0 = \Delta X_2^0 - \Delta X_1^0 \ (X = G,\ H,\ \text{or}\ S,\ \text{respectively}) \tag{1.22}$$

It is usually assumed that quantities $\Delta\Delta X^0$ are free from the contribution of achiral interactions; thus, analysis of these characteristics and their dependences on experimental conditions and on the nature of a solute/chiral selector system would provide insight on the mechanisms of chiral interaction. As a matter of fact, this is not the case. Combination of Equations 1.9 and 1.16 gives

$$\Delta\Delta G^0 = \Delta G_2^0 - \Delta G_1^0 = RT \ln \frac{k'_{es,2} + k'_{ns}}{k'_{es,1} + k'_{ns}} \tag{1.23}$$

that is, a difference in adsorption Gibbs energies for a pair of enantiomers is a lumped function depending both on enantioselective and nonselective retention mechanisms. The same holds true for $\Delta\Delta H^0$ and $\Delta\Delta S^0$. Keeping this in mind, let us consider what kind of information can be retrieved form observation on $\Delta\Delta X^0$ functions.

By analogy to Equation 1.7, one can write

$$\Delta\Delta G^0 = \Delta\Delta H^0 - T\Delta\Delta S^0 \tag{1.24}$$

dividing $\Delta\Delta G^0$ into the enthalpic and entropic constituents. It was predicted [67] and then observed, first in gas chromatography [68,69] and later in liquid chromatography [70,71], that at a high enough temperature T_{iso}, the elution order of a pair of enantiomers can be inverted. Since at the inversion (isoelution) point $\Delta\Delta G^0 = 0$, $T_{iso} = \Delta\Delta H^0 / \Delta\Delta S^0$. For the isoelution point to exist, the sign of the enthalpy difference must be equal to the sign of the entropy difference. A majority of reported separations occur in systems with existing isoelution point at temperatures much lower than this point (Figure 1.5a). Such separations are enthalpically controlled as for them $|\Delta\Delta H^0| > |T\Delta\Delta S^0|$. Commonly, this class of separations is attributed to one enantiomer interacting stronger with the chiral selector than the other one [22]. In view of the above said, we cannot exclude an essential contribution to $|\Delta\Delta H^0|$ from desolvation and solvent desorption processes. Indeed, in a situation when one enantiomer enters a chiral cavity and the other enantiomer does not, the number of solvent molecules released from both the chiral selector and either solute will be different, resulting in different solvent desorption enthalpies. So, Guillaume et al. [72] reported different numbers of water molecules excluded upon adsorption of D- and L-dansyl amino acids on a β-cyclodextrin (CD) CSP.

Processes at temperatures higher than T_{iso} (Figure 1.5b) constitute another class of enantioseparations. For them, both $\Delta\Delta H^0$ and $\Delta\Delta S^0$ are positive and $|\Delta\Delta H^0| < |T\Delta\Delta S^0|$, often $\Delta\Delta H^0 \approx 0$. Such processes are controlled by entropic effects. Examples of entropy-driven enantioseparations, although not numerous, are found among CSPs with all the various types of chiral selectors: crown ethers [65,73], antibiotics [74,75], zwitterionic *Chincona*-based alkaloids [62], polysaccharides [22,76], and in ligand-exchange chromatography [77]. Interestingly, there were no reports in the available literature of entropy-driven enantioseparations on CD-based CSPs. This is probably because an energy difference between inclusion-type complexes of CDs with favorably and unfavorably bound enantiomers results in an enthalpy gap that is too high to allow the entropic effects to dominate.

The thermodynamic regime of separation is influenced by experimental parameters and also depends for a given CSP on the nature of a solute. Toribio et al. [76] and Stringham and Blackwell [78] have induced transition from an enthalpy-driven to entropy-driven enantioseparation by changing the composition of the mobile phase in supercritical fluid chromatography (SFC). Reshetova and Asnin observed the change of a thermodynamic regime as a function of the mobile phase pH while separating enantiomers of weak acids on an antibiotic-based CSP [74]. More information can be obtained from studying $\Delta\Delta H^0 - \Delta\Delta S^0$ correlations in a series of chiral solutes. Choi et al. [73] found that racemates of arylaminoalcohols with the hydroxyl group attached to the chiral carbon atom were resolved

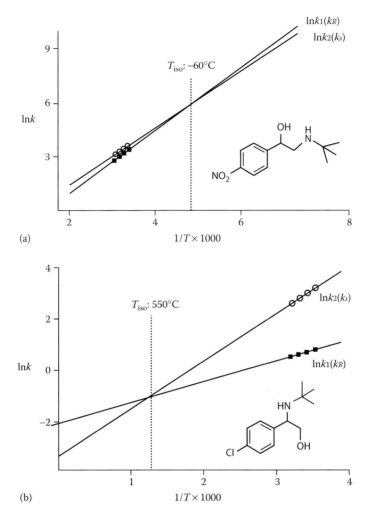

(a)

(b)

FIGURE 1.5 van't Hoff plots of lnk' vs. $1/T$ demonstrating the position of the isoelution point below (a) and above (b) of a sampled temperature interval. The formulas of the analytes are shown in the graphs. CSP: (+)-(18-crown-6)-2,3,11,12-tetracarboxylic acid immobilized on silica. (Reprinted with permission from Choi, H. J. et al., *Journal of Chromatography A*, 1164, 235, 2007.)

on a crown ether CSP under entropic control, whereas arylaminoalcohols with the amino group attached to the chiral carbon atom underwent enantioseparation under enthalpic control (see Figure 1.5). This obviously indicates different mechanisms of solute–selector binding for these two groups of aminoalcohols. Ilisz and colleagues, who have recently investigated enantioseparation of stereoisomers of multichiral monoterpene-based β-amino acids on zwitterionic Lindner phases [62] and on teicoplanin and teicoplanin aglycon phases [75], observed enthalpically controlled separation mechanism for some enantiomeric pairs and entropically

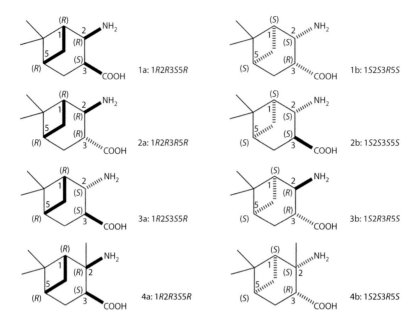

FIGURE 1.6 Chemical structures of stereoisomers of 2-amino-6,6-dimethylbicyclo [3.1.1] heptane-3-carboxylic acid (1 and 2), 3-amino-6,6-dimethylbicyclo[3.1.1]heptane-2-carboxylic acid (3), and 2-aminopinane-3-carboxylic acid (4).

controlled that for other pairs, the regime being dependent also on the mobile phase composition. Of special interest is the work [75], where of three pairs of enantiomers of a β-amino acid with four chiral centers (Figure 1.6), only one pair was resolved under an entropic control, thus manifesting a different enantioseparation mechanism. Other examples of analyte structure influencing the thermodynamic regime of enantioseparation are given in [22,70].

1.3.4 CONCENTRATION DEPENDENCE OF ADSORPTION ENTHALPY

The standard adsorption enthalpy relates to the infinite dilution conditions, therefore, cannot reveal a heterogeneous structure of a stationary phase. The measurement of adsorption isotherms at different temperatures allows determination of the so-called isosteric heat of adsorption (Q_{st}), a function of adsorbed amount characterizing the energy of a particular group of the adsorption sites being covered at a particular value of solute's solid phase concentration q in equilibrium with the liquid phase concentration c. By definition,

$$Q_{st}(q) = RT^2 \left(\frac{\partial \ln c}{\partial T} \right)_q \tag{1.25}$$

This function mimics the differential molar enthalpy of adsorption ($\Delta \bar{H}_{st}$), a thermodynamic quantity of a strict physical meaning; however, it does not coincide with the latter except at $q \to 0$ [34]. It is assumed that $Q_{st}(q)$ changes somewhat similar to $-\Delta \bar{H}_{st}(q)$, but unlike $\Delta \bar{H}_{st}$, it can be determined in a common chromatographic experiment.

There are a number of examples of using this quantity to elucidate retention mechanisms in chiral chromatography [34,79–84]. In general, interpretation of $Q_{st}(q)$ plots is not difficult. Invariance of the isosteric heat as in [80] proves the energetic homogeneity of the surface. Decrease in Q_{st} with increasing q reveals heterogeneity in adsorption sites [34,79,81,85]. A monotonous growth of Q_{st} or its nonmonotonous behavior are indicators of lateral interactions or caused by secondary equilibria such as ionization or association of solute molecules. Seidel-Morgenstern and Guiochon [82] have observed an increasing $Q_{st}(q)$ dependence for (+)-Tröger base (TB) adsorbed on microcrystalline cellulose triacetate, whereas the isosteric heat of adsorption of (−)-TB was almost invariable, decreasing only slightly with q. Such a behavior of (+)-TB is explained by its multilayer adsorption on polysaccharide CSPs [86]. The adsorption of (−)-enantiomer obeyed to the Langmuir model that agreed well with found lack of noticeable influence of q on Q_{st} for this enantiomer.

Comparison of $Q_{st}–q$ plots for optical antipodes provides additional information on the properties of a CSP and retention mechanisms. In Figure 1.7, isosteric heats of adsorption of the enantiomers of 3-chloro-1-phenylpropanol on a brash-type CSP are shown [85]. Two decreasing plots are approaching each other as adsorption progresses, meaning the exhaustion of strong enantioselective sites. When all the enantioselective sites are covered and only weak nonselective sites are left to adsorb, the plots will finally coincide. Another interesting case relating to the

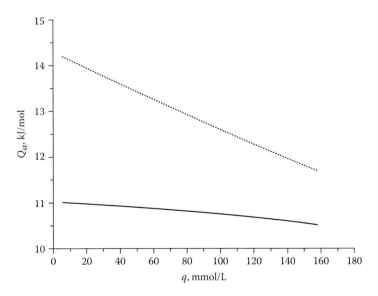

FIGURE 1.7 Isosteric adsorption heat as a function of adsorbed amount. Dotted line (*S*-3-chloro-1-phenylpropanol); solid line (*R*-3-chloro-1-phenylpropanol). (Reprinted with permission from Asnin, L. et al., *Journal of Chromatography A*, 1101, 158, 2006.)

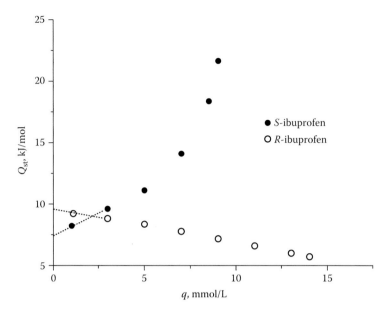

FIGURE 1.8 Isosteric heat of adsorption (Q_{st}) of S- and R-ibuprofen as a function of the adsorbed amount (q). Dashed lines are the extrapolations of $Q_{st}(q)$ plots to the infinite dilution conditions made using the last two points of the plots. Column: Nautilus-E, Mobile phase: acetic buffer/ethanol (40:60, v/v). (Reprinted with permission from Reshetova, E. N. and Asnin, L. D., *Russian Journal of Physical Chemistry*, 89, 275, 2015.)

adsorption of ibuprofen enantiomers on an antibiotic-based CSP is represented in Figure 1.8 [84]. As seen, at infinite dilution, Q_{st} of R-ibuprofen is larger than that of the S-isomer. The same relationship was found between the standard adsorption enthalpies for these compounds, with a quantitative agreement between the values of $Q_{st}(0)$ and ΔH^0 for each enantiomer. As q increases, the counter-directional $Q_{st}(q)$ dependences lead to an inversion of a relationship between heat effects of adsorption.

1.4 RETENTION MECHANISMS IN CHIRAL CHROMATOGRAPHY

Adsorption of a solute on a column packing is the immediate reason of chromatographic retention. Equation 1.2 connecting the function $k'(c)$ with the derivative of an adsorption isotherm is a mathematical reflection of this relation. Using this equation, the experimentally measured function $k'(c)$ can be converted to an adsorption isotherm. The latter in some aspects is a more convenient tool of representing and analyzing retention mechanisms. Most importantly, it has a clear physical meaning as expression of material balance in a two-phase system that can be further related to the concentration of the adsorption sites and the strength of interaction between an adsorption site and a solute molecule. In this section, we attempt at discussing retention mechanisms in terms of adsorption equilibrium and respective isotherm models.

1.4.1   ENANTIOSELECTIVE AND NONSELECTIVE RETENTION

It has become customary now to consider the adsorption of a chiral solute by a chiral solid as consisting of two parts: nonselective and enantioselective. Correspondingly, the total uptake (q) of enantiomer i in equilibrium with a bulk concentration c is given by

$$q(c) = q_{ns}(c) + q_{es,i}(c) \tag{1.26}$$

where the enantioselective term $q_{es,i}$ depends on the enantiomer, whereas the nonselective term is the same for either enantiomer, and both kinds of adsorption behavior are independent from each other. But is it really so? Consider a patch of the stationary phase surface bearing three chiral selectors as shown in Figure 1.9. The chiral selector contains a positive charge, say, a protonated amino group. Let us assume that a chiral analyte containing a negative charge binds to the chiral selector in the middle, and this interaction is enantioselective. Obviously, there will be ion–ion interaction between the charge of the analyte species and the charges of the other selectors. Generalizing this situation to a large enough patch of a surface, one can assert that any analyte that comes into an enantioselective interaction with a particular chiral selector will also interact with the electrostatic field of the local environment, and this later contribution

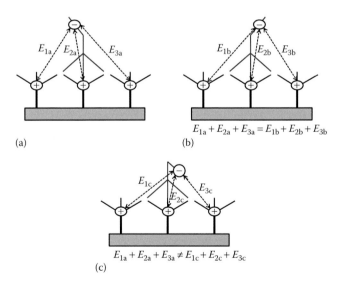

FIGURE 1.9 Illustration of the interaction of a chiral analyte with the total electric field of chiral selectors. Inserts (a) and (b) show two enantiomers in which the negative charge remains on the same height with respect to the surface. The total energy of electrostatic interaction is the same in both cases due to the principle of symmetry. Insert (c) shows an enantiomer configuration in which the negative charge is positioned at a different height over the surface as compared to the case in insert (a). The total energies of electrostatic interaction in cases (a) and (c) are not equal.

may be enantioselective or nonselective depending on the positioning of the adsorbed enantiomers with respect to the surface (Figure 1.9). Likewise, a localized interaction of an analyte with a particular nonselective site will be supplemented by the interaction of this analyte with the local electrostatic field. This circumstance is directly taken into account in molecular simulations of solute–stationary phase interactions. An expression for calculation of the energy of an interfacial system includes the Lennard–Jones term (van der Waals forces) and the electrostatic (Coulomb) term accounting for the interaction of the solute with the environment [87,88]. The separation of the contributions from the site–solute interaction and the solute–environment interaction is not so straightforward in macroscopic studies. The adsorption equilibrium constant for adsorption sites of type s (s = ns, es) relates to the difference between the free energies of a solute on the surface site ($G_{s,i}$) and in the bulk phase ($G_{0,i}$) through the Gibbs–Helmholtz equation:

$$b_{s,i} = \exp\left(-\frac{G_{s,i} - G_{0,i}}{RT}\right) \tag{1.27}$$

The energy contribution from the interaction of enantiomer i with the local electrostatic field $U_{s,i}^{\text{electrost}}$ can be singled out of the total free energy so that $G_{s,i} = G'_{s,i} + U_{s,i}^{\text{electrost}}$ leading to

$$b_{s,i} = \exp\left(-\frac{G'_{s,i} - G_{0,i}}{RT}\right) \cdot \exp\left(-\frac{U_{s,i}^{\text{electrost}}}{RT}\right) = b'_{s,i} \cdot b_{s,i}^{\text{electrost}} \tag{1.28}$$

where the adsorption constant b'_i explains the interaction of an enantiomer with the adsorption site given that the electrostatic field induced by the rest of the surface is null. The separation of these two factors has been attempted in the solvophobic theory developed for reversed-phase liquid chromatography (RP–LC) by Horváth et al. [89,90]. Although conditions of chiral chromatography differ from these of RP–LC, the mentioned studies by Horváth et al. show a principal possibility to analyze the "site" and "out-of-site" contributions to retention using extrathermodynamic considerations.

Thus, there is no neat solute–adsorption site interaction. A localized adsorption on a specific adsorption site always involves a delocalized component. Omitting this circumstance from consideration is vindicated by the fact that the net adsorption and delocalized components are lumped together into the experimentally measured adsorption constant according to Equation 1.28. In this context, the meaning of the adsorption isotherm (Equation 1.26) is not in that a certain amount of a solute is adsorbed due to interaction with a particular group of sites, which is only part true because of the influence of local environment, but in that one can count the number of molecules bound to sites of a particular sort.

1.4.1.1 Nonselective Interactions

Any interaction that is not stereochemically dependent is nonselective. This is because such interactions result in the formation of the adsorption complexes of enantiomers with a retention site that do not differ in their stability, consequently do not result in enantioseparation. Any intermolecular or interionic force or a combination of thereof can be involved in a nonselective retention. This is not a particular sort of intermolecular or Columbian interactions that induces enantioselective binding but a specific spatial arrangements of these as it will be explained in the next section. If such an arrangement is not possible, the interaction between a solute and an adsorption site will be nonselective regardless of types and combinations of forces involved. Usually, weak van der Waals forces and, occasionally, isolated polar and hydrogen bond interactions cause nonselective adsorption. As a result, the strength of nonselective binding is lower compared to stereospecific binding that typically involves an ensemble of coordinated interactions including the formation of hydrogen and/or donor–acceptor bonds, dipole–dipole stacking, ion–ion, and π–π interactions [8]. So far, only one example of the nonselective sites being the strongest over the whole population of the adsorption sites has been reported [39,40], the case described in detail in Table 1.3.

Nonselective adsorption can be localized or nonlocalized. The former category describes a situation when a solute species (molecule or ion) is retained over a particular adsorption site and its mobility is restricted in any dimension. The nonlocalized adsorption supposes that the mobility of an adsorbate is restricted only in one dimension, normally to the surface, but it can more or less freely travel along the surface. This sort of adsorption behavior requires the field of adsorption forces to be uniform in a plane parallel to the surface and typically arises where only weak nonspecific dispersive and electroinductive interactions are involved. The structure of CSPs, whether brash-type or coated-type, containing a multitude of diverse functional groups and/or structural elements of complex (on a molecular scale) geometry, for example, helical twisted cellulose chains, makes the nonlocalized adsorption on the surface unlikely. Only after all strong adsorption sites are occupied by adsorbate at a high degree of coverage, one may expect a certain contribution from this adsorption mechanism. The nonuniformity of the surface potential can be shielded by a layer of solvent molecules. If the influence of the surface forces extends farther than this monolayer, all solute molecules accumulated above this plane yet within the border of the adsorbed phase contribute to the nonselective adsorption and cannot be associated to any adsorption site. This retention mechanism combining adsorption on a surface and the accumulation of solute in a thin layer adjacent to the surface is called the adsorption-partitioning mechanism [91] and will be discussed in Section 1.4.4.

1.4.1.2 Enantioselective Interactions

For enantioseparation to take place, two enantiomers must form with a chiral selector transient complexes of different stability. As physical properties of enantiomers except their stereochemical configuration are identical, that distinction can be caused only by a specific spatial arrangement of each enantiomer with respect to the chiral selector. Simple geometrical consideration has led early researchers to the development of the "three-point interaction rule" that, as formulated by Pirkle and Pochapski, states that "Chiral recognition requires a minimum of three simultaneous interactions between the CSP and at least one of the enantiomers, with at least one of these interactions being stereochemically dependent" [92]. (A fascinating history of this rule that was started in 1933 by Easson and Steadman is recounted in excellent reviews by Lämmerhofer [6] and by Bentley [93].) Indeed, if a chiral selector has three binding points and one enantiomer of a select and has three complementary groups interacting with these points that the other enantiomer will fail at least at one of the interactions in virtue of its antipodal (with respect to the former enantiomer) configuration as shown in Figure 1.10. Later, Davankov clarified that two of the three interactions can be repulsive if the third interaction is strong enough to ensure the formation of one of two possible diastereomeric adducts [94]. Moreover, an achiral matrix to which the chiral selector is attached may play the role of one of the active points [95] (Figure 1.11). Another important commentary concerns the fact that such types of binding as dipole–dipole or π–π stacking are multipoint interactions in nature, and if one of these is present in a transient complex, the "formal" number of interactions needed for enantiorecognition can be less than three [92]. Also structural rigidity involving an asymmetric center of an analyte molecule facilitates differentiation between its optical isomers [6,95,96]. As it was noticed by Davankov [95], in a rigid chiral species the arrangement of any three arbitrarily taken points on the surface depends on the configuration of that species. Therefore, the requirement for one of the interactions to be stereochemically

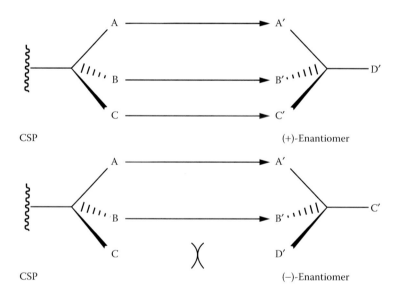

FIGURE 1.10 Illustration of the three-point interaction rule. The (+)-enantiomer (top) of the analyte is shown to be capable of three simultaneous interactions with the CSP (A-A′, B-B′, and C-C′), whereas the (−)-enantiomer (bottom) is capable of only two simultaneous interactions. If all three interactions are free-energy-lowering, the (−)-enantiomer will be less retained by the CSP. Alternatively, one interaction might be steric, in which case the enantiomer that affords the free-energy-lowering interactions with the least degree of steric interaction will be most retained. (Reprinted with permission from Pirkle, W. H. and Pochapsky, T. C., *Chemical Reviews* 89, 347, 1989.)

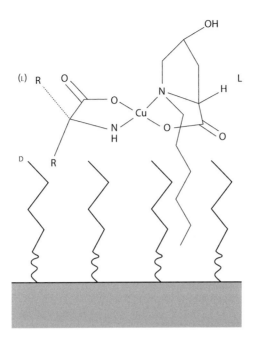

FIGURE 1.11 Chiral recognition induced by an achiral surface. The surface is a C18-type stationary phase. The chiral selector, *N*-alkyl-L-hydroxyproline (right), is immobilized on the surface due to interaction of its alkyl radical and C18 alkyl chains of the stationary phase. The aqueous mobile phase contains copper ions as complexing agent. The alkyl chain of the selector prefers a hydrophobic surrounding which is given by the C18-type stationary phase. D amino acids (left) are preferably retained because their hydrophobic moiety R also dips between the C18 alkyl chains of the stationary phase. The three-point interaction is guaranteed by two attractive interactions in the copper complex and one additional interaction with the stationary phase. (Reprinted with permission from Davankov, V. et al., *Chirality*, 2, 208, 1990.)

dependent is fulfilled for rigid species more easily than for conformationally labile structures.

Topiol and Sabio [97], applying a formal mathematical approach, criticized the three-point interaction model, or six-center interaction model as they called it, because each interaction is established between a pair of centers. They asserted that an eight-center (four-point) interaction is the minimal requirement for enantiodiscrimination. This model was conveniently illustrated by Bentley who suggested placing contact points not in a plane as in Figure 1.10 but in three dimensions as shown in Figure 1.12 [93]. It is seen that both enantiomers of molecule Cab(–CH$_2$–c) d can bind to active centers A, C, and D of a chiral selector, thus failing at chiral recognition. However, with an additional point B these two enantiomers can be differentiated. Despite strong arguments in favor of the four-point model, it has not received many citations in chiral chromatography studies. Researchers still prefer the three-point rule in explaining chiral recognition because of its simplicity and illustrative character.

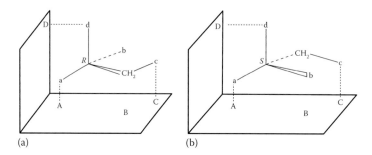

FIGURE 1.12 Bentley's binding model with a binding site D in a plane at right angles to that of sites A, B, and C. Note that the enantiomers of Cab(–CH2–c)d cannot be discriminated based on the three-point interaction a-A, c-C, and d-D. The fourth interaction between the CH_2 group and a site B is necessary to distinguish the *R*-enantiomer (a) from the *S*-enantiomer (b). (Reprinted with permission from Bentley, R., *Archives of Biochemistry and Biophysics*, 414, 1, 2003.)

The three(four)-point interaction model was developed for a chiral selector that, even if being three-dimensional itself, is located on the surface with no restrictions for an adsorbate species in the direction normal to that surface. These are not the conditions in a chiral pore or within a chiral matrix, where an adsorbate is surrounded by the walls of a pore or a cavity. It is logical to suppose that chiral recognition in such systems is steric in origin [92], meaning that the enantiomer that better fits a chiral pore is retained and its antipode should not enter the pore well and is therefore excluded. Experimental data, however, suggest that intermolecular interactions play their role in the enantioselectivity of chirally porous materials [19,96,98], so that Pirkle and Pochapsky's hypothesis that the three-point rule applies to this sort of CSPs too [92] seems sound. Of course, the number of contact points may readily exceed three or four as all peripheral atoms of an adsorbate species may be exposed to interaction with the pore/cavity walls. In order to comply with the three(four)-point rule, at least one of these interaction must be stereochemically dependent. Computer simulations of the inclusion of the enantiomers of 2-heptanol in the pores of a $Zn(C_6H_8O_8)$ homochiral MOF (HMOF 1) [99] illustrate this requirement. The pore structure of this adsorbent consists of pore channels running in the *z*-direction. Alcohol molecules lie along the pore channels. Table 1.5 summarizes energies of each structural moiety surrounding the chiral carbon of *R*- or *S*-2-heptanol in the pore (two distinct orientations of either enantiomer in the chiral channel were possible; only data for the favorable orientation are presented). The table demonstrates an enantioselective ability of the pore of 3.23 kJ/mol, mostly contributed by the interactions of the hydrogen atom and the OH group attached to the chiral carbon with the pore wall atoms. This happens because different stereoconfigurations of the enantiomers stipulate different positioning of these substituents with respect to the pore walls.

From mechanistic point of view, the formation of a strong diastereomeric complex between a chiral selector, either a surface ligand or a chiral pore, and an enantiomer depends on the fulfillment of a few conditions that were formulated by Lämmerhofer as a list of the following "fit rules" [6]:

TABLE 1.5
Average Total Energies (E_{total}), Intramolecular Energies (E_{intra}), and the Average Energies of each Structural Moiety around the Chiral Carbons of R- and S-2-Heptanol (E_{CH3}, E_H, E_{OH}, E_{C5H11}) Inside a Channel Pore of HMOF 1

Energy, kJ/mol	R-2-Heptanol	S-2-Heptanol	Difference (R–S)
E_{total}	−54.08	−50.84	−3.23
E_{intra}	76.52	76.78	−0.26
E_{CH3}	−17.95	−18.08	0.13
E_H	−17.82	−16.66	−1.16
E_{OH}	−16.97	−14.78	−2.19
E_{C5H11}	−77.85	−78.11	0.25

Source: Bao, X. et al., *Physical Chemistry Chemical Physics,* 12, 6466, 2010. With permission.

- *Steric fit* (i.e., size and shape complementarity) of the binding guest with regard to binding site of a selector
- *Electrostatic or functional fit* (i.e., a favorable geometric and spatial orientation of complementary functional groups that are amenable for electrostatic-type interactions such as ionic interaction, H-bond-mediated ionic interaction, hydrogen bonding, dipole–dipole interactions, π–π-interactions, cation–π-interactions, and anion–π interactions)
- *Hydrophobic fit* (i.e., mutual saturation of hydrophobic regions of both binding partners accompanied, in aqueous media, with destruction of their entropically unfavorable, structurally ordered water shell on the molecular surface. Such hydrophobic interactions in aqueous media are regarded as merely entropic in nature)
- *Dynamic fit and induced fit* (as to maximize binding interactions by dynamic and conformational adaptation in the course of complex formation)
- Mutual saturation of extended molecular surfaces by each other (i.e., of host and guest)

It should be kept in mind that fulfillment of these conditions does not necessarily result in enantioseparation. If both enantiomers of a pair are involved in strong binding with a chiral selector, enantioseparation may not happen. For instance, enantiomers of small molecules which fit entirely into chiral cavities of CDs are usually not separable [96]. A mandatory prerequisite for chiral recognition in accordance with the three-point rule is *ideal fit* for one enantiomer and *nonideal fit* for its antipode.

All types of intermolecular forces can be involved in selector–solute binding. Table 1.6 compares these forces in terms of bond strength and working distance. Of course, the figures given are only rough estimates that can be significantly altered by experimental conditions. A typical miscalculation consists in the underestimation of an importance of dispersive interactions. Any atom of any functional group

TABLE 1.6
Characteristics of Intermolecular Forces

Interaction Forces	Relative Strength,[a] kJ/mol	Range[b]
Electrostatic interactions		
Ionic interactions (salt bridge)		Medium ($1/r^2$)
via H-bond	40	
without H-bond	20	
Ion-dipole interactions	4–17	Short
H-bonds	4–17	Long
Van der Waals forces		
Orientation forces (permanent dipole—permanent dipole)	4–17	Short ($1/r^3$)
Induction forces (permanent dipole—induced dipole)		Very short ($1/r^6$)
Dispersion forces (induced dipole—instantaneous dipole)	2–4	Very short ($1/r^6$)
Aryl–aryl charge transfer (π–π interaction)	4–17	Medium
face-to-face		
face-to-edge		
Hydrophobic interactions	4	Very short

Source: Lämmerhofer, M., *Journal of Chromatography A,* 1217, 814, 2010; Berthod, A., *Analytical Chemistry*, 78, 2093, 2006. With permission.

[a] *In vacuo.*

[b] r = distance between the interacting species.

is capable of this sort of interaction if one can approach the interacting partner close enough. Thus, all the peripheral atoms of a solute molecule which are within the working distance of the dispersive forces of an adsorption site will be involved in dispersion attraction. Although a single atom–atom potential is low, the summation over all interacting atoms will give a noticeable contribution if their number is high. For instance, the energy contribution of a methyl group having three peripheral hydrogen atoms in adsorption on graphitized carbon is about 9 kJ/mol (in gas phase) [100], which is of order of magnitude of a hydrogen bond.

Mechanism of solute–selector binding strongly depends on the nature of the chiral selector. It is the formation of metal chelate complexes in ligand exchange chromatography, host–guest inclusion in the case of supramolecular selectors such as CDs or crown ethers, key and lock binding on protein-based CSPs, combination of donor–acceptor interaction, hydrogen bonding, dipole–dipole and π–π stacking with Pirkle selectors, and so on. There is a vast amount of literature on chiral recognition mechanisms on different types of CSPs. Some important reviews on this problem are listed in Table 1.1.

1.4.2 ENANTIOSELECTIVE SITES AND CHIRAL SELECTORS

The reader is already aware that there is distinction between chiral selector and enantioselective site. The former is a molecular or supramolecular unit including one or more asymmetric centers that form a chiral cavity capable of but not necessarily exhibiting stereospecific interaction. The enantioselective site is a small part of a CSP, either on the surface or in the bulk that interacts with stereoisomers differently in a stoichiometric ratio 1:1. By definition, q_{es}^* in Equation 1.13 is the concentration of the enantioselective sites, but not the chiral selectors. These latter may lose the enantiodiscriminating ability and become nonselective sites [8]. A chiral selector may interact selectively with one molecule via its chiral cavity and nonselectively with a few molecules via the peripheral atoms, achiral substituents, and tethering groups (for brush-type CSPs). Then this chiral selector will be a source of one enantioselective site and a few nonselective sites. The authors of [101] considering the mass balance of the adsorption sites of a Pirkle-type CSP (Whelk-O1) with regard to the enantiomers of naproxen, came to a conclusion that chiral ligands adsorb one molecule of the favorably retained enantiomer but in average total more than one molecule of its antipode. It was explained by supposing that the enantiomer that was bound strongly blocked other molecules from interacting with the selector. The other enantiomer, not able in the virtue of its spatial configuration to interact with all functional groups of the selector, left some of them intact, giving an opportunity for one more molecule to bind to this same ligand.

When two or more chiral selectors cooperate in the retention of a solute, the whole group constitutes a single site. Table 1.7 collects the data on the saturation capacities of CSPs with respect to different chiral probes. It is seen that the partial saturation capacities and the total concentration of adsorption sites are functions of an analyte. Although some of these data can be explained by AAD depending on the nature of an analyte, those data where the saturation capacity drops by a factor of five and more following the increase in a prober's size are to be explained by multisite adsorption.

On the same CSP, different chiral selectors or different parts of a chiral selector may play the role of an enantioselective site for different analytes. Matsunaga and Haginaka [45] have proved that β-blockers bind to chicken α_1-acid glycoprotein (AGP) at the primary enantioselective site located at the Trp26 residue, while ketoprofen and benzoin bind to this enzyme at the secondary less efficient enantioselective site. Multiplicity of enantioselective sites is also characteristic for human serum albumin-based CSPs [102]. Peyrin and coworkers have found two sorts of enantioselective, with respect to dansyl amino acids, sites associated with the vancomycin moiety bound to silica [103]. One site located in the aglycone pocket was inhibited by N-acetyl-D-ala. The second site was unaffected by this compound but exhibited enantioselectivity toward dansyl amino acids.

The increasing of the binding density of chiral ligands not always ensures a proportionally rising enantioselectivity. Figure 1.13a shows behaviors of apparent enantioselectivity and retention factors of both enantiomers of dichlorobenzene–leucine on a dihydroquinidine-based anion-exchanger as functions of the surface selector density. As seen, the enantioselectivity slowly decreases as selector concentration increases. According to Hetteger et al. [104], high selector loading leads to a limited

TABLE 1.7

Saturation Capacity of Several CSPs[a]

Analyte	q_{ns}^*, mmol/L	q_{es}^*, mmol/L	Q*(total), mmol/L	Reference
Chiralcel OJ				
Methyl mandelate	549	25	574	[231]
Ketoprofen	610	85	695	[152]
Chiralcel OB				
1-Indanol	586	93	679	[47]
1-Phenyl-1-propanol	724	53	777	[235]
Kromasil CHI-TBB				
2-Phenylbutiric acid	458	127	585	[236]
Ibuprofen, SFC	1392	293	1685	b
DACH-ACR				
Lorazepam	288	77	365	[237]
Temazepam	67	12	79	[237]
Chiris Chiral AX:QD1				
TFAE	296	7	303	[34]
3-Chloro-1-phenylpropanol	1366	125	1491	[85]
CHIRAL-AGP				
Alprenolol, pH = 5	9.7	3.2	12.9	[209]
1-(1-Naphthyl)ethylamine, pH = 5	50.6	1.3	51.9	[209]
Methyl mandelate, pH = 5	12.6	2.1	14.7	[209]

a   Unless otherwise mentioned, data refer to liquid chromatography.

b   Data of Reference [238] recalculated using the bi-Langmuir model.

conformational freedom of the immobilized selectors, and thus hindered interactions with analytes. A distinct pattern is shown in Figure 1.13b presenting plots of apparent enantioselectivity of three amino acid derivatives on a quinine-based CSP as functions of the selector loading on silica. The retention factors of both enantiomers in all three cases presented were linear functions of the selector loading [105]. After the selector concentration achieves a value of 80 µmol/g, α becomes constant, revealing a direct proportionality between q_{es}^* and the selector loading. This is an interesting finding meaning that in this particular CSP there is a constant ratio between the number of selector moieties retaining enantioseparation ability and the number of those losing enantioseparation ability.

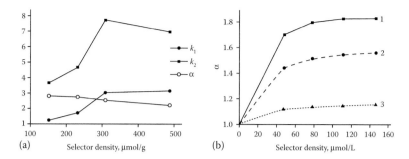

FIGURE 1.13 (a) Retention characteristics of racemic DCB-Leu resolved on a dihydro-quinidine-based CSP as functions of the concentration of the chiral selector on the silica surface. Mobile phase composition: MeOH/AcOH/NH$_4$OAc = 99/1/0.25 (v/v/w); temperature: 25°C. (Reprinted with permission from Hetteger, H. et al., *Journal of Chromatography A*, 1337, 85, 2014.) (b) Apparent enantioseparation factors α of racemic *N*-Fmoc-phenylalanine: (1) *N*-3,5-sipropoxybenzyloxycarbonyl-leucine, (2) Dichlorprop, and (3) resolved on silica grafted 6′-neopentoxy-cinchonidine-1-adamantylcarbamate as function of the concentration of the chiral selector on the silica surface. Mobile phase composition: MeOH/H2O/AcOH/ NH$_4$OAc = 580:400:20:5 (v/v/v/w); temperature 25°C. (Reprinted with permission from Levkin, P. et al., *Journal of Chromatography A*, 1269, 270, 2012.)

1.4.3 TWO-SITE MODEL

The concept of two contributions to the retention of optical isomers on CSPs has been formalized in the frameworks of the two-site model that attributes the nonselective and enantioselective parts of the total uptake of a chiral solute by the stationary phase (see Equation 1.26) to the nonselective and enantioselective adsorption sites, respectively [8,33]. Assuming independent behavior of these two groups of sites and Langmuirian adsorption on either group of sites, Guiochon et al. [27,83] derived the bi-Langmuir adsorption isotherm model that reads

$$q = q_{ns}^* \frac{b_{ns}c}{1 + b_{ns}c} + q_{es}^* \frac{b_{es,i}c}{1 + b_{es,i}c} \qquad (1.29)$$

where index *i* refers to a particular enantiomer and the adsorption constants b_{ns} and b_{es} have the same meaning as in Equations 1.10 and 1.11 above. For a long time, Equation 1.29 was a "work horse" in enantioselective adsorption research, which satisfactorily described a wide range of chiral adsorption systems. Gradually, examples were accumulated of this model failing, revealing its shortcomings [8]. Of these, the most important is the neglecting of a heterogeneous structure of the ensembles of the adsorption sites. In a number of works, multisite models [8] were used to account for the diversity of adsorption sites [36,38,106–108]. Another expressions for local isotherms taking into account the factor of heterogeneity, such as the Freundlich and Langmuir–Freundlich [109], Toth [47,110], or Jovanovich–Freundlich isotherm models [111], were proved to be useful.

In a general form, a two-site model isotherm equation is given by

$$q = q_{ns}^{*} T_{ns}\left(\mathbf{B}_{ns},c\right) + q_{es}^{*} T_{es,i}\left(\mathbf{B}_{es,i},c\right) \tag{1.30}$$

In this equation, term $q_{ns(es)}^{*} T_{ns(es)}\left(\mathbf{B}_{ns(es)},c\right)$ stands for a local isotherm. The factor $T(\mathbf{B}, c)$ symbolizes the concentration-dependent part of a local isotherm. The argument $\mathbf{B}$ of this function is the vector of isotherm parameters. Equation 1.30 presumes that the nonselective T_{ns} terms are the same for enantiomers, whereas the enantioselective terms are not. In general, local isotherm functions for the enantioselective sites can be different for optical antipodes ($T_{es,1} \neq T_{es,2}$), but a situation when only the parameters of the local isotherm differ ($T_{es,1} = T_{es,2} \wedge \mathbf{B}_{es,1} \neq \mathbf{B}_{es,2}$) is more frequent. In the case of a multichiral analyte, the nonselective terms must be the same for all stereoisomers. The enantioselective sites may in principle differentiate between some stereoisomeric forms of an analyte but not between the others. Experimental data do not seem to support this theoretical opportunity. As a rule, chiral columns, if display enantioselectivity with respect to a multichiral compound, separate or show tendency to separate all its stereoisomers [112–117].

Equation 1.30 assumes independent behavior of enantioselective and nonselective sites. But it is not impossible that two neighboring sites belonging to these different types will together retain a solute molecule, especially if this molecule is large enough. The experimental data by Brüger and Arm [118] suggested that surface silanol groups might interfere with the interactions between the analyte and the chiral selector. Cann and coworkers have proved through molecular simulations that the surface end-capped and free silanol groups can interact with the bound chiral selectors, affecting their conformation [88]. These authors have not observed a bridging interaction involving a chiral selector and a surface silanol. They, however, described bridging interactions between neighboring chiral selectors [87]. Since we know that chiral selectors can play the role of nonselective sites, one can imagine the formation of a selector–solute–selector complex in which one of the selectors or the part of one involved in the bridging interaction represents a group of nonselective sites. To the best of our knowledge, this phenomenon (cross-site interaction) has never been taken into account in macroscopic adsorption mechanisms on CSPs. Obviously, necessary changes to Equation 1.30 would give a complex and hardly operable mathematical expression.

Another assumption underlying Equation 1.30 requires constancy of the saturation capacities q_{ns}^{*} and q_{es}^{*}. This is not true in general. Kasat et al. [119] have shown that the adsorption of norephedrine on polysaccharides induces structural changes in the materials. This may result in a dependency of the saturation capacity on solute concentration and will definitely result in changes in the quality of the adsorption sites. The above discussed bridging interaction will lead to a decrease in the concentration of the adsorption sites. Research on the induction of enantioselectivity in achiral zeolites in the course of adsorption of racemic chiral solutes [120,121] delivers one more possible mechanism compromising the invariance of q^{*}, apt to microporous materials. In this case, a chiral molecule trapped in an achiral micropore converts it into an enantioselective site because the probability of entrapping a second molecule depends on the energies of $R–R$, $R–S$, and $S–S$ dimers, which are all different in the confined space of the pore. This mechanism also applies to chirally porous solids, in which case the adsorption of a chiral solute may convert nonselective sites to enantioselective ones and vice versa depending on probabilities of homo- and heterodimerization.

1.4.4   Adsorption-Partitioning Two-Site Model

If an adsorbed layer of a mobile phase, whose thickness is larger than one molecular diameter, is formed on the surface of a CSP, the solute amount accumulated in the part of this adsorbed phase located above the first monolayer covering the surface will constitute a part of the nonselective uptake. This contribution is equal to $(v_a–v_1)c_a$, where v_a and v_1 are the total volume of the adsorbed phase and the volume of the first monolayer, respectively, and c_a the solute concentration in the adsorbed phase beyond the first monolayer in equilibrium with the bulk concentration c. The amount retained by the surface nonselective sites is determined by a common adsorption isotherm and is proportional to the saturation capacity of the nonselective sites q_{ns}^*. The solute molecules in the "upper" layer of the adsorbed phase are not in direct contact with the surface. Nevertheless their standard chemical potential is not equal to that of the bulk phase solute because the composition of the solvent is different in the adsorbed layer and in the bulk phase. By analogy with liquid/liquid extraction, one can assume a direct proportionality between c_a and c, with the partition coefficient E. Then the final isotherm equation for the adsorption–partition model reads

$$q = (v_a - v_1)Ec + q_{ns}^* T_{ns}(\mathbf{B}_{ns}, c) + q_{es}^* T_{es,i}(\mathbf{B}_{es,i}, c) \qquad (1.31)$$

This "partition" contribution is expected to be low, an order of magnitude of a discrepancy between the total uptake q and the excess adsorption, a characteristic that is actually measured in a chromatographic experiment [91,122]. Only if E is very high that is possible when the solute solubility strongly depends on solvent composition, this part of the nonselective adsorption will be noticeable.

1.4.5   Lateral Interactions

When the surface concentration of solute increases, the probability of solute molecules to encounter increases too. At saturation, the surface is uniformly covered by adsorbed solute and probability of an adsorbed solute molecule to neighbor within a distance of interaction another adsorbed solute molecule(s) is high. It should be understood, however, that saturation does not necessary mean the formation of a closest surface package of solute. Rather, it means the occupation of all adsorption sites. If these are located distantly from each other, there may be no contact between solute molecules occupying neighboring sites. Even in this case adsorbed molecules may still interact with each other due to surface mobility, that is, collide during their migration on the surface. (As physical adsorption is a reversible process, an adsorbate molecule does not occupy an adsorption site for infinitely long time.

Eventually, it will desorb and either go to the bulk phase or travel over the surface till it is retained by another adsorption site.) Micro- and small mesopores provide one more opportunity for solute–solute interactions in the adsorbed phase. Indeed, a solute molecule entering a small pore whose walls are already covered by solute molecules may find itself in interaction with these adsorbed species. An attractive interaction would stabilize this molecule inside the pore even without a direct contact with the pore walls. The concept of volume filling of micropores well known in gas adsorption [123] did not find a wide use in liquid/solid processes in spite of developed mathematical formalism [124–126]. At the same time, the volume filling is likely to occur in periodically ordered mesoporous materials. Actually, it cannot be excluded in any micro- and mesoporous adsorbents provided the pore/solute size ratio is less than 5.

Gradual increase of the bulk solute concentration beyond the saturation level can result, given strong cooperative adsorbate–adsorbate interaction, in appearance of an inflection point on an adsorption isotherm. Several authors did observe an S-shaped isotherm in chiral systems [127,128]. In the region of intermediate concentrations before the inflection point, lateral interactions can still manifest themselves in deviation of an adsorption isotherm from the Langmuir behavior [48,101]. The general Equation 1.30 can take lateral interactions into account through a proper choice of the functions $T(\mathbf{B},c)$. Different versions of rational functions originated in statistical thermodynamics [127–130] or obtained based on other principles [131] can be used. The Moreau isotherm [130] is particularly useful in describing adsorbate–adsorbate interactions in chiral chromatography [48,101]. This is a simple equation with three adjustable parameters, the parameter I being responsible for intermolecular interactions:

$$q = q^* \frac{b_i c + I\left(b_i c\right)^2}{1 + 2b_i c + I\left(b_i c\right)^2} \tag{1.32}$$

The implementation of the two-site model requires using a Moreau term for each group of the adsorption sites that would result in the bi-Moreau isotherm equation with two sets of the parameters b and I for the enantioselective and nonselective sites, respectively [48]. Implicitly, it would mean that a molecule adsorbed on an enantioselective site interacts laterally with those occupying only the enantioselective sites, the same principle holding for the nonselective sites. Even if a physically reasonable model accounting for the lack of cross-site interaction can be developed [101], this is an unlikely situation. On the other hand, a model that would take into account interaction between different terms of an adsorption isotherm would result in a multiparameter complex expression that hardly could be used in practice. Szabelski and Talbot did not neglect the cross-site influences in their computer simulation of lateral interactions on chiral surfaces, but they used another simplification to avoid excessive mathematical difficulties, assuming that the strength of interactions does not depend on the type of adsorption site (es or ns) [132]. Their results suggest an important effect of lateral interactions on enantioselectivity and adsorption kinetics.

1.4.6 SECONDARY EQUILIBRIA

In liquid chromatography, many processes besides the interaction of a solute with the stationary phase take place in a column, caused by the presence in the system of the liquid phase. Those include the solvation of a solute and the stationary phase, ionization of an ionogenic solute and ionogenic chiral selectors, dissociation of mobile phase electrolytes, solute–solute and solute–solvent association, and so on. All equilibrium processes that are not the solute adsorption proper are called secondary equilibria. These may influence essentially enantiomer retention and separation.

1.4.6.1 Solvation

The solute is solvated by the liquid phase and in order for an adsorption event to occur this solvate shell must be destroyed. The formed transient adsorption complex is resolvated, yet the composition of the newly created solvation shell does not necessarily match that of a solute species in the mobile phase. The heat of solvation ranges from 90 to 130 kJ/mol when strong polar and ionic interactions are involved. In nonpolar solvents, it is less than 90 kJ/mol. The Gibbs free energy of solvation is usually essentially lower than the enthalpy due to the entropy term. Yet, this is the energy balance of desolvation/resolvation processes rather than only desolvation that contributes to the overall adsorption thermodynamics. This balance depends on the degree of the "pulling" of an adsorbate species inside of a chiral cavity. If the species is included completely, the most part of the energy needed to break down the solute's solvation shell will be consumed without tantamount compensation from resolvation. In the case of noninclusive adsorption, the most part of the adsorbate's external surface will be exposed to the solvent, and the overall enthalpy balance is expected to be low to moderate.

Thermodynamic study of the solvation effect in liquid/solid adsorption is complicated by the lack of information about the contribution of the resolution process, which is hardly to be determined. Nevertheless, some tentative estimates have been made [64,65] to show that the enthalpy and free energy of solvation (modulo) exceeds or comparable to those of net adsorption. Considering that other contributions in Equation 1.21, from the desorption of eluent and its dilution, are usually relatively small, this is the energy balance of desolvation, resolvation, and net adsorption that controls adsorption thermodynamics, other processes only altering it to a degree. Chervenak and Toon, who studied protein–substrate binding, contended that solvation reorganization accounted for 25%–100% of the measured enthalpy of binding [133].

For comprehending the effect of solvation on enantioseparation, let us rewrite Equation 1.21 for the free energy, grouping the terms by their importance to enantioselectivity:

$$\Delta G^0 = \underbrace{\Delta G^0_{desorp} + \Delta G^0_{netads} + \Delta G^0_{resolv}}_{\text{Strong influence}} + \underbrace{\Delta G^0_{dil}}_{\text{Weak influence}} + \underbrace{\Delta G^0_{desolv}}_{\text{No influence}} \quad (1.33)$$

The first group of the terms determines enantioselectivity as each term depends on the enantiomer. Even the desorption term will be different for optical antipodes if those displace different number of solvent molecules. For this reason, the dilution contribution is also enantiomer dependent, although its importance is much lower since the absolute value of this term is low. The only component that is indifferent with regard to enantiomer configuration is the Gibbs energy of desolvation as this process occurs in the liquid phase, therefore nonchiral. For the $\Delta\Delta G^0$ value one can write

$$\Delta\Delta G^0 = \Delta\Delta G^0_{\text{desorp}} + \Delta\Delta G^0_{\text{netads}} + \Delta\Delta G^0_{\text{resolv}} + \left(\Delta\Delta G^0_{\text{dil}} \approx 0\right) \tag{1.34}$$

Of the two solvation processes, only the resolution equilibrium affects enantioselectivity. This influence depends on a difference between the external surfaces of adsorbed enantiomer species, both in terms of size and quality. When one enantiomer is deeply inserted into the chiral cavity while its antipode is not, one may expect a relatively high value of the $\Delta\Delta G^0_{\text{resolv}}$ term. If both isomers after having been adsorbed expose similar parts of their molecules to solvent, this term will be negligible. It is worth to note that occasionally reported cases of the reversal of elution order caused by variation in the mobile phase composition [134–136] can be attributed to the influence of solvation only if $\Delta\Delta G^0_{\text{resolv}}$ becomes positive as a result of this variation and exceeding the absolute value of the sum $\left(\Delta\Delta G^0_{\text{desorp}} + \Delta\Delta G^0_{\text{netads}}\right)$. Otherwise, the reversal is either due to the influence of the mobile phase composition on the structure of adsorption sites, that is, on the $\Delta\Delta G^0_{\text{netads}}$ summand, or because of the competitive adsorption between the solute and mobile phase modifiers, that is, due to the $\Delta\Delta G^0_{\text{desorp}}$ summand.

The application of the stoichiometric approach to the study of the solvation effects is also complicated because the composition of the solvation shell is rarely known. Besides, it changes upon the transfer of solute molecule from the liquid to the adsorption phase. A few semi-empirical retention models taking into account solvation have been proposed, relating the retention factor and the mobile phase composition [6,137]. Common shortcomings of these models consist in neglecting the heterogeneous structure of the CSP and applying other gross approximations obscuring the physical meaning of the model parameters. Therefore, the analysis of solvation effects is limited to a qualitative consideration both in linear [71,138,139] and in nonlinear [64,108,140] chromatography.

Recent achievements in computer simulation allow researchers to investigate solute–solvent interactions at the molecular level. Zhao and Cann have proved by means of MD calculations that 2-propanol in the hexane-based solvent affects the separation of the enantiomers of styrene oxide on Whelk-O1 in several ways, one of which is the stabilization of the solute in the bulk phase through hydrogen bonding [87]. The alcohol may enter the chiral cleft thus reducing the solute docking probability. The alcohol may also compete with the solute for the interaction with the chiral selector. Finally, the solvent may stabilize or destabilize a docked analyte Whelk-O1 complex. All these options could of course be predicted without the MD technique. What is important here is that the simulation predicts frequencies of these events and through this relative contributions of each mechanism to retention.

1.4.6.2 Electrolyte Dissociation

Ionogenic compounds dissociate in solution. As a result, two types of species are present in the liquid phase, neutral molecules and ions, each able to interact with the stationary phase. An excellent account of dissociation equilibrium in chromatography on uncharged adsorbents was given by Kazakevich [141]. The first essential theory in this area was that by Horvath, Melander, and Molnar (HMM) who developed an equation relating retention factor of an electrolyte solute to the proton concentration. For monoprotic acid analyte HA, the HMM equation reads [90]

$$k' = \frac{k'_{HA} + k'_A \dfrac{K_a}{[H^+]}}{1 + \dfrac{K_a}{[H^+]}} \tag{1.35}$$

where:

k'_{HA} and k'_A are hypothetical retention factors for the neutral and anionic forms of the analyte

K_a the respective dissociation constant

Similar equation can be developed for multiprotic acids, basic analytes, zwitterions, and so on [90]. Figure 1.14 demonstrates typical HMM k'-pH dependences for anionic and cationic solutes. Their characteristic S-shape can be distorted if other than adsorption/partition retention mechanisms interfere, such as ion-exchange [142] or ion-pairing [141] (see Section 1.5.3).

At high solute concentrations, consideration in terms of retention factor is not more adequate. An adsorption isotherm should be used instead. The general Equation 1.30 of the two-site model can be easily transformed to deal with the dissociation of an analyte. The concentrations of the ionized (I) and neutral (N) forms of an 1:1 electrolyte, taken as a simple example, relate to the total analyte concentration c as $c_I = \beta c$ and $c_N = c[1-\beta]$, where β is the dissociation degree. A local adsorption isotherm for the sites of type s ($s = $ ns, es) is then given by

$$q_s = q^*_{s,I} T_{s,I} \left(B_{s,I}, c_I(c) \right) + q^*_{s,N} T_{s,N} \left(B_{s,N}, c_N(c) \right) \tag{1.36}$$

Note that the saturation capacity q^*_s is divided into two subgroups: the sites that adsorb ionized species ($q^*_{s,I}$) and the sites that adsorb neutral molecules ($q^*_{s,N}$). It is possible that a group of adsorption sites will be active with regard only to a certain sort of analyte species, either ionized or neutral. Then Equation 1.36 will be simplified by eliminating a respective term. One cannot exclude a possibility of the competition between neutral and ionic species for binding to an adsorption site. This problem was considered by Gritti and Guiochon in [144,145].

The dissociation degree is a function of the mobile phase pH. The latter, in its turn, depends on the solute concentration in the mass transfer zone if solute is an acid or a base [146,147]. This dependence is determined by the mobile phase composition, especially by its buffer capacity. Figure 1.15a shows $\beta(c)$ curves for naproxen

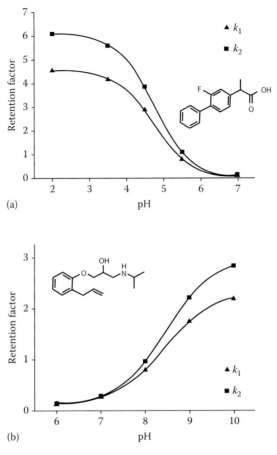

FIGURE 1.14   Effect of pH on the retention factors of enantiomers of an acidic (a) and basic (b) analytes. The formulas of analytes are shown in the graphs. (Reprinted with permission from Tachibana, K. and Ohnishi, A., *Journal of Chromatography A*, 906, 127, 2001.)

in three mobile phases composed of an acetic buffer in an acetonitrile/water solvent. β is almost constant in the solution of 0.01 M acetic acid, it decreases as a function of naproxen concentration in buffer solutions, the steeper, the lower the buffer concentration. Figure 1.15b compares pH profiles and chromatograms of naproxen on a Whelk-O1 column measured with the mentioned mobile phases. As seen, the elution of an acid does disturb the pH profile, the perturbation being stronger for the eluent with lower buffer capacity.

1.4.7   COMPETITIVE ADSORPTION

Enantioseparation supposes that a mixture of enantiomers or, in the case of multichiral compounds, a mixture of stereoisomers contact with a CSP. These can compete with each other for the interaction with adsorption sites. In the simple case of the

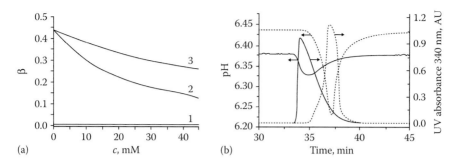

FIGURE 1.15   (a) Dissociation degree of naproxen (β) as a function of its concentration in mobile phase A1 (1), AN1 (2), and AN3 (3). (b) pH profiles (left axis) and detector signal profiles (right axis) of *S*-naproxen recorded with mobile phases AN1 (dashed lines) and AN3 (solid lines). Sample volume 100 μL; sample concentration 43.4 mM. A1: 0.01 M CH$_3$COOH in MeOH-H$_2$O (80:20, v/v) (1); AN1: 0.01 M CH$_3$COOH + 0.01 M CH$_3$COONa in MeOH-H$_2$O (80:20, v/v); AN3: 0.03 M CH$_3$COOH + 0.03 M CH$_3$COONa in MeOH-H$_2$O (80:20, v/v). (Reprinted with permission from Asnin, L. et al., *Journal of Chromatography A*, 1217, 7055, 2010.)

bi-Langmuir model, the competitive adsorption isotherm for a pair of enantiomers designated by indices 1 and 2 is a set of two partial isotherm equations:

$$q_1 = q_{ns}^* \frac{b_{ns}c_1}{1 + b_{ns}(c_1 + c_2)} + q_{es}^* \frac{b_{es,1}c_1}{1 + b_{es,1}c_1 + b_{es,2}c_2} \tag{1.37}$$

$$q_2 = q_{ns}^* \frac{b_{ns}c_2}{1 + b_{ns}(c_1 + c_2)} + q_{es}^* \frac{b_{es,2}c_2}{1 + b_{es,1}c_1 + b_{es,2}c_2} \tag{1.38}$$

Total adsorbed amount $= q_1 + q_2$

Note that the saturation capacities q_{ns}^* and q_{es}^* are the same for both enantiomers. This is a mandatory requirement both of thermodynamics [148] and of material balance. If a CSP contains enantioselective sites that retain one enantiomer but do not retain another one, the adsorption constant (not the saturation capacity) for the unretained enantiomer will be equal to zero and the enantioselective term in the respective partial adsorption isotherm will be nullified. This is, however, a hypothetical situation. More frequently one observes $b_{es,2} > b_{es,1} > 0$.

More elaborated competitive models taking into account different aspects of surface heterogeneity [131,149] and lateral interactions [150–152] have been developed. The real adsorption solution model [153,154] dealing with nonideal behaviors in the liquid and the stationary phases has been applied once in chiral chromatography [155]. The reader is advised to refer to the reviews [156] and [8] for more information.

As a rule, the adsorption of enantiomers (and stereoisomers [157]) is competitive. There are, however, a few exceptions from this rule. According to Mihlbachler et al. [128], the adsorption of TB on the Chiralpak AD CSP is cooperative. MD simulations have shown that the presence of the (−)-TB promotes a more efficient interaction between the (+)-TB and the stationary phase [86]. Kaczmarski et al. [131] have reported a small but statistically significant cooperative effect in the adsorption of the enantiomers of 1-indanol on a cellulose tribenzoate CSP.

Besides analytes, mobile phase additives and the solvent itself compete for the interaction with adsorption sites. Usually, the influence of the eluent is neglected. This is justified when (1) the adsorption coefficients of mobile phase components are much smaller than those for analytes and (2) the concentration of mobile phase components is much higher than that of the analytes. These conditions are almost always met for the solvent. It is not so as to mobile phase additives, which frequently are strong adsorbates and whose concentration may be quite comparable with that of target solutes, especially in preparative chromatography. Even under such conditions, one can avoid explicit inclusion of an additive into the total adsorption model if this compound is eluted far from the enantiomers to be separated. Only if an additive is eluted close to or along with the peaks of the target analytes and if its concentration is comparable with that of the target analytes, its effect on the elution profile can be dramatic [84,158,159]. This phenomenon in chiral chromatography was considered by Arnell et al. [160,161]. From mathematical point of view, it requires extension of a competitive adsorption model for one more component, the additive.

1.4.8 BRIDGING MICROSCOPIC AND MACROSCOPIC APPROACHES

The microscopic approach reveals fine details of the interaction of analytes with stationary phase on the molecular level and allows prediction of enantioselectivity of CSPs through comparison of the interaction energies of a chiral selector with each enantiomer. The state-of-the-art method consists in even a more realistic representation of the stationary phase surface as an ensemble of chiral selectors and achiral ligands, end-caped and free silanols (Figure 1.16) [29,88]. Moreover, the fact of adsorption occurring inside a pore channel can also be taken into account [5,29]. Following this strategy, the apparent enantioselectivity is evaluated, a characteristic that includes the contribution of nonchiral interactions (see Equation 1.16). This

FIGURE 1.16 A side view of the simulation cell showing two interfaces with Whelk-O1 selectors, end-caps, and silanol groups. The solvent consists of *n*-hexane (thin gray lines), 2-propanol (green), *R*-styrene oxide (yellow), and *S*-styrene oxide (purple). (Reprinted with permission from Zhao, C. F. and Cann, N. M., *Analytical Chemistry,* 80, 2426, 2008.)

technique is a clear progress compared to an old approach of considering only chiral selectors that delivers an estimate of the true enantioselectivity (in sense of Equation 1.17). The "molecular simulation" apparent enantioselectivity can be directly compared with an experimental α value. A convincing coincidence was achieved in some cases [29], revealing the potential of the method. Nevertheless, computer simulation still cannot predict the retention times of analytes. This is because the estimation of retention time (factor) through molecular simulation requires not only knowledge of the chemical structure of the solid surface but also of the pore structure of an adsorbent particle. This latter requirement is an insuperable obstacle at present. The best information that can be obtained today about the pore structure is the pore size distribution and an average characteristic of pore curvature called tortuosity coefficient. No reliable information concerning pore length and shape or a pore network type is available. Moreover, the distribution of the chiral selector within a particle is not known, and this distribution is likely to be irregular, with uneven accessibility of selectors in different parts of the particle, for example, in narrow and wide pores. All these difficulties make the prediction of chromatographic retention on the basis of molecular simulation an issue of a distant future with the only exception of MOFs and similar materials for which the pore structure is well defined and pore parameters are known and are constant over the entire particle.

1.5 EFFECT OF EXPERIMENTAL VARIABLES ON RETENTION AND RESOLUTION OF ENANTIOMERS

1.5.1 TEMPERATURE

The influence of temperature can be exerted on two levels, thermodynamic and structural. The effect of temperature on retention thermodynamics was comprehensively described in Section 1.3. It obeys the Le Chatelier rule: increase of temperature shifts exothermic adsorption equilibrium toward the bulk phase, resulting in decreasing retention, and shifts endothermic adsorption toward the adsorbed complex, promoting retention. Deviations from this rule find their explanation in structural changes in the stationary phase [28,162,163]. In SFC, another (thermodynamic) factor can play a role in atypical behavior of the retention factor when temperature decreases below the critical point of the mobile phase. For a relevant discussion see [164].

Unless the van't Hoff plots for separable optical antipodes are parallel, which is a rare phenomenon, variation of temperature will eventually result in the reversal of elution order. Frequently, the point of intersection of the van't Hoff plots is lower [71,133] or higher [112,136,164–166] than a conventional temperature interval of 20°C–50°C. Sometimes, the isoelution point is found within this interval [59,70]. In this case, temperature is a convenient parameter to control the elution order. An interesting example of temperature influence on the elution order of stereoisomers was delivered by Schlauch and Frahm [60]. They studied the resolution of four-stereoisomer mixture of 1-amino-2-hydroxycyclohexanecarboxylic acids by ligand-exchange chromatography. As seen in Figure 1.17, increasing temperature results in the reversal of elution order of the enantiomers of the *cis*-acid, but it does not affect relative positions of the *trans*-acid stereoisomers.

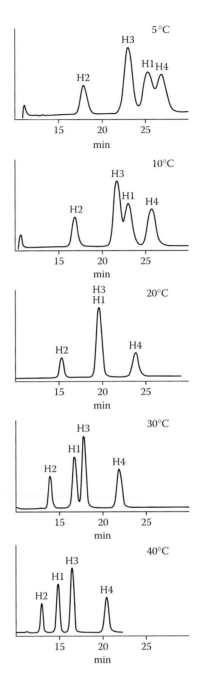

FIGURE 1.17 Chromatograms of 1-amino-2-hydroxycyclohexanecarboxylic acids on the copper(II)-D-penicillamine column with 2 mM $CuSO_4$ in acetonitrile-water (7.5:92.5, v/v) mobile phase at different temperatures. (1) *cis*-(1R,2S), (2) *trans*-(1R,2R), (3) *cis*-(1S,2R), and (4) *trans*-(1S,2S). (Reprinted with permission from Schlauch, M. and Frahm, A. W., *Analytical Chemistry*, 73, 262, 2001.)

TABLE 1.8

Isoelution Temperature of Cyclic β-Aminocarboxylic Acids on Chiralpak ZWIX(+) Column

Analyte[a]	Resolved Enantiomers	Mobile Phase[b]	T_{iso}, °C
1	1*R*2*R*3*S*5*R*/1*S*2*S*3*R*5*S*	A	105
	Same	B	395
4	1*R*2*R*3*S*5*R*/1*S*2*S*3*R*5*S*	A	−10
	Same	B	15

Source: Ilisz, I. et al., *Journal of Chromatography A*, 1334, 44, 2014. With permission.

[a] Formulas of the analytes are shown in Figure 1.6.

[b] Mobile phases: (A) MeOH/MeCN (50/50v/v) containing 25 mM triethylamine and 50 mM HCOOH and (B) MeOH/MeCN (50/50v/v) containing 25 mM triethylamine and 50 mM AcOH.

No correlation between the nature of the CSP and of the analyte and the isoelution temperature has been found. Moreover, minor, seeming unessential, modifications in the structure of an analyte may result in a drastic change in T_{iso}. This is illustrated by data in Table 1.8 comparing values of the isoelution point for enantiomer pairs of two cyclic β-amino acids (see compounds No. 1 and 4 in Figure 1.6) differing only by a methyl group at C2. The isoelution temperature has been shown to be affected by the mobile phase composition [59,70,71,78,166]. Indeed, when the acidic modifier of the mobile phase was changed from acetic (mobile phase B in the table) to formic acid (mobile phase A), the value of T_{iso} dropped from 395°C to 105°C. Balmér et al. [70] resolved the racemate of a metoprolol-related amino alcohol on a Chiralcel OD column in the normal-phase mode, and they observed an essential increase in the isoelution point following the addition of water to the mobile phase. The authors explained this finding by the ability of water to affect the affinity of one (of two existing) type of the binding sites.

Temperature may influence the structure of the stationary phase and through this retention and enantioselectivity. An indication that this happened is the strong distortion or bending of the van't Hoff plots. O'Brien et al. [28] proved that tris(4-methylbenzoate) cellulose coated on silica, the CSP Chiralcel OJ, underwent a structural transformation at 18°C in contact with hexane-alcohol solvents. A corresponding break of the van't Hoff plots for a pair of enantiomers was observed as shown in Figure 1.18a. This phenomenon is especially well pronounced in a ln α versus 1/T graph (Figure 1.18b). The graph encompasses two regions of this plot, one corresponding to an entropy-driven separation at $T < 18$°C and another corresponding to an enthalpy-driven separation above this point. Hence, the authors conclude that at low temperatures the enantioselectivity is explained by entropy demanding inclusion between cellulose chains. Beyond 18°C, the enantioseparation occurred mainly due to selective hydrogen-bonding interactions in wide open, that is, excluding the inclusion mechanism, chiral cavities [28]. Later, O'Brien and coworkers [163] reported about an irreversible temperature-induced structural transformation of the amylose-based

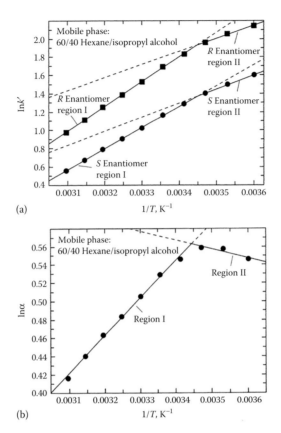

FIGURE 1.18 van't Hoff plots for k' (a) and α (b) of the enantiomers of a chiral diol (see [28] for structure) with hexane/2-propanol (60:40, v/v). (Reprinted with permission from O'Brien, T. et al., *Analytical Chemistry*, 69, 1999, 1997.)

CSP Chiralpak AD, influencing its chiral recognition properties (Figure 1.19). This effect was modulated by the polar component of the mobile phase. Interestingly, the analogous cellulose-based CSP Chiralcel OD did not demonstrate such an atypical behavior; the van't Hoff plots of α obtained on this CSP were independent from the direction of temperature change.

Pirkle [162] has observed an unusual wave-type dependence of $\ln k'$ on $1/T$ for optical isomers of a spirolactam on a phenylglycine-based CSP with 2.5% 2-propanol in hexane as the mobile phase. This manifests a combination of oppositely directing factors, some of them being dominating at a certain temperature range. Pirkle himself emphasized the importance of the solvation and desolvation of the binding sites by 2-propanol. As temperature increases, the alcohol molecules solvating the binding sites are desorbed, thus rendering their displacement by the solute unnecessary. The spared "displacement" enthalpy contributes to an upturn of k' with temperature yet to a certain point at which the factors determining a regular trend of the van't Hoff plot become dominating.

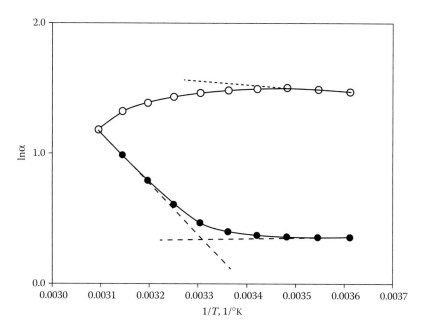

FIGURE 1.19 van't Hoff plots of α for dihydropyrimidinone acid on a Chiralpak AD column. Mobile phase: 15% EtOH in Hexane with 0.1% TFA. Procedures: heating (filled circle); cooling (hollow circle). (Reprinted with permission from Wang, F. et al., *Journal of Chromatography A*, 958, 69, 2002.)

Measurements of adsorption isotherms revealed somewhat contradictory results regarding the effect of temperature on the concentration of the binding sites. Study of a Lindner AX:QD1-type CSP conducted in the laboratory of G. Guiochon showed that the saturation capacity of the enantioselective and nonselective sites increases with temperature if 2,2,2-trifluoro-1-(9-anthryl)ethanol is used as a prober [34]. When the saturation capacity was measured with respect to 3-chloro-1-phenylpropanol, q_{es}^* decreased when temperature grew while q_{ns}^* remained constant, around ~1350 mmol/L, throughout almost the entire temperature interval studied (15°C–40°C) with the only exception at 15°C where this value was ~1700 mmol/L [85]. The binding capacity of the enantioselective sites of the antibiotic-based adsorbents Chirobiotic T and TAG declined with temperature [66]. In another study with Chirobiotic TAG [167], the concentration of the enantioselective sites fluctuated as a function of temperature around an average value within ±10%; the concentration of the nonselective sites increased in correlation with temperature.

Coherent results were obtained in two independent studies with protein-based CSPs. Götmar et al. [63], who investigated silica-bound amyloglucosidase and used alprenolol as a prober, and Jacobson et al. [83], who experimented with a Resolvosil-BSA column and *N*-benzoyl-alanine as a prober, both observed the concentration of the selective sites decreasing and that of the nonselective sites growing with temperature. The reproducibility of the results suggests exposing an intrinsic property of protein CSPs. One can suppose that the fraction of protein molecules whose chiral pocket loses its chiral recognition ability due to the temperature-induced distortion of the configuration increases as temperature grows.

1.5.2 Mobile Phase Composition

Mobile phase in nongaseous methods, not only in liquid but also in SFC, influences chromatographic process in many ways. It solvates the analyte(s) in the bulk phase and on the surface, solvates the binding sites, and modulates the affinity of the binding sites by affecting their conformation and through adsorption inside of the chiral cavities. Usually, mobile phases are two, three, or even four component mixtures, each component playing its particular role. One component, with the lowest elution power, can be conventionally called diluent. This is an alkane or another low polar solvent in NPRL, water in RP–LC, and pressurized CO_2 in SFC. Of course, the diluent is not obligatorily an inert solvent. It participates in the solvation of analytes, can interact with the binding sites, but it alone cannot elute the analytes out of a column. In order to increase the elution power of the mobile phase, the diluent is mixed with a stronger eluting solvent, the modifier, frequently alcohol, ester, or chloroorganics in NPLC; methanol or acetonitrile in RP–LC; and methanol in SFC. The percentage of the modifier controls the retention of analytes and, to a lesser degree, enantioselectivity. The latter property depends more on the nature of the modifier than on its concentration [70,71,136,168]. Sometimes, however, the effect of the modifier percentage can be as strong as resulting in the reversal of elution order of enantiomers [135,169,170].

The modifier influences retention through (1) regulating the solvation of a solute in an eluent, (2) displacing solute molecules from the binding sites by mechanism of competitive adsorption, and (3) altering the properties of the stationary phase, the last two mechanisms also influencing chiral recognition. The addition of a modifier may improve or, on the contrary, reduce solvation of a solute. The sign of the effect depends on the nature of a solute and the solvent constituents. For instance, the addition of an alcohol to water will improve the solvation of lipophilic compounds but impair the solvation of hydrophilic solutes. More complicated situations are also possible. Figure 1.20 represents the solubility, a parameter correlating with solvation, of D-, L-, and racemic tryptophan in water/MeOH mixtures [171]. It is seen that the solubility curve has two regions of diminishing solubility connected by a segment of increasing solubility. This figure demonstrates one more important phenomenon: difference in the solvation of racemate and pure enantiomers. If a true racemate is to be resolved, the products of its dissociation, that is, pure enantiomers, will have different solvation status in the bulk phase comparing to the dimer.

The desorption ability (mechanism (2)) of the modifier depends on its affinity toward binding sites and also on its concentration. To illustrate this, consider a simple competitive Langmuir isotherm for an analyte (a) in the presence of a modifier (m):

$$q_a = \frac{q^* b_a c_a}{1 + b_a c_c + b_m c_m} \tag{1.39}$$

As it is seen, the effect of the modifier is determined by the product $b_m c_m$ rather than just by b_m. Therefore, the growth of modifier concentration will always result in

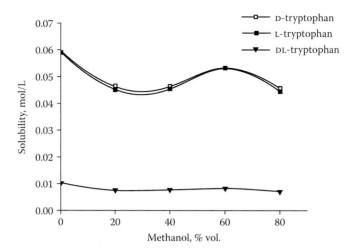

FIGURE 1.20   Solubility of tryptophan in water–methanol at 20°C. (Reprinted with permission from Jandera, P. et al., *Journal of Chromatography A*, 917, 123, 2001.)

increasing the displacement effect regardless of the nature of compounds involved. Despite abundant records of this phenomenon in the chromatographic literature, there are few studies considering this effect in terms of competitive adsorption isotherm. The adsorption of racemic temazepam on a hybrid polymer CSP from a CH_2Cl_2/MeOH mobile phase was investigated to confirm that the theory of competitive adsorption as it is described in Section 1.4.7 does account for and predicts the displacement effect in a proper way [50]. Interestingly, the adsorption of methanol, the modifier, obeyed to the competitive Langmuir model as in Equation 1.39, whereas that of temazepam followed the bi-Langmuir model as in Equations 1.37 and 1.38. This is an important finding as it indicates that the modifier does not distinguish between the nonselective and enantioselective sites, probably because it interacts with small moieties in these sites, which are similar in both types of the sites. A close behavior was characteristic for 2-propanol on the same CSP (hexane the diluent) but not for ethyl acetate, which adsorbed according to the bi-Langmuir model [51].

Admittedly, the modifier may alter the properties of binding sites (mechanism (3)) via stabilizing particular conformations of selectors [31,88,172], solvating selector's functional groups or influencing local electrostatic field in general [173], or by means of adsorption inside a chiral cavity. CD selectors are known to accommodate nonpolar molecules inside of the hydrophobic inner cavity. In contact with nonpolar or aprotic solvents, these cavities will be occupied by strongly retained solvent molecules that makes the inclusion adsorption mechanism impossible. In aqueous solvents, the inner cavity is occupied by a hydro-organic mixture, which can be partly or completely displaced by a solute molecule, thus allowing the formation of an inclusion complex [6,174]. Accordingly, two types of retention mechanisms on bound CD CSPs are distinguished, depending on the mobile phase used (Figure 1.21). Another example is provided by Ching and coauthors [175]. Studying the separation of propranolol enantiomers on a perphenylcarbamate β-CD-bonded CSP, they

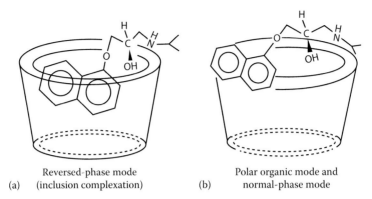

	Reversed-phase mode		Polar organic mode and
(a)	(inclusion complexation)	(b)	normal-phase mode

FIGURE 1.21   Preferred molecular recognition mechanisms of CDs according to Armstrong et al. [174] in reversed-phase mode (a) as well as polar organic and normal-phase mode (b), respectively. In the aqueous RP mode lipophilic portions of the guest molecule bind in the cavity by inclusion complexation (driven by hydrophobic interactions) (a) while in organic media complexation is driven by hydrophilic interactions at the rim surfaces (b). (Reprinted with permission from Lämmerhofer, M., *Journal of Chromatography A*, 1217, 814, 2010.)

observed that the addition of triethylammonium acetate to the mobile phase reduced retention dramatically yet without loss of enantioselectivity. The authors interpreted their results in terms of an additive species blocking the entrance into the CD cavity. The phenomenon of the inclusion of mobile phase components into the matrix of an adsorbent was extensively studied on polysaccharide CSPs [168,176,177]. These materials being prone to swelling absorb lipophilic mobile phase components that affects their adsorption properties. First, it was recognized that nonswollen cellulose triacetate showed only a weak chiral recognition ability but swollen in a range of solvents demonstrated increased enantioselectivity [178]. Later, the dependence of the enantioselective ability on preadsorption of organic compounds was observed in other polysaccharides coated on silica [28,168,179–181]. It was suggested that solvent molecules penetrate the matrix of a polysaccharide, thus opening chiral cavities. The size and shape of the cavities depend on the structure of the penetrating molecules; therefore, the performance of polysaccharide CSPs is a function of the modifier. This hypothesis, even if being an oversimplified version of reality, in its core—the alteration of the steric environment of chiral cavities by modifier molecules—is corroborated by the whole body of chromatographic and spectroscopic evidence [135,168,176,177].

These three discussed mechanisms can be correlated. Alternatively, they can be anti-correlated when, for example, the increase of modifier concentration improves the solvation power of the mobile phase but diminishes its desorption ability or *vice versa*. The concerted action of all three factors will results in a monotonous change of k' as a function of the mobile phase composition, whereas oppositely influencing factors should lead to a k'-plot with the increasing and decreasing branches. Figure 1.22 illustrates these two types of behavior. The plots represent the retention factors of the enantiomers of N'-benzylnornicotine on a β-CD-bonded CSP [170]. The mobile phase is aqueous buffer/MeOH in insert A and aqueous buffer/MeCN

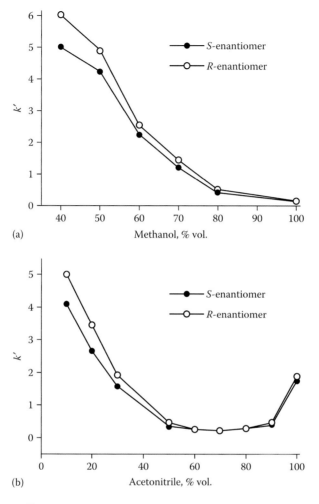

(a)

(b)

FIGURE 1.22 Effect of mobile phase composition on the enantiomeric resolution of race-mic N'-benzylnornicotine on a β-CD-based CSP. Mobile phase: (a) aqueous buffer/metha-nol; (b) aqueous buffer/acetonitrile. (Reprinted with permission from Seeman, J. I. et al., *Analytical Chemistry*, 60, 2120, 1988.)

in insert B. It is interesting to note that an exothermic water/methanol mixture [182] gives a monotonous plot while an endothermic water–acetonitrile mixture [182] pro-duces a quasi-parabolic plot.

Besides the diluent and the modifier, an eluent can contain other minor additives. These can be buffer components that are mixed with a mobile phase to maintain a desired pH or additives serving other purposes. Frequently, minor concentra-tions of acids or bases are added in order to suppress dissociation of analytes or to reduce undesired activity of strong nonselective adsorption sites. Some neutral polar compounds can also be used for this purpose. Such additives improve peak shapes

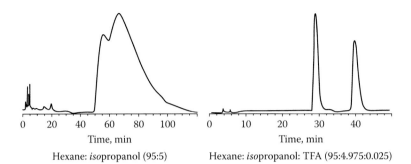

0   20   40   60   80   100   0   10   20   30   40
              Time, min   Time, min
      Hexane: *iso*propanol (95:5)   Hexane: *iso*propanol: TFA (95:4.975:0.025)

FIGURE 1.23   Chiral separation of *RS*-ketoprofen. Comparison of chromatograms with and without trifluoroacetic acid (TFA) addition. Column: Chiralcel OJ (250 × 4.6 mm i.d.). (Reprinted with permission from Tang, Y., *Chirality*, 8, 136, 1996.)

(Figure 1.23) and not rarely affect enantioseparation dramatically both in HPLC [183–185] and in SFC [76,186,187]. Tang [183] and Stringham [188] have reported on the improvement in the enantioseparation of chiral amines obtained with strong acid additives, TFA [183] and alkylsulfonic acids [188], in the mobile phase. This surprising result was attributed to the formation of ammonium cation–acid anion ion pairs. The effect of ion-pairing additive on chromatography of tryptophan enantiomers on a Chirobiotic T column was investigated by Loukili et al. [189]. One more additive (in NPLC and in polar mode) sometimes introduced unintentionally as a contaminant of major mobile phase components is water. Balmér and coworkers ascertained that the addition of water to hexane/1-propanol/diethylamine eluents is decisive for the enantiomer separation of the several chiral amino alcohols on a Chiracel OD column [70,190]. The influence was so strong that resulted in the reversal of elution order as water content increased. The authors rationalized their findings within the framework of the two-site model assuming that water affects the binding properties of only one of the chiral sites [70]. Arnell et al. [160,161] have investigated the effect of a strongly adsorbed additive on the concentration profiles of target analytes under overloaded conditions. The presence of an additive resulted in strong distortion of the chromatograms especially when the concentration of enantiomers in the mobile phase became comparable to that of the additive.

As the methods of nonlinear chromatography were gaining popularity [191], researchers turned to apply this technique in combination with the two-site model in order to scrutinize the effect of mobile phase composition on retention and enantioselectivity in chiral systems [8,56]. Data collected by different research groups show that mobile phase composition strongly affect the concentration of the adsorption sites, besides the expected influence on their adsorption affinity [37,40,64,107,108,167,171,192]. Although not always definite correlations between the mobile phase composition and the saturation capacity of the stationary phase are revealed [193,194], in general an increase in the solvation power of the mobile phase reduces the saturation capacity. This can be rationalized if one remembers that the adsorption sites are distributed over a range of binding energies and assume that an increased energy cost of desolvation (see Equation 1.33) makes

TABLE 1.9
Retention Parameters of L- and D-Methinonine on Chirobiotic TAG for Different MeOH Concentrations in the Mobile Phase

MeOH, %	q_l^*, g/L	$b_l{}^a$, L/g	q_h^*, g/L	$b_{h,L}$, L/g	$b_{h,D}$, L/g	K_l	$K_{h,L}$	$K_{h,D}$	K_L	K_D	α
15	105	0.010	1.42	0.17	0.92	1.05	0.24	1.31	1.29	2.36	1.82
25	79	0.015	1.09	0.19	1.31	1.19	0.21	1.43	1.39	2.61	1.88
35	64	0.025	1.30	0.14	1.26	1.60	0.18	1.64	1.78	3.24	1.82
45	57	0.025	1.44	0.12	1.24	1.43	0.17	1.79	1.60	3.21	2.01
55	54	0.030	1.62	0.11	1.20	1.62	0.18	1.94	1.80	3.56	1.98

Source: Fuereder, M. et al., *Journal of Chromatography A,* 1346, 34, 2014. With permission.

a   The averaged value for L- and D-methinonine (the b_l values for the enantiomers do not differ within the experimental error).

adsorption thermodynamically impossible for a part of the surface sites at the low-energy end of this distribution.

The contribution of a group of sites to the retention factor depends actually not on their concentration q^* but on the product q^*b according to Equation 1.13. Further consideration shows that for many systems a change in the mobile phase composition differently affects the nonselective and enantioselective contributions to retention. An illustrative example is adopted from the work [167] describing the behavior of the enantiomers of methionine on the Chirobiotic TAG CSP, eluted by water/methanol buffer solutions. Table 1.9 presents the values of q^* and b for the low-energy sites (supposedly nonselective) and high-energy sites (selective), the respective partial partition coefficients $K_l = q_l^*b_l$ and $K_h = q_h^*b_h$, and the total partition coefficient $K = K_l + K_h$ as functions of the methanol percentage. It becomes clear that in spite of fluctuations in the concentration of the high-energy sites, probably originated in the experimental error, the retention potential of this group of sites with respect to the D-enantiomer increases while with respect to the L-enantiomer decreases, and the contribution of the low-energy sites increases as the mobile phase becomes more lipophilic. This example of linear dependence of the partial retention contributions on the modifier percentage should not be, of course, taken as a general case. These dependencies may be strongly nonlinear if measured over a wide range of mobile phase compositions as it was demonstrated in [193]. Such would indicate oppositely influencing forces, each becoming dominant at a certain range of mobile phase compositions, as explained above.

1.5.3 pH and Ionic Strength

The effect of the pH and ionic strength on elution characteristics is a part of the mobile phase influence. It is, however, convenient to discuss this issue in a separate section because the mechanism of this effect is quite different from how the modifier and nonbuffer additives act. One can distinguish four types of solute–stationary phase systems with respect to the ionic status of the solute and the stationary phase:

1. *Ionizable solute*: neutral stationary phase
2. *Neutral solute*: neutral stationary phase
3. *Ionizable solute*: ionizable stationary phase
4. *Non-ionogenic solute*: ionizable stationary phase

Each of these systems differently respond to the change of pH and ionic strength of the mobile phase, the rate of this response ranging from strong for ionizable solutes on any type of CSP to weak in the case of neutral molecules on neutral adsorbents.

1.5.3.1 Neutral Stationary Phase

The underlying mechanisms for the retention of ions on neutral adsorbents are considered in [141,191]. The effect of the pH on the retention factor of acidic/basic solutes is well described by the HMM theory explained in Section 1.4.6.2. Some experimental evidences obtained on polysaccharide [143] and MIP [195] CSPs have been reported. In this theory, the effect of pH consists in controlling the dissociation of an ionizable analyte. Accordingly, the strongest effect on k' is observed in the pH domain around the pK_a of the analyte ($pK_a \pm 2$) where the ratio between analyte's charged and neutral forms varies strongly as a function of the proton concentration. Outside this domain, "the retention of ionizable compounds is nearly independent of the pH and depends only on the nature of the buffer used when the ionic form of the compound forms an ion-pair with one of the buffer ions" [191]. The difference between retention factors of optical antipodes on neutral stationary phases changes with the pH of mobile phases. For basic compounds, it increases with pH increasing beyond the pK_a value; for acidic compounds, it increases with pH decreasing below the pK_a value [143]. Correspondingly, apparent enantioselectivity α changes from 1 to a certain constant value or vice versa depending on whether the analyte is anion or cation.

The ionic strength has an effect on many processes determining retention and peak profile, such as solute and buffer's dissociation, solute–solute and solute–solvent interactions, and ion pairing. Ståhlberg has noticed one more influencing factor, the formation of the double electric layer on the surface due to the adsorption of mobile phase electrolytes [196]. This layer causes an electrostatic attraction or repulsion between the surface and the charged solute. Since the potential and the length of the double layer depends on the ionic strength, the Ståhlberg theory predicts an additional contribution to the overall effect of this parameter. Because it is difficult to rationalize such a multifactorial phenomenon, empirical explanations dominate when relationships between the ionic strength and elution characteristics are discussed. Ishikava and Shibata [197] demonstrated a profound influence of the supporting salt's concentration on the retention, enantiomer separation, and the shape of the band profile of cationic solutes (Figure 1.24). The authors attributed this finding to ion pairing. Kazakevich specified that this effect should be pH dependent [141]. That is, the retention increases with the salt concentration when the solute exists in an ionic form. When a change of pH shifts the dissociation equilibrium toward the neutral form, the

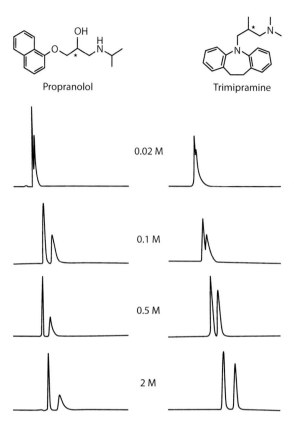

FIGURE 1.24 Effect of concentration of NaClO$_4$ on the separation of propranolol and trimipramine. Column: Chiralcel OD-R (250 × 4.6 mm i.d.). Mobile phase: aq. NaClO$_4$/ acetonitrile (60:40, v/v). (Reprinted with permission from Ishikawa, A. and Shibata, T., *Journal of Liquid Chromatography,* 16, 859, 1993.)

effect abates. More than salt concentration, the valence of the buffer affects the shape of overloaded (i.e., at high solute concentration) band profiles. Probably, the buffer ion with more than one charge plays the role of a bridging species for analyte molecules, creating conditions for adsorbate–adsorbate interactions [191].

The nature of buffer is also important. Occasionally, different retention factors and enantioselectivities are observed with different buffer systems [198]. Lee et al. [199] separated enantiomers of dansyl amino acids on β-CD with triethylamine acetate (TEAA) buffer (pH 4.1). They observed a dramatic retention decrease when the TEAA concentration increased, which was accompanied by loss of enantioselectivity, on the one hand, and increase in chromatographic efficiency, on the other hand. The latter fact indicates improvement in mass transfer kinetics, and all these facts together prove that TEAA blocks the chiral cavity of the CD, preventing the formation of a solute–CD inclusion complex. A quite different observation was reported by Fujimura and coauthors [200], who also separated enantiomers of dansyl amino acids on β-CD but used a phosphate buffer (pH 6.5). In this case, the retention factor and resolution increased with the buffer concentration. The underlying mechanism is not clear, but it is obvious that phosphate species facilitate the inclusion of the analyte into the CD cavity.

It should be kept in mind that CSPs obtained by immobilization of a chiral substance on silica or functionalized silica bear surface groups which can be ionized in aqueous buffers [201]. In experiments by Hou et al. [202] with a cellulose derivative coated on an aminopropyl silica, the retention factors of optical antipodes of acidic and neutral analytes decreased as a function of the sodium perchlorate concentration up to a concentration value of 0.2 M. At higher concentrations, the retention of either type of compounds changed negligibly. The author explained their findings by interaction of perchlorate anion with protonated surface amino groups. When all surface groups were masked at a salt concentration of 0.2 M, no further change in elution characteristics was observed. On the other hand, when basic analytes were examined, their retention increased and the enantioselectivity coefficient grew monotonously with increasing salt concentration. Hence, the authors supposed the involvement of the ion-pairing mechanism [202].

1.5.3.2 Ionizable Stationary Phase

Many classes of CSP selectors bear ionizable functionalities. Those are proteins, glycopeptide macrocyclic antibiotics, *Cinchona* alkaloids, amino acids and amino acid derivatives, and so on. This gives rise, when in contact with solutions allowing ionization, to ion-exchange mechanism interfering with molecular adsorption mechanisms. The pH of mobile phase influences enantioseparation under such conditions in two ways. First, it controls dissociation equilibria in the mobile phase. Second, the pH controls the ionization of selector functionalities and affects the conformation of chiral selectors [32,203]. A visible indication of a mixed—adsorption/partition and ion exchange—retention mechanism is a deviation of k' versus pH dependence from an HMM S-shaped trajectory. This phenomenon is illustrated in Figure 1.25 that shows k'–pH plots for the enantiomers of the weak acid ketoprofen eluted from an eremomycin-based column with water–ethanol solvents [142]. The plots have maxima that correlate with pK_a of the analyte in a given mobile phase. As eremomycin is an antibiotic with pI in a basic pH domain [204], it plays the role of an anion exchanger at pH < 7. As the pH increases, the fraction of ketoprofen anions capable of binding to strong anion-exchange sites grows that explains the increasing branch of the plots. As the pH traverses the pK_a point, the analyte species are mostly ionized; on the other hand, the surface concentration of positively charged ligands decreases rapidly. Thus, the contribution from the ion-exchange retention mechanism declines, and that of the (neutral) adsorption mechanisms is low. A combination of these two factors stipulate the decreasing branch of the plots.

Figure 1.25 represents one of two main types of k'–pH dependencies for ionizable CSPs: a curve with a maximum. The other frequently found type is a monotonous (decreasing for acids and increasing for bases) curve; representative examples are given in Figure 1.26. Occasionally, this curve has an S-shaped form, which does not necessarily mean a dominating influence of the dissociation factor as it would mean for the neutral adsorbent. The experimental data show that the retention of nonelectrolytes also depends on pH. Nonlinear dependences with a broad maximum

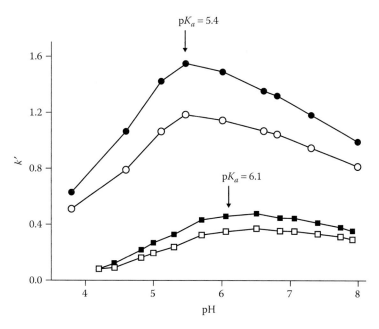

FIGURE 1.25 Effect of mobile phase pH on the retention of *R*-ketoprofen (filled symbols) and *S*-ketoprofen (open symbols) on a Nautilus-E column with water/ethanol buffers with 40% (circles) and 60% (squares) of EtOH. Arrows show the pK_a values of ketoprofen in the respective mobile phase. (Reprinted in a slightly modified form with permission from Reshetova, E. N. and Asnin, L. D., *Russian Journal of Physical Chemistry*, 85, 1434, 2011.)

were observed for benzoin on an ovomucoid column [205] and for 3-phenylphthalide on a teicoplanin column [206]. Enantioselectivities were also functions of pH. This behavior displays the effect of acidity on the properties and, possibly, concentration of the binding sites. Survey of the literature data did not reveal general correlations between the nature of analyte and selector and the behavior of the k'(pH) function. It is clear, however, that pK_a of the analyte and p*I* of the selector are the parameters that influence the shape of this plot the most. Nearby corresponding pH values, one can expect the occurrence of singular points on a k'–pH plot, an inflection point, a maximum, or a curvature.

An important insight into the impact of pH was gained from nonlinear chromatographic measurements. In a series of works on the adsorption of β-blockers on the CBH I CSP, Guiochon and coauthors revealed that the pH controls the concentration and affinity of the enantioselective and nonselective binding sites [61,207,208]. The nonselective interactions were concluded to be mostly hydrophobic and only slightly ionic, whereas the contributions of hydrophobic and ionic interactions to the binding of solute molecules to the enantioselective sites were comparable. Hence, the acidity of the eluent had a different effect on these groups of sites and on the interaction of the enantioselective sites with enantiomers that resulted in different adsorption behaviors of optical antipodes as functions of pH, depending also on the hydrophobicity of an analyte. Later, Götmar et al. [209] compared retentions of basic, acidic, and neutral

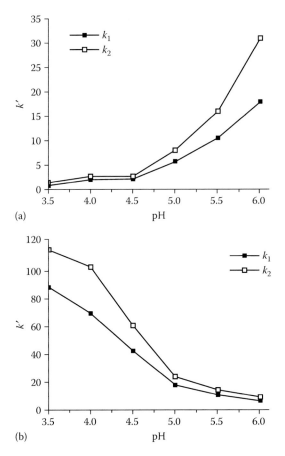

FIGURE 1.26 Effect of mobile phase pH on the retention on the ovomucoid CSP of the enantiomers of the basic compound verapamil (a) and the acidic compound benoxaprofen (b). (Reprinted with permission from Iredale, J. et al., *Chromatographia*, 31, 329, 1991.)

chiral analytes on the silica-immobilized AGP protein at different pH. Again, it was found that the mobile phase pH differently influenced the nonselective and enantioselective sites. Moreover, different compounds differently responded to change in pH. Table 1.10 taken from the cited work provides the details. Chen et al. [111] have observed strong influence of the pH on the behavior of an imprinted polymeric stationary phase, using the enantiomers of phenylalanine anilide as probe analytes. This was explained by a combination of factors relating to the dissociation of the solute and to the dissociation of the carboxylic groups of the stationary phase. As the pH grows, the concentration of strong ionized binding sites grows too. The binding sites are distributed between the enantioselective and nonselective sites in an approximate ratio 1:10. This ratio remains nearly constant when the pH changes. The nonselective part of the surface was reported to be homogeneous, whereas

TABLE 1.10
Effect of pH on Apparent α and True α_{true} Enantioselectivities and on Relative Enantioselectivity Contribution to Linear Retention of the Analytes[a]

Analyte	pH	α	α_{true}	Relative Enantioselectivity Contribution, %	
				First Isomer	Second Isomer
Alprenolol	4.04	1.20	1.25	82	85
	4.47	1.40	1.44	90	93
	4.97	1.47	1.50	95	96
1-(1-Naphthyl)ethylamine	4.98	1.35	1.81	43	58
	5.48	1.32	1.62	52	63
	5.97	1.18	1.29	62	68
	6.52	1.26	1.40	67	74
2-Phenylbutiric acid	3.98	1.32	1.38	84	88
	4.47	1.40	1.45	88	92
	4.97	1.57	1.66	87	92
Methyl mandelate	4.49	1.30	1.42	72	79
	4.96	1.34	1.43	78	83
	5.99	1.35	1.41	86	90

Source: Götmar, G. et al., *Analytical Chemistry,* 74, 2950, 2002. With permission.

[a] Stationary phase, immobilized R1-acid glycoprotein on silica; eluent, acetic buffer at $I = 0.050$ M and 0.25% 2-propanol.

the enantioselective part of the surface was not so. The degree of heterogeneity decreased with increasing pH.

Because ionizable compounds, for example, 1:1 electrolytes, exist in two forms, ionic and neutral, and each form can adsorb to the stationary phase differently, these species develop in ion-exchange columns two adsorption fronts moving with different velocities. Then a shift of the pH toward suppressing dissociation will lead to a leveling of the band profiles. Figure 1.27 confirms this supposition. It shows the elution profiles of *S*-naproxen obtained in the frontal analysis and elution modes on a weak anion-exchanger at two pH values, 5.9 and 7.3. At the former pH, the dissociation of naproxen is suppressed as its $pK_a = 7.4$ at the given mobile phase [106].

The effect of the ionic strength is multifactorial. One aspect of this influence is explained by electrochemistry and consists in dependence of the activity coefficients of ions in the mobile phase on this parameter. It also determines the characteristics of the double electric layer formed on the surface of charged stationary phases as noticed by Ståhlberg [210] and Cantwell and Puon [211]. This "double layer" effect can be quantitatively predicted and when compared with experimental data (unfortunately, not in chiral chromatography) failed to explain satisfactorily chromatographic behavior of charged analytes especially at high sample loadings [212]. Many more factors have to be taken into account to provide a fair

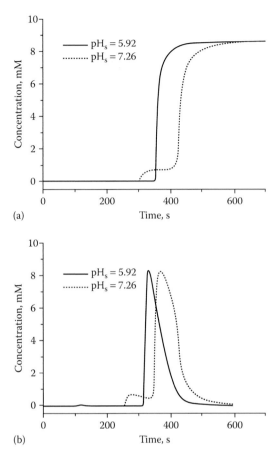

(a)

(b)

FIGURE 1.27 Breakthrough curves (a) and elution profiles (b) of *S*-naproxen with 90% methanol + 10% acetic buffer mobile phase at different pH. Column: Whelk-O1. (Reprinted with permission from Asnin, L. et al., *Journal of Chromatography A,* 1192, 62, 2008.)

explanation. The ionic strength can affect differently even by different mechanisms the strong and the weak adsorption sites, respectively [101,212,213]. The buffer ions compete with analyte ions for interaction with charged binding sites, thus interfering in the ion-exchange retention mechanisms [214]. So was explained a decrease in retention of chiral amino acid-type hormones on a carbamylated quinine CSP caused by an increase of ammonium acetate buffer concentration, the enantioselectivity remaining constant. A similar effect of ammonium acetate buffer content was reported in the work [142]. In this study concerning the separation of profen enantiomers on an antibiotic-type CSP, it was shown that it was not the concentration of ammonium acetate that influenced retention but the concentration of acetic acid added as a constituent of this buffer. Hence, the authors concluded that the ion-exchange mechanism was not the only important factor determining retention in this system.

1.6 MULTICHIRAL COMPOUNDS

So far, we mostly discussed monochiral solutes, that is, compounds which have one asymmetric center. Many substances subjected to analysis or preparative separation in chromatography contain more than one chiral atom. For a long time, separation scientists have considered this problem from one single viewpoint: the more chiral atoms are present in a molecule, the more stereoisomers are to be separated [113]. Surprisingly, it was not noticed that mechanisms of chiral recognition could be somewhat different for mono- and multichiral analytes. In principle, there are two ways in which a multichiral molecule can interact with a chiral selector. The chiral centers of a molecule can form a configuration that will interact with a chiral selector as a whole unit, that is, as a quasimonochiral compound. This situation is similar to the adsorption of monochiral substances. The second way is when chiral centers act independently and each of them form a separate binding center. These binding centers may compete with each other for interaction with an adsorption site or they may bind to different types of adsorption sites. These options are illustrated in Figure 1.28. A respective mathematical description has been recently given in the short communication [215]. For the sake of brevity, it is reproduced here by example of a bichiral compound AB, where A and B symbolize different asymmetric centers. Consider first the case of competition between chiral centers A and B for binding to the sole chiral selector S. Then two equilibria are possible with respective equilibrium constants b_{ABS} and b_{BAS}:

$$AB + S \leftrightarrow ABS \quad b_{ABS} = \frac{[ABS]}{[AB] \cdot [S]} \tag{1.40}$$

$$AB + S \leftrightarrow BAS \quad b_{BAS} = \frac{[BAS]}{[AB] \cdot [S]} \tag{1.41}$$

A respective adsorption isotherm reads

$$q = \frac{q_s^*(b_{ABS} + b_{BAS})c}{1 + (b_{ABS} + b_{BAS})c} \tag{1.42}$$

where $c = [AB]$. As seen, this is the Langmuir isotherm with the absorption coefficient being the sum of the partial coefficients for the AB-S and BA-S binding.

When a solid bears two types of chiral selectors S_A and S_B, the former able to adsorb molecule AB via the center A and the latter adsorbing this molecule via the center B, the conditions of the bi-Langmuir model met and the adsorption isotherm equation reads

$$q = \frac{q_{SA}^* b_{SA} c}{1 + b_{SA} c} + \frac{q_{SB}^* b_{SB} c}{1 + b_{SB} c} \tag{1.43}$$

where q_{SA}^* and b_{SA} are, respectively, the surface concentration and adsorption coefficient of the S_A-type selector, and q_{SB}^* and b_{SA} are the respective parameters for the S_B-type selector.

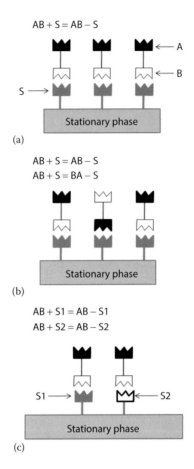

FIGURE 1.28 Illustration of three scenarios of binding equilibria: (a) a single binding mode selectand—a single selector; (b) a multiple binding mode selectand—a single selector; (c) a single binding mode selectand—multiple selectors. The molecule AB is a selectand, symbols S, S1, and S2 denote selectors. S1 and S2 may designate different selectors or different conformations of the same selector. (Reprinted with modification with permission from Asnin, L. D. and Nikitina, Y. K., *Journal of Separation Science, 37*, 390, 2014.)

Equations 1.42 and 1.43 show that study of the adsorption of multichiral substances within the frame of the macroscopic approach is a complicated task because the adsorption models involved are not distinguishable from those for the monochiral compound. On the other hand, a set of the retention factors of all stereoisomers of a multichiral compound provide sufficient information to distinguish the binding mode in which the chiral centers of a molecule behave independently from that in which the chiral centers behave correlatively. Again, consider for the sake of simplicity a bichiral analyte. It is easy to show that a linear combination of the retention factors of the enantiomers LL and DD taken with the sign "plus" and those for the enantiomers LD and DL taken with the sign "minus" has to be equal to zero if the chiral centers behave independently. That is,

$$k'_{LL} + k'_{DD} - k'_{LD} - k'_{DL} = 0 \qquad (1.44)$$

Table 1.11 summarizes data on the application of this criterion to some dipeptide derivatives chromatographed on three different Pirkle phases (Figure 1.29) [239].

TABLE 1.11
Retention Factors of Dipeptide Derivatives on Three Pirkle-Type CSPs

DNB-a.a.-a.a.-CO$_2$CH$_3$	Retention Factor				$\Sigma_\pm$[a]	$\Sigma_\pm$/max(k'),[b] %
	LL	DD	LD	DL		
			CSP1[c]			
ala-ala	Not reported					
ala-leu	2.8	6.86	4.5	7.52	−2.36	−31.4
ala-met	11.4	25.76	19.3	32.23	−14.37	−44.6
ala-phe	9.8	17.64	15.7	23.71	−11.97	−50.5
leu-ala	3.7	5.25	6.2	9.55	−6.8	−71.2
leu-leu	4.6	11.09	8	11.2	−3.51	−31.3
leu-met	3.9	8.81	6.7	10.25	−4.24	−41.4
leu-phe	7.1	12.07	15.2	20.06	−16.09	−80.2
phenylgly-ala	10.3	12.88	30.75	21.5	−29.07	−94.5
phenylgly-leu	12.6	17.14	46.97	30.7	−47.93	−102.0
phenylgly-met	11.7	13.1	31.76	23.7	−30.66	−96.5
phenylgly-phe	22.7	22.7	69.31	47.8	−71.71	−103.5
			CSP2[c]			
ala-ala	7.5	22.8	11.8	40.71	−22.21	−54.6
ala-leu	3.8	11.4	5	16.65	−6.45	−38.7
ala-met	6.3	19.908	9.7	32.689	−16.181	−49.5
ala-phe	6.5	13.195	7.9	18.486	−6.691	−36.2
leu-ala	4	19.32	5.4	23.706	−5.786	−24.4
leu-leu	3.7	24.235	4.6	16.422	6.913	28.5
leu-met	4.2	19.824	6	26.82	−8.796	−32.8
leu-phe	6.5	20.67	10.1	40.097	−23.027	−57.4
phenylgly-ala	8.3	22.41	12.4	35.34	−17.03	−48.2
phenylgly-leu	4.3	11.567	6.2	15.562	−5.895	−37.9
phenylgly-met	7.3	19.345	10.5	25.83	−9.685	−37.5
phenylgly-phe	7.3	15.403	10.2	22.44	−9.937	−44.3
			CSP3[c]			
ala-ala	1.3	22.386	3.7	22.052	−2.066	−9.2
ala-leu	Not reported					
ala-met	2.8	28.868	4.1	26.732	0.836	2.9
ala-phe	2.9	25.839	5.4	27	−3.661	−13.6

(Continued)

TABLE 1.11 (*Continued*)

Retention Factors of Dipeptide Derivatives on Three Pirkle-Type CSPs

DNB-a.a.-a.a.-	Retention Factor					
CO$_2$CH$_3$	LL	DD	LD	DL	$\Sigma_\pm$ [a]	$\Sigma_\pm$/max(k'),[b] %
leu-ala	2.4	25.464	2.4	26.904	−1.44	−5.4
leu-leu	1.4	21.994	1.6	20.816	0.978	4.4
leu-met	2.7	49.977	3.8	45.258	3.619	7.2
leu-phe	2	34.62	3.6	40.356	−7.336	−18.2
phgly-ala	3.7	18.87	5.2	19.916	−2.546	−12.8
phgly-leu	2.7	17.847	4.1	16.687	−0.24	−1.3
phgly-met	4.9	30.478	7.9	28.598	−1.12	−3.7
phgly-phe	5.1	30.753	8	29.2	−1.347	−4.4

Source: Pirkle, W. H. et al., *Journal of Chromatography,* 398, 203, 1987. With permission.

[a] $\Sigma_\pm = k'_{LL} + k'_{DD} - k'_{LD} - k'_{DL}$.

[b] max(k'), the retention factor of the last eluted stereoisomer.

[c] See Figure 1.29.

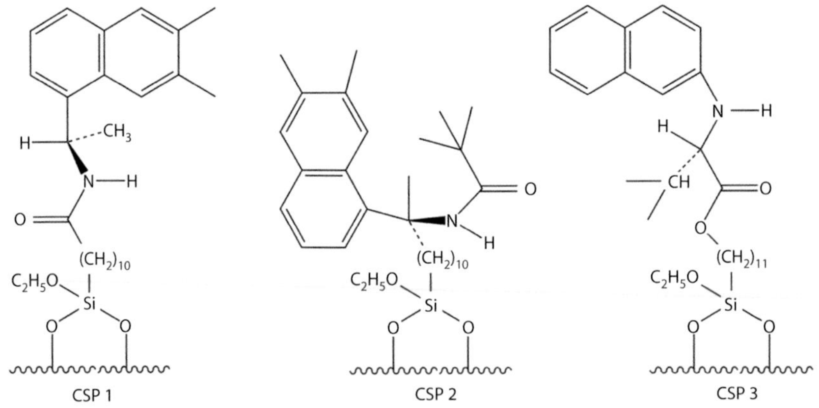

FIGURE 1.29   Structures of Pirkle-type CSPs. An illustration to data in Table 1.11.

It is interesting to see that all analytes examined on the CSP 1 and CSP 2 do not pass this test, whereas for most analytes on the CSP 3 a deviation of criterion (Equation 1.44) from zero is low. One can suggest based on this result that the chiral selector of this CSP, *N*-(2-naphthyl)valine, despite its simple structure has two binding sites, one complementary to the N-terminus and the other complementary to the C-terminus of dipeptide. Table 1.12 compares the retention data for two series of N-derivatized ala-ala dipeptides obtained on the same *tert*-butylcarbamoylquinine CSP under different elution conditions [116]. One can see that the condition (Equation 1.44) depends both on the elution mode and on the derivatizing group. This criterion was applied

TABLE 1.12

Retention Factors of *N*-Protected ala-ala Measured on a *tert*-Butylcarbamoylquinine CSP under Reversed-Phase and Polar Organic Mode Conditions

N-Protecting Group[a]	Retention Factor				$\Sigma_{\pm}$[b]	$\Sigma_{\pm}$/max(k'),[c] %
	LL	DD	LD	Dl		
	Reversed-Phase Mode					
DNB	19.44	3.4	13.7	4.41	4.73	24.33
DNP	9.07	5.37	6.23	14.91	−6.7	−44.94
DNZ	4.02	2.34	3.58	2.76	0.02	0.50
CC	4.87	3.67	4.87	4.87	−1.2	−24.64
Fmoc	3.84	2.68	3.47	3.1	−0.05	−1.30
	Polar Organic Mode					
DNB	20.16	4.91	20.88	5.83	−1.64	−7.85
DNP	8.47	6.07	7.94	24.01	−17.41	−72.51
DNZ	6.04	3.87	6.98	4.49	−1.56	−22.35
CC	6.19	5.15	6.9	7.35	−2.91	−39.59
Fmoc	4.89	3.53	5.2	4.32	−1.1	−21.15

Source: Czerwenka, Ch. et al., *Journal of Pharmaceutical and Biomedical Analysis,* 30, 1789, 2003. With permission.

[a] DNB, 3,5-dinitrobenzoyl; DNP, 2,4-dinitrophenyl; DNZ, 3,5-dinitrobenzyloxycarbonyl, CC, carbazole-9-carbonyl; FMOC, 9-fluorenylmethoxycarbonyl.

[b] $\Sigma_{\pm} = k'_{LL} + k'_{DD} - k'_{LD} - k'_{DL}$.

[c] max(k'), the retention factor of the last eluted stereoisomer.

to many other data on bichiral compounds found in the literature (not reported data). The condition (Equation 1.44) was revealed to fail more frequently than to be satisfied. One can speculate that independent behavior of chiral centers of multichiral substances in the formation of the binding complex is not common.

1.7 CONCLUSIONS

Since the late 1980s, when it was recognized that the interaction of solute with the stationary phase is not confined to enantioselective interactions alone, that nonselective interactions are to play an essential role in the retention and also in resolution of enantiomers, there began a search for methods to separate these two interaction mechanisms in order to study their partial influences that led eventually to the adjustment of nonlinear chromatography techniques for the needs of the said research. A key step was the development of the two-site model which allowed estimation of the enantioselective and nonselective contributions to retention. A modern understanding of this model supposes that the "CSP comprises ability to enantioselective and nonselective interactions, and that the sources of these interactions have complex structures <...>. Each source may combine several types of adsorption sites with very different adsorption affinities and even with a different chemical nature. In fact, the very term 'site' should be considered in a broad sense as in the case of nonlocalized adsorption there are no binding sites that would retain solute molecules" [8]. Moreover, the partial enantioselective and nonselective contributions are not fully independent. Besides possible mechanisms of mutual interdependence caused by direct site–site or bridging site–solute–site interactions, they are correlated via the general electrostatic field formed as a result of interference of local electrostatic fields of all surface atoms belonging to both enantioselective and nonselective sites. This subtle moment seems to have never been considered in chiral chromatography and is still waiting to be investigated.

The application of the two-site model extended to take into account secondary equilibria in the mobile phase allows us to explain chromatographic behavior of chiral solutes including the shape of chromatographic profile, degree of enantiomer resolution (but not the elution order), and effect of experimental parameters on retention. Some clues concerning chiral recognition mechanism may be derived from the macroscopic data by the inquisitive mind; however, its unambiguous explanation is a task for molecular methods. These, being able to provide extensive information on the selector–selectand binding, are not yet developed enough to give a comprehensive explanation, let alone quantitative predictions, for the interaction of solute with the column packing. Although modern computer methods are elaborated enough to simulate the surface of a CSP including nonchiral entities, it is still not possible to elucidate on the molecular level the effect of the porous structure on retention and enantioselectivity. This is not only because neither the pore structure nor the distribution of the chiral selector within the pores is known with necessary details, but also because the behavior of a solute molecule inside the pore is not fully understood.

At present, there is a general understanding of adsorption in chiral systems expressed in the form of the two-site model and the four-point (formerly known as the three-point) interaction rule. This by no means implies that there is no more room for research in this area. An important assumption of the two-site model of independent behavior of the enantioselective and nonselective sites has not been seriously questioned yet, although there are sound reasons for challenging this assumption. The mechanism of the transition between the enantioselective and nonselective states of a chiral selector is still obscure. Is it a dynamic equilibrium or it is an irreversible process in a given mobile phase? Can the ratio between the enantioselective and nonselective forms of a selector be influenced by the adsorption of a solute? These questions important for the understanding of retention mechanisms and for the optimization of enantioseparation are rarely questioned and so far have never been answered. A little is known about the thermodynamic of enantioselective adsorption in HPLC. It is true that there are hundreds of publications reporting the van't Hoff analysis of retention data. On the other hand, there is not a single work, to the best of our knowledge, in which all components of Equations 1.21 and 1.33 would be independently measured. Only when every contributing process is estimated, we can declare that the thermodynamic picture of adsorption is clear. It is not an idle curiosity. The key to the explanation of the effect of solvent on enantioselectivity lies in the separation of the solvation, desorption, and net adsorption contributions to $\Delta\Delta G^0$ (see Equation 1.34). A class of multichiral compounds so far poorly studied needs more attention. As it was shown, there should be no apparent difference in the adsorption isotherms of monochiral and multichiral analytes. Therefore, it will require ingenuity of researchers to devise experimental and theoretical approaches to investigate retention mechanisms for such compounds. A new emerging sort of CSPs, the chirally porous adsorbent, put new questions concerning the nature of adsorption sites and the mechanisms of chiral recognition in these materials. In particular, one may expect an essential influence of solute–solute and solute–solvent interactions inside the pore on both enantioselective and nonselective interactions. The described problems are not an exhaustive list of issues to be resolved to obtain a thorough understanding of solute–stationary phase interactions in chiral chromatography. There are many more, and with every new type of a CSP and every new class of chiral solutes, there emerges new questions. Therefore, a researcher who would decide to work in this area will not be disappointed by a lack of research tasks.

REFERENCES

1. G. M. Henderson, H. G. Rule, *Journal of Chemical Society* (1939) 1568.
2. L. Zechmeister, *Annals of the New York Academy of Sciences* 49 (1948) 220.
3. C. E. Dalglish, *Journal of Chemical Society* (1952) 3940.
4. G. Folkers, M. Yarim, P. Pospisil, in *Chirality in Drug Research*, eds. E. Francotte and W. Lindner, Wiley-VCH Verlag GmbH, Weinheim, Germany, 2006, pp. 323–340.
5. R. K. Lindsey, J. L. Rafferty, B. L. Eggimann, J. I. Siepmann, M. R. Schure, *Journal of Chromatography A* 1287 (2013) 60.
6. M. Lämmerhofer, *Journal of Chromatography A* 1217 (2010) 814.
7. P. Szabelski, K. Kaczmarski, *Acta Chromatographica* 20 (2008) 513.
8. L. D. Asnin, *Journal of Chromatography A* 1269 (2012) 3.
9. R. Tsuchida, M. Kobayashi, A. Nakamura, *Journal of the Chemical Society of Japan* 56 (1935) 1339.
10. G. Karagunis, G. Coumoulos, *Nature* 141 (1938) 162.
11. G. M. Henderson, H. G. Rule, *Nature* 141 (1938) 917.
12. M. Kotake, T. Sakan, N. Nakamura, S. Senoh, *Journal of the American Chemical Society* 73 (1951) 2973.
13. G. Hesse, R. Hagel, *Chromatographia* 6 (1973) 277.
14. H. Koller, K.-H. Rimbock, A. Mannschreck, *Journal of Chromatography* 282 (1983) 89.
15. G. Nickerl, A. Henschel, R. Grünker, K. Gedrich, S. Kaskel, *Chemie Ingenieur Technik* 83 (2011) 90.
16. P. Paik, A. Gedanken, Y. Mastai, *Applied Materials & Interfaces* 1 (2009) 1834.
17. P. Peluso, V. Mamane, S. Cossu, *Journal of Chromatography A* 1363 (2014) 11.
18. I. J. da Silva Jr., M. A. G. dos Santos, V. de Veredas, C. C. Santana, *Separation and Purification Technology* 43 (2005) 103.
19. M. Zhang, X.-D. Xue, J.-H. Zhang, S.-M. Xie, Y. Zhang, L.-M. Yuan, *Analytical Methods* 6 (2014) 341.
20. M. Zhang, J.-H. Zhang, Y. Zhang, B.-J. Wang, S.-M. Xie, L.-M. Yuan, *Journal of Chromatography A* 1325 (2014) 163.
21. S.-M. Xie, Z.-J. Zhang, Z.-Y. Wang, L.-M. Yuan, *Journal of American Chemical Society* 133 (2011) 11892.
22. C. B. Castells, P. W. Carr, *Chromatographia* 52 (2000) 535.

23. F. H. Ling, V. Lu, F. Svec, J. M. J. Fréchet, *Journal of Organic Chemistry* 67 (2002) 1993.
24. F. Ling, E. Brahmachary, M. Xu, F. Svec, J. M. J. Fréchet, *Journal of Separation Science* 26 (2003) 1337.
25. V. Schurig, R. Weber, *Journal of Chromatography* 217 (1981) 51.
26. R. Däppen, V. Meyer, H. Arm, *Journal of Chromatography* 464 (1989) 39.
27. S. Jacobson, S. Golshan-Shirazi, G. Guiochon, *Journal of the American Chemical Society* 112 (1990) 6492.
28. T. O'Brien, L. Crocker, R. Thompson, K. Thompson, P. H. Toma, D. A. Conlon, B. Feibush, C. Moeder, G. Bicker, N. Grinberg, *Analytical Chemistry* 69 (1997) 1999.
29. C. F. Zhao, N. M. Cann, *Analytical Chemistry* 80 (2008) 2426.
30. J. Reeder, P. P. Castro, C. B. Knobler, *Journal of Organic Chemistry* 59 (1994) 3151.
31. T. Bürgi, A. Baiker, *Journal of the American Chemical Society* 120 (1998) 12920.
32. Ch. Czerwenka, M. M. Zhang, H. Kählig, N. M. Maier, K. B. Lipkowitz, W. Lindner, *Journal of Organic Chemistry* 68 (2003) 8315.
33. T. Fornstedt, P. Sajonz, G. Guiochon, *Chirality* 10 (1998) 375.
34. G. Götmar, L. Asnin, G. Guiochon, *Journal of Chromatography A* 1059 (2004) 43.
35. H. Kim, K. Kaczmarski, G. Guiochon, *Chemical Engineering Science* 60 (2005) 5425.
36. H. Kim, G. Guiochon, *Analytical Chemistry* 77 (2005) 93.
37. H. Kim, G. Guiochon, *Analytical Chemistry* 77 (2005) 1708.
38. H. Kim, K. Kaczmarski, G. Guiochon, *Journal of Chromatography A* 1101 (2006) 136.
39. L. Asnin, K. Horvath, G. Guiochon, *Journal of Chromatography A* 1217 (2010) 1320.
40. L. Asnin, G. Guiochon, *Journal of Chromatography A* 1217 (2010) 2871.
41. P. Spégel, L. Schweitz, S. Nilsson, *Analytical Chemistry* 75 (2003) 6608.
42. B. Loun, D. S. Hage, *Journal of Chromatography* 579 (1992) 225.
43. I. Marle, S. Jönsson, R. Isaksson, C. Pettersson, G. Pettersson, *Journal of Chromatography* 648 (1993) 333.
44. J. Ghuman, P. A. Zunszain, I. Petitpas, A. A. Bhattacharya, M. Otagiri, S. Curry, *Journal of Molecular Biology* 353 (2005) 38.
45. H. Matsunaga, J. Haginaka, *Journal of Chromatography A* 1106 (2006) 124.
46. G. Götmar, D. Zhou, B. J. Stanley, G. Guiochon, *Analytical Chemistry* 76 (2004) 197.
47. A. Felinger, D. Zhou, G. Guiochon, *Journal of Chromatography A* 1005 (2003) 35.
48. M. Enmark, J. Samuelsson, T. Undin, T. Fornstedt, *Journal of Chromatography A* 1218 (2011) 6688.
49. C. F. Zhao, N. M. Cann, *Journal of Chromatography A* 1131 (2006) 110.
50. A. Cavazzini, G. Nadalini, V. Malanchin, V. Costa, F. Dondi, F. Gasparrini, *Analytical Chemistry* 79 (2007) 3802.
51. A. Cavazzini, G. Nadalini, V. Costa, F. Dondi, *Journal of Chromatography A* 1143 (2007) 134.
52. Y. K. Nikitina, I. Ali, L. D. Asnin, *Journal of Chromatography A* 1363 (2014) 71.
53. S. Allenmark, *Chirality* 5 (1993) 295.
54. G. Götmar, T. Fornstedt, G. Guiochon, *Chirality* 12 (2000) 558.
55. T. Fornstedt, G. Guiochon, *Analytical Chemistry* 73 (2001) 608A.
56. J. Samuelsson, R. Arnell, T. Fornstedt, *Journal of Separation Science* 32 (2009) 1491.
57. S. Jönsson, A. Schön, R. Isaksson, C. Pettersson, C. Pettersson, *Chirality* 4 (1992) 505.
58. T. Fornstedt, P. Sajonz, G. Guiochon, *Journal of the American Chemical Society* 119 (1997) 1254.
59. K. Fulde, A. W. Frahm, *Journal of Chromatography A* 858 (1999) 33.
60. M. Schlauch, A. W. Frahm, *Analytical Chemistry* 73 (2001) 262.
61. G. Götmar, T. Fornstedt, M. Andersson, G. Guiochon, *Journal of Chromatography A* 905 (2001) 3.
62. I. Ilisz, Z. Pataj, Z. Gecse, Z. Szakonyi, F. Fülöp, W. Lindner, A. Peter, *Chirality* 26 (2014) 385.

63. G. Götmar, J. Samuelsson, A. Karlsson, T. Fornstedt, *Journal of Chromatography A* 1156 (2007) 3.
64. L. D. Asnin, K. Kaczmarski, E. N. Reshetova, *Russian Chemical Bulletin* 58 (2009) 1731.
65. L. Asnin, K. Sharma, S. W. Park, *Journal of Separation Science* 34 (2011) 3136.
66. M. Haroun, C. Ravelet, A. Ravel, C. Grosset, A. Villet, E. Peyrin, *Journal of Separation Science* 28 (2005) 409.
67. B. Koppenhoeffer, E. Bayer, *Chromatographia* 19 (1984) 123.
68. V. Schurig, J. Ossig, R. Link, *Angewandte Chemie* 101 (1989) 197.
69. K. Watabe, R. Charles, E. Gil-Av, *Angewandte Chemie, International Edition* 28 (1989) 192.
70. K. Balmér, P.-O. Lagerstrom, B.-A. Persson, G. Schill, *Journal of Chromatography* 592 (1992) 331.
71. W. H. Pirkle, P. G. Murray, *Journal of High Resolution Chromatography* 16 (1993) 285.
72. Y. C. Guillaume, J.-F. Robert, C. Guinchard, *Talanta* 55 (2001) 263.
73. H. J. Choi, Y. J. Park, M. H. Hyun, *Journal Chromatography A* 1164 (2007) 235.
74. E. N. Reshetova, L. D. Asnin, *Russian Journal of Physical Chemistry* 83 (2009) 547.
75. L. Sipos, I. Illisz, Z. Pataj, Z. Szakonyi, F. Fülöp, D. W. Armstrong, A. Péter, *Journal of Chromatography A* 1217 (2010) 6956.
76. L. Toribio, M. J. del Nozal, J. L. Bernal, C. Cristofol, C. Alonso, *Journal of Chromatography A* 1121 (2006) 268.
77. Z. Gecse, I. Ilisz, M. Nonn, N. Grecsó, F. Fülöp, R. Agneeswari, M. H. Hyun, A. Péter, *Journal of Separation Science* 36 (2013) 1335.
78. R. W. Stringham, J. A. Blackwell, *Analytical Chemistry* 69 (1997) 1414.
79. L. Yang, H. Zhang, T. Tan, A. U. Rahman, *Journal of Chemical Technology and Biotechnology* 84 (2009) 611.
80. S.-M. Lai, Z.-C. Lin, *Separation Science and Technology* 34 (1999) 3173.
81. P. Szabelski, K. Kaczmarski, A. Cavazzini, Y.-B. Chen, B. Sellergren, G. Guiochon, *Journal of Chromatography A* 964 (2002) 99.
82. A. Seidel-Morgenstern, G. Guiochon, *Journal of Chromatography* 631 (1993) 37.
83. S. Jacobson, S. Golshan-Shirazi, G. Guiochon, *Journal of Chromatography* 522 (1990) 23.
84. E. N. Reshetova, L. D. Asnin, *Russian Journal of Physical Chemistry* 89 (2015) 275.
85. L. Asnin, K. Kaczmarski, A. Felinger, F. Gritti, G. Guiochon, *Journal of Chromatography A* 1101 (2006) 158.
86. K. Mihlbachler, M. A. De Jesús, K. Kaczmarski, M. J. Sepaniak, A. Seidel-Morgenstern, G. Guiochon, *Journal of Chromatography A* 1113 (2006) 148.
87. C. F. Zhao, N. M. Cann, *Journal of Chromatography A* 1149 (2007) 197.
88. M. Ashtari, N. M. Cann, *Journal of Chromatography A* 1218 (2011) 6331.
89. C. Horvath, W. Melander, I. Molnar, *Journal of Chromatography* 125 (1976) 129.
90. C. Horvath, W. Melander, I. Molnar, *Analytical Chemistry* 49 (1977) 142.
91. Y. V. Kazakevich, R. LoBrutto, F. Chan, T. Patel, *Journal of Chromatography A* 913 (2001) 75.
92. W. H. Pirkle, T. C. Pochapski, *Chemical Reviews* 89 (1989) 347.
93. R. Bentley, *Archives of Biochemistry and Biophysics* 414 (2003) 1.
94. V. A. Davankov, *Chirality* 9 (1997) 99.
95. V. Davankov, V. R. Meyer, M. Rais, *Chirality* 2 (1990) 208.
96. D. R. Taylor, K. Maher, *Journal of Chromatographic Science* 30 (1992) 67.
97. S. Topiol, M. Sabio, *Journal of the American Chemical Society* 111 (1989) 4109.
98. X. Bao, L. J. Broadbelt, R. Q. Snurr, *Molecular Simulation* 35 (2009) 50.
99. X. Bao, L. J. Broadbelt, R. Q. Snurr, *Physical Chemistry Chemical Physics* 12 (2010) 6466.

100. A. V. Kiselev, *Discussion of Faraday Society* 40 (1965) 205.
101. L. Asnin, K. Kaczmarski, G. Guiochon, *Journal of Chromatography A* 1217 (2010) 7055.
102. J. Haginaka, *Journal of Chromatography B* 875 (2008) 12.
103. I. Slama, C. Ravelet, A. Villet, A. Ravel, C. Grosset, E. Peyrin, *Journal of Chromatographic Science* 40 (2002) 83.
104. H. Hetteger, M. Kohout, V. Mimini, W. Lindner, *Journal of Chromatography A* 1337 (2014) 85.
105. P. Levkin, N. M. Maier, W. Lindner, V. Schurig, *Journal of Chromatography A* 1269 (2012) 270.
106. L. Asnin, K. Kaczmarski, G. Guiochon, *Journal of Chromatography A* 1192 (2008) 62.
107. L. Asnin, G. Guiochon, *Journal of Chromatography A* 1217 (2010) 1709.
108. H. Kim, G. Guiochon, *Analytical Chemistry* 77 (2005) 2496.
109. F. L. X. Liu, M. Wei, X. Duan, *AIChE Journal* 53 (2007) 1591.
110. A. Cavazzini, K. Kaczmarski, P. Szabelski, D. Zhou, X. Liu, G. Guiochon, *Analytical Chemistry* 73 (2001) 5704.
111. Y. Chen, M. Kele, I. Quiñones, B. Sellergren, G. Guiochon, *Journal of Chromatography A* 927 (2001) 1.
112. A. Péter, G. Török, F. Fülöp, *Journal of Chromatographic Science* 36 (1998) 311.
113. H. Y. Aboul-Enein, *Journal of Chromatography A* 906 (2001) 185–193.
114. H. Y. Aboul-Enein, I. A. Al-Duraibi, *Journal of Liquid Chromatography & Related Technologies* 21 (1998) 1817.
115. Ch. Czerwenka, W. Lindner, *Analytical and Bioanalytical Chemistry* 382 (2005) 599.
116. Ch. Czerwenka, M. Lämmerhofer, W. Lindner, *Journal of Pharmaceutical and Biomedical Analysis* 30 (2003) 1789.
117. S. Akapo, C. McCrea, J. Gupta, M. Roach, W. Skinner, *Journal of Pharmaceutical and Biomedical Analysis* 49 (2009) 632.
118. R. Brüger, H. Arm, *Journal of Chromatography* 592 (1992) 309.
119. R. B. Kasat, N.-H. L. Wang, E. I. Franses, *Journal of Chromatography A* 1190 (2008) 110.
120. T. S. van Erp, T. P. Caremans, D. Dubbeldam, A. Martin-Calvo, S. Calero, J. A. Martens, *Angewandte Chemie* 122 (2010) 3074.
121. A. Martin-Calvo, S. Calero, J. A. Martens, T. S. van Erp, *Journal of Physical Chemistry C* 117 (2013) 1524.
122. H. L. Wang, U. Duda, C. J. Radke, *Journal of Colloid and Interface Science* 66 (1978) 152.
123. F. Rouquerol, J. Rouquerol, K. Sing, *Adsorption by Powders and Porous Solids: Principles, Methodology and Applications.* Academic Press, New York, 1999, pp. 110–112.
124. V. K. Dobruskin, *Langmuir* 12 (1996) 5606.
125. J. K. Garbacz, S. Furmaniak, A. P. Terzyk, M. Grabiec, *Journal of Colloid and Interface Science* 359 (2011) 512.
126. F. Stoeckli, *Russian Chemical Bulletin* 50 (2001) 2265.
127. A. Seidel-Morgenstern, G. Guiochon, *Chemical Engineering Science* 48 (1993) 2787.
128. K. Mihlbachler, K. Kaczmarski, A. Seidel-Morgenstern, G. Guiochon, *Journal of Chromatography A* 955 (2002) 35.
129. L. Asnin, K. Kaczmarski, G. Guiochon, *Journal of Chromatography A* 1138 (2007) 158.
130. M. Moreau, P. Valentin, C. Vidal-Madjar, B. C. Lin, G. Guiochon, *Journal of Colloid and Interface Science* 141 (1991) 127.
131. K. Kaczmarski, D. Zhou, M. Gubernak, G. Guiochon, *Biotechnology Progress* 19 (2003) 455.
132. P. Szabelski, J. Talbot, *Journal of Computational Chemistry* 25 (2004) 1779.
133. M. C. Chervenak, E. J. Toon, *Journal of the American Chemical Society* 116 (1994) 10533.

134. P. Macaudiére, M. Lienne, M. Caude, R. Rosset, A. Tambuté, *Journal of Chromatography* 467 (1989) 357.
135. S. Ma, S. Shen, H. Lee, M. Eriksson, X. Zeng, J. Xu, K. Fandrick, N. Yee, C. Senanayake, N. Grinberg, *Journal of Chromatography A* 1216 (2009) 3784.
136. O. Gyllenhaal, M. Stefansson, *Journal of Pharmaceutical and Biomedical Analysis* 46 (2008) 860.
137. K. A. Dill, *Journal of Physical Chemistry* 91 (1987) 1980.
138. W. H. Pirkle, C. J. Welch, *Journal of Liquid Chromatography* 14 (1991) 2027.
139. C. V. Hoffmann, R. Reischl, N. M. Maier, M. Lämmerhofer, W. Lindner, *Journal of Chromatography A* 1216 (2009) 1157.
140. M. Guillaume, A. Jaulmes, B. Sebille, N. Thuaud, C. Vidal-Madjar, *Journal of Chromatography B* 753 (2001) 131.
141. Y. V. Kazakevich, *Journal of Chromatography A* 1126 (2006) 232.
142. E. N. Reshetova, L. D. Asnin, *Russian Journal of Physical Chemistry* 85 (2011) 1434.
143. K. Tachibana, A. Ohnishi, *Journal of Chromatography A* 906 (2001) 127.
144. F. Gritti, G. Guiochon, *Journal of Separation Science* 31 (2008) 3657.
145. F. Gritti, G. Guiochon, *Journal of Chromatography A* 1216 (2009) 1776.
146. F. Gritti, G. Guiochon, *Journal of Chromatography A* 1216 (2009) 8874.
147. T. Pabst, G. Carta, *Journal of Chromatography A* 1142 (2007) 19.
148. M. D. LeVan, T. Vermeulen, *Journal of Physical Chemistry* 85 (1981) 3247.
149. G. Guiochon, A. Felinger, A. M. Katti, S. Golshan-Shirazi, *Fundamentals of Preparative and Nonlinear Chromatography,* 2nd ed., Academic Press, Boston, MA, 2006.
150. B.-G. Lim, C.-B. Ching, R. B. H. Tan, *Separations Technology* 5 (1995) 213.
151. I. Quiñones, G. Guiochon, *Langmuir* 12 (1996) 5433.
152. F. James, M. Sepúlveda, F. Charton, I. Quiñones, G. Guiochon, *Chemical Engineering Science* 54 (1999) 1677.
153. C. J. Radke, J. M. Prausnitz, *AIChE Journal* 18 (1972) 761.
154. A. L. Myers, *AIChE Journal* 29 (1983) 691.
155. C. Migliorini, M. Mazzotti, G. Zenoni, M. Pedeferri, M. Morbidelli, *AIChE Journal* 46 (2000) 1530.
156. V. M. Gun'ko, *Theoretical and Experimental Chemistry* 43 (2007) 139.
157. X. Wang, C. B. Ching, *Industrial & Engineering Chemistry Research* 42 (2003) 6171.
158. S. Golshan-Shirazi, G. Guiochon, *Analytical Chemistry* 61 (1989) 2373.
159. T. Fornstedt, G. Guiochon, *Analytical Chemistry* 66 (1994) 2116.
160. R. Arnell, P. Forssén, T. Fornstedt, *Analytical Chemistry* 79 (2007) 5838.
161. R. Arnell, P. Forssén, T. Fornstedt, R. Sardella, M. Lämmerhofer, W. Lindner, *Journal of Chromatography A* 1216 (2009) 3480.
162. W. H. Pirkle, *Journal of Chromatography* 558 (1991) 1.
163. F. Wang, T. O'Brien, T. Dowling, G. Bicker, J. Wyvratt, *Journal of Chromatography A* 958 (2002) 69.
164. R. W. Stringham, J. A. Blackwell, *Analytical Chemistry* 68 (1996) 2179.
165. A. Berthod, W. Li, D. W. Armstrong, *Analytical Chemistry* 64 (1992) 873.
166. I. Ilisz, N. Grecsó, A. Aranyi, P. Suchotin, D. Tymecka, B. Wilenska, A. Misicka, F. Fülöp, W. Lindner, A. Péter, *Journal of Chromatography A* 1334 (2014) 44.
167. M. Fuereder, I. N. Majeed, S. Panke, M. Bechtold, *Journal of Chromatography A* 1346 (2014) 34.
168. F. Zhan, G. Yu, B. Yao, X. Guo, T. Liang, M. Yu, Q. Zeng, W. Weng, *Journal of Chromatography A* 1217 (2010) 4278.
169. L. Chankvetadze, N. Ghibradze, M. Karchkhadze, L. Peng, T. Farkas, B. Chankvetadze, *Journal of Chromatography A* 1218 (2011) 6554.
170. J. I. Seeman, H. V. Secor, D. W. Armstrong, K. D. Timmons, T. J. Ward, *Analytical Chemistry* 60 (1988) 2120.

171. P. Jandera, M. Škavrada, K. Klemmová, V. Bačkovská, G. Guiochon, *Journal of Chromatography A* 917 (2001) 123.
172. G. D. H. Dijkstra, R. M. Kellogg, H. Wynberg, J. S. Svendsen, I. Marko, K. B. Sharpless, *Journal of the American Chemical Society* 111 (1989) 8069.
173. N. M. Maier, S. Schefzick, G. M. Lombardo, M. Feliz, K. Rissanen, W. Lindner, K. B. Lipkowitz, *Journal of the American Chemical Society* 124 (2002) 8611.
174. D. W. Armstrong, L. W. Chang, S. C. Chang, X. Wang, H. Ibrahim, G. R. Reid III, T. E. Beesley, *Journal of Liquid Chromatography & Related Technologies* 20 (1997) 3279.
175. C. B. Ching, P. Fu, S. C. Ng, Y. K. Xu, *Journal of Chromatography A* 898 (2000) 53.
176. R. M. Wenslow Jr., T. Wang, *Analytical Chemistry* 73 (2001) 4190.
177. T. Wang, R. M. Wenslow, Jr., *Journal of Chromatography A* 1015 (2003) 99.
178. R. Isaksson, P. Erlandsson, L. Hansson, A. Holmberg, S. Berner, *Journal of Chromatography* 498 (1990) 257.
179. M. Maftouh, Ch. Granier-Loyaux, E. Chavana, J. Marini, A. Pradines, Y. V. Heyden, C. Picard, *Journal of Chromatography A* 1088 (2005) 67.
180. M. Okamoto, H. Nakazawa, *Journal of Chromatography* 588 (1991) 177.
181. I. Matarashvili, L. Chankvetadze, S. Fanali, T. Farkas, B. Chankvetadze, *Journal of Separation Science* 36 (2013) 140.
182. A. Wakisaka, H. Abdoul-Carime, Y. Yamamoto, Y. Kiyozumi, *Journal of the Chemical Society, Faraday Transactions* 94 (1998) 369.
183. Y. Tang, *Chirality* 8 (1996) 136.
184. Y. K. Ye, R. Stringham, *Journal of Chromatography A* 927 (2001) 53.
185. Y. Liu, A. Berthod, C. R. Mitchell, T. L. Xiao, B. Zhang, D. W. Armstrong, *Journal of Chromatography A* 978 (2002) 185.
186. C. Hamman, D. E. Schmidt Jr., M. Wong, M. Hayes, *Journal of Chromatography A* 1218 (2011) 7886.
187. R. Pell, W. Lindner, *Journal of Chromatography A* 1245 (2012) 175.
188. R. W. Stringham, *Journal of Chromatography A* 1070 (2005) 163.
189. B. Loukili, Ch. Dufresne, E. Jourdan, C. Grosset, A. Ravel, A. Villet, E. Peyrin, *Journal of Chromatography A* 986 (2003) 45.
190. K. Balmér, P.-O. Lagerström, S. Larsson, B.-A. Persson, *Journal of Chromatography* 631 (1993) 191.
191. F. Gritti, G. Guiochon, *Journal of Chromatography A* 1099 (2005) 1.
192. S. Ottiger, J. Kluge, A. Rajendran, M. Mazzotti, *Journal of Chromatography A* 1162 (2007) 74.
193. I. Poplewska, R. Kramarz, W. Piątkowski, A. Seidel-Morgenstern, D. Antos, *Journal of Chromatography A* 1173 (2007) 58.
194. C. Wenda, A. Rajendran, *Journal of Chromatography A* 1216 (2009) 8750.
195. X. Huang, H. Zou, X. Chen, Q. Luo, L. Kong, *Journal of Chromatography A* 984 (2003) 273.
196. J. Ståhlberg, *Chromatographia* 24 (1987) 820.
197. A. Ishikawa, T. Shibata, *Journal of Liquid Chromatography* 16 (1993) 859.
198. C. R. Mitchell, D. W. Armstrong, in *Methods in Molecular Biology–Chiral Separations: Methods and Protocols*, Vol. 243, eds. G. Gübitz and M. G. Schmid, Humana Press, Totowa, NJ, 2004, pp. 61–112.
199. S. H. Lee, A. Berthod, D. W. Armstrong, *Journal of Chromatography* 603 (1992) 83.
200. K. Fujimura, S. Suzuki, K. Hayashi, S. Masuda, *Analytical Chemistry* 62 (1990) 2198.
201. A. Méndez, E. Bosch, M. Roses, U. D. Neue, *Journal of Chromatography A* 986 (2003) 33.
202. J. G. Hou, Y. L. Wang, C. X. Li, X. Q. Han, J. Z. Gao, J. W. Kang, *Chromatographia* 50 (1999) 89.
203. R. A. Olsen, D. Borchardt, L. Mink, A. Agarwal, L. J. Mueller, F. Zaera, *Journal of the American Chemical Society* 128 (2006) 15594.

204. A. F. Prokhorova, E. N. Shapovalova, A. V. Shpak, S. M. Staroverov, O. A. Shpigun, *Journal of Chromatography A* 1216 (2009) 1216.
205. J. Iredale, A.-F. Aubry, I. Wainer, *Chromatographia* 31 (1991) 329.
206. D. W. Armstrong, Y. Liu, H. Ekborgott, *Chirality* 7 (1995) 474.
207. T. Fornstedt, G. Götmar, M. Andersson, G. Guiochon, *Journal of the American Chemical Society* 121 (1999) 1164.
208. G. Götmar, T. Fornstedt, G. Guiochon, *Analytical Chemistry* 72 (2000) 3908.
209. G. Götmar, N. R. Albareda, T. Fornstedt, *Analytical Chemistry* 74 (2002) 2950.
210. J. Ståhlberg, *Journal of Chromatography A* 855 (1999) 3.
211. F. F. Cantwell, S. Puon, *Analytical Chemistry* 51 (1979) 623.
212. F. Gritti, G. Guiochon, *Journal of Chromatography A* 1282 (2013) 46.
213. F. Gritti, G. Guiochon, *Journal of Chromatography A* 1217 (2010) 5584.
214. H. Gika, M. Lämmerhofer, I. Papadoyannis, W. Lindner, *Journal of Chromatography B* 800 (2004) 193.
215. L. D. Asnin, Y. K. Nikitina, *Journal of Separation Science* 37 (2014) 390.
216. B. L. He, in *Chiral recognition in separation methods*, ed. A. Berthod, Springer-Verlag, Berlin/Heidelberg, Germany, 2010, pp. 153–201.
217. A. Cavazzini, L. Pasti, A. Massi, N. Marchetti, F. Dondi, *Analytica Chimica Acta* 706 (2011) 205.
218. G. K. E. Scriba, *Chromatographia* 75 (2012) 815.
219. B. Natalini, R. Sardella, A. Macchiarulo, M. Marinozzi, E. Camaioni, R. Pellicciari, in *Advances in Chromatography*, Vol. 49, eds. E. Grushka and N. Grinberg, CRC Press, Boca Raton, FL, 2011, pp. 71–134.
220. A. Ciogli, D. Kotoni, F. Gasparrini, M. Pierini, C. Villani, *Topics in Current Chemistry* 340 (2013) 73.
221. K. Harata, *Chemical Reviews* 98 (1998) 1803.
222. T. Ikai, Y. Okamoto, *Chemical Reviews* 109 (2009) 6077.
223. B. Chankvetadze, *Journal of Chromatography A* 1269 (2012) 26.
224. J. Haginaka, in *Advances in Chromatography*, Vol. 49, eds. E. Grushka and N. Grinberg, CRC Press, Boca Raton, FL, 2011, pp. 37–70.
225. W. J. Cheong, F. Ali, J. H. Choi, J. L. Kim, Y. Sung, *Talanta* 106 (2013) 45.
226. T. Takeuchi, J. Haginaka, *Journal of Chromatography B* 728 (1999) 1.
227. I. D'Acquarica, F. Gasparrini, D. Misiti, M. Pierini, C. Villani, in *Advances in Chromatography*, Vol. 46, eds. E. Grushka and N. Grinberg, CRC Press, Boca Raton, FL, 2008, pp. 109–174.
228. I. Ilisz, Z. Pataj, A. Aranyi, A. Péter, *Separation and Purification Reviews* 41 (2012) 207.
229. C. Fernandes, M. E. Tiritan, M. Pinto, *Chromatographia* 76 (2013) 871.
230. M. Lämmerhofer, W. Lindner, in *Advances in Chromatography*, Vol. 46, eds. E. Grushka and N. Grinberg, CRC Press, Boca Raton, FL, 2008, pp. 1–107.
231. F. Charton, S. C. Jacobson, G. Guiochon, *Journal of Chromatography* 630 (1993) 21.
232. F. Charton, M. Bailly, G. Guiochon, *Journal of Chromatography A* 687 (1994) 13.
233. S. Jacobson, G. Guiochon, *Journal of Chromatography* 600 (1992) 37.
234. A. Berthod, *Analytical Chemistry* 78 (2006) 2093.
235. A. Felinger, A. Cavazzini, G. Guiochon, *Journal of Chromatography A* 986 (2003) 207.
236. J. Lindholm, T. Fornstedt, *Journal of Chromatography A* 1095 (2005) 50.
237. A. Cavazzini, F. Dondi, S. Marmai, E. Minghini, A. Massi, C. Villani, R. Rompietti, F. Gasparinni, *Analytical Chemistry* 77 (2005) 3113.
238. S. Peper, M. Lubbert, M. Johannsen, G. Brunner, *Separation Science and Technology* 37 (2002) 2545.
239. W. H. Pirkle, D. M. Alessi, M. H. Hyun, T. C. Pochapsky, *Journal of Chromatography* 398 (1987) 203.

2 The Role of Chromatography in Alzheimer's Disease Drug Discovery

Jessica Fiori, Angela De Simone, Marina Naldi, and Vincenza Andrisano

CONTENTS

2.1   INTRODUCTION

Alzheimer's disease (AD), the most common form of dementia in adults, is a neurodegenerative disorder whose physiopathological events include progressive cognitive impairment and memory loss, associated with a deficit in cholinergic neurotransmission (Davies and Maloney, 1976). Histological hallmarks that characterize this disorder comprise plaques of β-amyloid (Aβ) peptide, neurofibrillary tangles (NFTs), a dramatic loss of synapses and neurons, and a decreased level of choline acetyltransferase that correlates with a decline in mental status scores (Bartus et al., 1982). This disease is still a serious burden for society because no effective pharmacological treatment has yet been found. The only symptomatic drugs available on the market can just partially restore the cholinergic deficit in the first stages of the disease. Therefore, the research is strongly focused on the discovery of effective new drugs, which can halt and reverse the disease. Owing to the multifactorial nature of AD, one of the most promising drug discovery approaches for treatment is addressed to compounds with a multitarget biological profile, the so-called multitarget-directed ligands (MTDLs) (Cavalli et al., 2008). MTDLs developed so far include derivatives that can simultaneously restore brain acetylcholine (ACh) levels, decrease oxidative stress, inhibit Aβ aggregation and formation, decrease tau protein hyperphosphorylation, and protect neuronal cells against toxic insults (Prati et al., 2015).

The research efforts in the field of drug discovery are based on the knowledge of the molecular aspects of the disease and on the development of new techniques necessary to investigate the biological systems at molecular level. The selection of new leads to enter clinical trials is therefore a challenging task involving various essential steps, the first being the selection of molecules able to bind to AD validated target(s), and then the study of the effects of hitting the target at molecular, cellular, whole animal, and human level. In AD, acetylcholinesterase (AChE) has been the first target for the development of new drugs since the discovery of the cholinergic deficit in the central nervous system (CNS). However, basic research showed that cognitive impairment could be due not only to a cholinergic deficit but also to a cascade of biochemical events leading to the accumulation in the brain of proteins such as Aβ and hyperphosphorylated tau protein. Important targets are amyloid fibrillogenesis, beta-secretase amyloid precursor protein (APP) cleaving enzyme (BACE1), one of the enzymes which cleave APP, and GSK3-β, a tau protein phosphorylating kinase. On the other hand, another noncholinergic role of AChE in the AD has been discovered: Some evidences suggest that the AChE peripheral binding site may play a key role in the development of senile plaques, accelerating Aβ deposition. Once the disease targets have been selected, the determination of the new compounds' activity must be carried out quickly and in a way that allows the verification of the design hypothesis. Drug activity is in fact mediated by different types of interactions with specific biological targets and the esteem of these interactions may elucidate the mechanism of action. To this end, in first instance, high-throughput screening (HTS) methods of a large number of compounds for the selection of few lead compounds are required. Second, specific methods, which elucidate the selected compound mechanism of action *in vitro*, have to be employed before the ultimate and most advanced tools, transgenic animal models of the disease, can be used to study the effects of single compounds on the disease phenotype *in vivo*, followed by clinical trials on real patients. Toward all these aims, separation science is a demand and a resource for solving the urgent analytical problems associated with all the steps in the long pathways which go from the selection of active compounds to the complete development of new drugs.

Here, we review the purposely designed chromatographic methodologies which have contributed to defining the most important steps toward the discovery of new drugs for AD. To begin, an important contribution is given by *in vitro* assessment of the activity of chemical libraries on isolated targets by the affinity chromatography on HPLC immobilized-enzyme column (or immobilized enzyme reactors, IMERs). A further step is the *in vivo* verification of activity of the lead selected compound by monitoring the appropriate biomarkers in animal biological fluids, which are any isolated and characterized molecules capable of probing the activity/toxicity of the lead compound. Then, the verification of absorption, distribution, metabolism, and excretion (ADME) profile in humans is essential to prove that the potential new AD drugs reach the CNS, hitting the AD targets.

2.2   HITS AND LEADS SELECTION FOR AD DRUG DISCOVERY BY IMERs

In the past 15 years, big efforts have been directed toward the improvements of bioanalytical tools for exploring molecular recognition phenomena, such as the interaction of small molecules with target proteins, by exploiting the hyphenation of separation systems with selective detectors. In this context, immobilized target proteins, such as enzymes- and receptors-based analytical methods for drug discovery, facilitated automated HTS of new drug candidate molecules. In fact, the importance of using a bioanalytical platform for HTS in order to select enzyme inhibitors from a large library of compounds in the early drug discovery phase is well known.

We are here reviewing the work published in the field of AD, which involves the use of immobilized target enzymes inserted in separation systems for the screening of new enzyme inhibitors. The possibility of performing a mild immobilization strategy on purposely dedicated polymeric materials gave rise to stable IMERs, which are used to follow the enzymatic reaction, to determine enzyme kinetics, and to screen for new inhibitors after their insertion into a chromatographic system coupled to a suitable detector, selective for monitoring either substrate or product of the catalyzed reaction.

In one of the most used *operandi*, the substrate solution is injected and transported by the mobile phase, usually an aqueous buffer with a small percentage of organic solvent, to the IMER, where the enzymatic reaction takes place with the formation of products. Then, unprocessed substrate and newly formed product are eluted and selectively identified by the detector (UV-DAD; fluorescence; or mass spectrometry, MS). The resulting substrate- and/or product-specific peaks in the chromatogram can be used to determine loss of substrate, formation of new product and, by the decrease of product peak, to screen for inhibitors. Usually, IMERs maintain high specificity and catalytic efficiency; moreover, they result in more stable forms under unfavorable conditions such as heat, pH, and organic solvents. After immobilization, enzymes were found to preserve their activity for longer periods. Thereby, advantages in their use are minimized cost, short analysis time, and operating in a continuous mode. IMERs can be widely used in many fields of drug discovery (hit identification, lead optimization, ADMET studies, commercial drug quality control) (Bertucci et al., 2003; Girelli and Mattei, 2005). In the AD drug discovery process, IMERs were used for screening acetylcholinesterase, butyrylcholinesterase (BuChE), and b-site APP cleaving enzyme inhibitors from synthetic or natural compound libraries, since these enzymes represent the major targets of this neurodegenerative pathology. They were also applied for the finest characterization of kinetics and thermodynamic parameters of the enzymatic reactions and mode of inhibition. Principal steps for IMERs preparation and validation are the choice of support materials, immobilization strategy, the determination of enzyme activity after immobilization, and the screening of enzyme inhibitors.

2.2.1 Choice of Support Material

The most important features of polymeric matrix are good biocompatibility, known porosities, structure, and chemical group capable of interacting with the selected enzyme. All these properties can influence mechanical strength and enzyme stability in a broad range of temperature, pH, ionic strength, and organic solvents. Other physical properties, such as form, shape, porosity, pore size distribution, swelling capability, and charges affect the kinetics of immobilization process. Most promising chromatographic matrices to be used for enzyme immobilization are monoliths, a novel generation of stationary phase that have been widely used because of their major features such as fast separation and enzymatic conversion due to lack of diffusion resistance during mass transfer (Josic et al., 2001). In particular, the macroporous matrices allow high-speed analysis and lower back pressure. The most popular polymeric macroporous material is known under the trademark Convective Interaction Media (CIM®). The dimensions of CIM disk are very small (12 mm in diameter and 3 mm in thickness) and are placed in a plastic housing (www.biaseparations.com).

The choice of matrix is considered of critical importance, since enzyme activity and stability could be altered due to the specific microenvironment in which it is placed. In fact, the degree at which matrix, covalent bond, and steric constraints influence enzyme physical properties will depend on the chemistry of the matrix and on the method of immobilization.

A detailed study was carried out for the choice of best support material for AChE immobilization since it represents the first enzyme involved in AD that was immobilized on IMER support. AChE catalyses the hydrolysis of the neurotransmitter ACh, therefore inhibitors of this enzyme proved to be useful to improve the cholinergic deficit in AD.

Two polymeric monolithic disks with different reactive groups (epoxy and ethylenediamine) and a packed silica column were compared. The different matrices have different chemistries (epoxy or aldehydic active groups), structural features (polymeric or silica-based material), and format (loose material or monolithic column). The advantages and disadvantages of different supports were established comparing immobilization rate, immobilized enzyme stability, conditioning time for HPLC analyses, optimum mobile phase and peaks shape, aspecific interaction, and costs. The first AChE immobilization was performed on a monolithic ethylenediamino CIM disk (12 mm × 3 mm i.d.) by *in situ* covalent immobilization after matrix activation. The proposed immobilization procedure gave a 3.0% yield of active immobilized enzyme and was found appropriate to covalently retain 0.22 ± 0.01 U of AChE (Bartolini et al., 2004). The same amount of enzyme corresponding to 12.4 μg of protein was used for AChE immobilization on epoxy-CIM disk. Yield was found around 3.0% and 0.18 ± 0.01 U were retained on the epoxy matrix. The amount of active

AChE-immobilized units on Glut-P-IMER, determined in flow-through conditions after insertion into a HPLC system resulted in 4.35 ± 0.01 U (immobilization yield: 29%) (Bartolini et al., 2005). This last approach is based on the application of a wide pore silica particles commercially available as bulk material that offers the possibility of performing a large number of parallel studies batch-wise (offline studies). In that case, batch-wise experiments could be previously performed in eppendorf tubes in order to optimize the immobilization conditions (i.e., optimum pH of immobilization, enzyme/matrix ratio, and time of incubation). An aliquot of 30 mg of glutaraldehyde-P (Glut-P), pH 5.0, and 8 h of incubation were found to be suitable conditions for AChE immobilization. The obtained AChE-IMERs were characterized by offline studies. Loose particles were packed into an empty column (Amersham Bioscience, UK) to obtain the Glut-P-IMER. The higher percent of immobilized enzyme active units reported for AChE-GlutP was found related to the wider surface contact area between the small particles of silica matrix and the enzyme. This facilitates the interaction between the reactive lysines' primary amino groups of enzyme and glutaraldehyde groups on silica matrix. Glut-P-AChE-IMER showed a slower catalysis rate since the pore geometry can produce mobile phase stagnation and low substrate diffusion. On the other hand, epoxy and ethylendiamino CIM disks did not show this inconvenience due to their monolithic structure, high porosity, shorter length which facilitates conditioning time, easy access of substrate, and fast recovery of enzymatic activity. The least stable IMER in terms of immobilized enzyme activity was obtained for AChE-epoxy-CIM, probably due to the lacking spacer chain between solid support and enzyme. In fact, keeping the enzyme at a favorable distance from matrix avoids enzyme adsorption and denaturation due to the loss of suitable conformation. Despite Glut-P-IMER presents a lot of advantages in terms of chromatographic support costs, allows offline the optimization of the immobilization conditions, offers the possibility of varying the amount of matrix, and the final IMER dimension, the EDA–CIM monolithic support was found to be the best choice, because immobilized enzyme activity was retained for a much longer time.

Under storage conditions, over 80% of the initial activity was retained on EDA and Glut-P IMERs up to 2 months. Epoxy-CIM-IMER showed an higher instability and 70% of activity was lost after 2 months of use due to the lack of spacer chain. Moreover, after 15 months of daily use and more than 2000 injections, 12% of enzyme initial activity was still retained by immobilized enzyme on EDA–CIM disk. The AChE-epoxy-CIM was found to be inactivated in an even shorter time. Moreover, comparison of functional groups on matrices showed that both types of polymer and surface modification might affect stability after the immobilization reaction (Bartolini et al., 2005). So, for all these reason the EDA–CIM disks are preferred for AChE and other enzyme immobilization as reported by studies carried out with BuChE and BACE-1.

In fact, different matrices where also used for BuChE immobilization. Luckarift et al. (2006) investigated the influence of different supports of stationary phase and immobilization methodology on this enzyme. The effect of two flow-through systems was initially investigated to evaluate silica-entrapped BuChE: a fluidized-bed system and a packed-bed system. In both systems, a limitation was represented by BuChE low stability and other problems related to silica packaging or channeling of particles. So the possibility of using IMERs prepared with silica encapsulation *in situ* via His-tag attachment or the silica-immobilized enzyme to metal ion affinity resin was investigated. Four different columns were prepared: (1) soluble BuChE (soluble-BuChE–IMER); (2) BuChE immobilized in silica (Si-BuChE–IMER); (3) BuChE immobilized in silica, with N-terminal His6-peptide (N-His6-BuChE–IMER); and (4) BuChE immobilized in silica, with C-terminal His6-peptide (C-His6-BuChE–IMER). The amount of protein immobilized was determined for each system. The columns with silica were demonstrated to retain much more total protein (>90%) than the soluble-BuChE–IMER and exhibited high substrate conversion efficiency (~60%). What differs from Si-BuChE–IMER and C-His6-BuChE–IMER or N-His6-BuChE–IMER is the loss of activity over time due to the gradual elution of silica particles from the column during continuous flow. So, physical attachment of the silica particles via the His-tag resulted in more stable IMER preparations, in terms of stability and retention of enzyme activity than the soluble-BuChE–IMER (Luckarift et al., 2006).

By following the AChE positive results, BuChE enzyme was covalently immobilized on EDA–CIM disk (Andrisano et al., 2001; Bartolini et al., 2005). The concentration of active units, determined after inserting IMER into an HPLC system connect with an UV–vis detector, was found in agreement with that found in healthy brains (0.99 ± 0.01); the immobilization yield was found to be 8.3%.

In agreement with the previously described immobilizing procedure, and the advantages of EDA–CIM disk, β-secretase micro-IMER (hrBACE-1-micro-IMER) was prepared under the same experimental conditions. The immobilization yield was found to be 43.17 ± 0.43% and 6.18% of initial active units were retained on the EDA matrix (Mancini et al., 2007).

2.2.2 ENZYME IMMOBILIZATION STRATEGIES

Many techniques could be used for enzyme immobilization, which can be distinguished into enzyme entrapment (Vilanova et al., 1984; Luckarift et al., 2006), cross-linking (Hu et al., 2008; Freije et al., 2008; Wu et al., 2011), adsorption, covalent bonding, and a combination of these methods. Among these, covalent immobilization was found to be the best immobilization strategy for AD target enzymes.

Each AChE, BChE, and BACE-1 IMER was prepared by using an *in situ* covalent immobilization procedure on an ethylenediamine monolithic CIM (EDA–CIM) disk according to enzyme features. The covalent immobilization was accomplished by two different techniques:

1. By insertion of the monolithic column in a flow-through system and the passage of solution containing target enzyme (Bertucci et al., 2003)
2. By dipping the monolithic column into a solution containing the target enzyme (Mallik and Hage, 2006)

The immobilization on CIM disk is based on initial enzyme adsorption on the support followed by a covalent linkage between the nucleophilic groups of enzyme as amine, thiol, or hydroxy groups and the epoxy groups on the matrix (Wheatley and Schmidt, 1999). Since enzymes are very large molecules, their active center could not be totally accessible. So, the insertion of a spacer is fundamental to improve the amount of immobilized enzyme (Bartolini et al., 2005). Moreover, it is important to keep the enzyme away from matrix in order to preserve it from adsorption phenomena or loss of suitable conformation. The chemistry at the basis of enzyme coupling on EDA–CIM disk and Glut-P is based on a Shiff base reaction followed by one additional step to reduce and stabilize the imine groups. A previous activation with a bifunctional agent as glutaric dialdehyde is required for EDA–CIM, since the former matrix is originally a weak ionic exchange polymeric column. Other important considerations have to be made about the reactive groups on the matrix to influence enzyme stability. AChE-epoxy-CIM is less stable than AChE-EDA-CIM because residual hydroxyl vicinal group can give some instability to the newly formed C–N bond. So, even for this reason EDA–CIM disk is the favorite immobilization matrix. The experimental conditions adopted for all enzymes are similar. EDA–CIM disk is required to be inserted into the HPLC system approximately for 30 min and conditioned with mobile phase consisting of phosphate buffer (50 mM, pH 6.0). Then, the CIM disk is removed, placed in a glass beaker covered with 10 ml of 10% glutaraldehyde solution in phosphate buffer (50 mM, pH 6.0) for derivatization and kept under stirring for 6 h, in the dark. The reacted matrix was then washed with phosphate buffer (50 mM, pH 6.0). After that, less than 1 mL of enzyme solution (15 U/mL) in phosphate buffer is added to the matrix and left to react overnight under gentle stirring (Bartolini et al., 2005, 2009). Some differences consisting of flushing glutaraldehyde solution into HPLC system are reported for BACE-1 immobilization procedure. The enzyme is diluted in 20 mM Hepes buffer at pH 7.4 containing 125 mM NaCl and left to react with matrix under mild stirring in a cold room (~4°C) for 22 h (Mancini and Andrisano, 2010).

2.2.3 Determination of Enzyme Activity after Immobilization

The reaction rates of the immobilized enzyme depend on the enzymatic intrinsic activity, the amount of loaded enzyme, the substrate accessibility to the active sites and its concentration, and diffusivity. After each immobilization, the amount of immobilized enzyme, its activity, and stability have to be determined. In fact, the amount of immobilized active enzyme doesn't always correspond to that of active enzyme. The offline quantification of immobilization is carried out usually by colorimetric assay or using fluorescent or radiolabeled ligands (Lowry et al., 1951; Bradford, 1976; Besanger et al., 2006). Differently, the determination of active units when IMER is inserted into a HPLC apparatus is carried out by injecting increasing concentration of the specific substrate into the chromatographic system and monitoring the resulting product by UV–visible, fluorescence, or MS detector (Hodgson et al., 2005) according to spectroscopic features. Concerning cholinesterase-IMERs, thiocholine is obtained as product from acetylthiocholine (ATCh) or butyrylthiocholine (BTCh) hydrolysis. Since choline molecule doesn't present any significant chromophore for UV detection, its thio-analogue is used and Ellman's reagent (Ellmann et al., 1961) is added to the mobile phase. Reaction of thiocholine with Ellman reagent gives rise to a colored anion that can be monitored at 412 nm. In the first reported AChE-IMER application, the amount of resulting adduct was found proportional to ATCh concentration, injected into the chromatographic apparatus. The chromatographic conditions such as pH, mobile phase composition, temperature, and percentage of organic solvent influence not only chromatographic parameters, but also the enzymatic activity. For AChE-IMERs, the best chromatographic condition were well established in terms of type of buffer, saline concentration, and Ellman's reagent content. The experiments were carried out conditioning the hrAChE-CIM disk with 0.1 M phosphate buffer solution at pH 7.4, and then with a mobile phase consisting of Tris–HCl buffer (pH 8.0) containing Ellman's reagent, $MgSO_4$, and $KClO_3$. The evaluation of the ChE active units retained was obtained after injecting aliquots of substrate at increasing concentration. Aliquots of 10 µL ATCh aqueous solution at increasing concentration (range comprised between 3.1 and 250 mM), were injected in the HPLC system, at a flow rate of 1.0 mL/min and UV detection at 412–450 nm. (Bartolini et al., 2004, 2005).

The described chromatographic conditions allowed yellow anion to be eluted in 2 min and the substrate saturating concentration was also determined by Michaelis–Menten equation, by plotting the area of the resulting product versus substrate concentration.

The chromatographic conditions adopted for BuChE-IMER studies were similar. Potassium chlorate was maintained as a selective anion exchanger competitor for the protonated amine groups on the matrix, while the presence of magnesium salt proved to be not as detrimental for activity as is the case for AChE. Another modification consisted in the addition of organic modifier in the mobile phase. The importance of this parameter is due to its capability to suppress some nonspecific interactions that could arise between more hydrophobic tested inhibitors (donepezil, propidium) and chromatographic support. The organic modifier can also influence the enzyme activity and chromatographic separation. The best compromise between optimized chromatographic separation and high enzyme activity was obtained for this application by using 2-propanol. The addition of 1% (v/v) of this organic solvent provided an optimal elution profile and increased product peak area. Flow rate was set at 1.0 mL/min and UV detection at 480 nm. Aliquots of 20 µL BTCh aqueous solution at increasing concentrations (range comprised between 3.10 and 300 mM) were injected.

In order to account for micromoles of both ATCh and BTCh hydrolyzed, eluates for each substrate injection were collected in 5 mL volumetric flasks during 5 min of chromatographic elution. The absorbance at 412 nm of each eluate was spectrophotometrically acquired, using the mobile phase as a blank. Micromoles of hydrolyzed substrate were calculated by applying Lambert–Beer law. As previously reported, the catalysis rates (µmol/min) were then derived by dividing the µmoles of substrate hydrolyzed by the contact time (0.34 min) (Andrisano et al., 2001). A linear correlation between absorbance values, relative chromatographic peak areas, and µmoles of product formed per minute was obtained and used in further analysis. A Michaelis–Menten plot was obtained by plotting the catalysis rates versus normalized substrate concentration, and K_{Mapp} and v_{maxapp} were derived. From the v_{maxapp} value, the immobilized active units were then determined. Normalized substrate concentration was calculated by the following formula:

$$[\text{substrate}]_{\text{normalized}} = (C_{\text{inj}} \times V_{\text{inj}})/BV$$

where:

C_{inj} is the injected substrate concentration
V_{inj} is the injected volume
BV is the bed volume of the BChE-IMER

Different approaches for the determination of enzyme active units on BACE-1-IMER were followed (Mancini et al., 2007). In fact, the classic in solution methods adopted for evaluation of BACE-1 inhibitors activity are based on fluorescence resonance energy transfer (FRET) technologies and different types of substrate are adopted for this assay according to spectroscopic features of tested molecules (Mancini et al., 2011). Three different substrates were employed to evaluate human recombinant (hr) BACE-1-micro-IMER activity: JMV2236, M-2420, and Substrate IV, all mimicking the peptidic sequence of the Swedish-mutated APP sequence, targeted by BACE-1 (Mancini et al., 2007; De Simone et al., 2014). These fluorogenic peptide substrates containing the β-secretase site of the Swedish mutation of APP were injected and cleaved by immobilized BACE-1, so their relative fluorescent products were detected (Figure 2.1).

In the first application, a column switching system was used. This system is applied when analytical column cannot be directly coupled to the IMER. After covalent linking of enzyme on the solid matrix the active units where determined by using JMV2236 as substrate (Andrau et al., 2003). In particular, the products of the enzymatic cleavage and the remaining noncleaved substrate were collected on a C18 column trap and switched to the liquid chromatography electrospray MS (LC–ESI–MS) system to determine the kinetic constants.

For all substrates the best chromatographic condition were established: mobile phase composition, flow rate, and fluorescence detector parameters. Kinetic

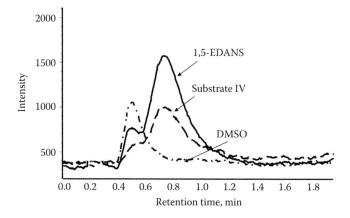

FIGURE 2.1 The overlaid chromatograms were obtained by injecting reference hydrolysis product [5-(2-aminoethylamino)-1-naphthalenesulfonic acid] 1,5 EDANS, Substrate IV, and DMSO onto hrBACE1-IMER. The peak obtained after substrate injection and enzymatic hydrolysis was found consistent with that of standard product in terms of retention time. (From De Simone, A. et al., *J. Chromatogr. B*, 953–954, 108–114, 2014. With permission.)

TABLE 2.1
Chromatographic Conditions Used for Different BACE-1 Substrates and Relative Kinetic Parameters Evaluated with Immobilized BACE1-IMERs

| Substrate Name | Chromatographic Conditions | | | Kinetic Parameters | | |
	Mobile Phase	Flow	Λ Detection	k_M (μM)	v_{max} (pmol min^{-1})	References
JMV2236	0.1 M sodium acetate, 10 mM MgCl$_2$ (pH 4.5)/ DMSO 95/5(v/v)	1 mL/min	320 nm	12.10 ± 2.00	1.24 ± 0.08	Mancini et al. (2007)
M-2420	25 mM sodium phosphate (pH 5.5) CHAPS0.1% w/v/ DMSO 95/5(v/v)	1 mL/min	320$_{ex}$/450$_{em}$	8.32 ± 0.49	3.48 ± 0.09	Mancini and Andrisano (2010)
Substrate IV	50 mM sodium phosphate (pH 5.0)	1 mL/min	355$_{ex}$/490$_{em}$	8.28 ± 0.53	1.47 ± 0.00	De Simone et al. (2014)

parameters in terms of k_m and v_{max} were calculated for each substrate as previously described for ChE-IMERs and are reported in Table 2.1 with chromatographic conditions.

A pH value, different from physiological one, was chosen for the mobile phase, the best pH value was found in the range of 4.5–5.5 according to literature data. In fact, BACE-1 is an aspartic protease showing an optimum catalytic efficiency at acidic pH, near to the pK_a of aspartate (Haniu et al., 2000; Andrau et al., 2003).

2.2.4 INHIBITION STUDIES AND SELECTION OF HITS AND LEAD COMPOUNDS

The small reaction volume, short analysis time, high sensitivity, and low cost are the major advantages of these methods compared to the in-solution conventional ones. Indeed, this strategy results are suitable for the HTS of potential AChE, BChE, and BACE1 inhibitors from synthetic compound libraries, natural medicinal plants, and other complex samples. In the online system, the enzyme substrate is processed very fast (a few minutes vs. hours) as the use of monolithic matrix enhances reaction rate due to the high local concentration of immobilized enzyme, together with a better accessibility of the active site for the substrate. In particular, enzyme inhibition is evaluated by simultaneously injecting a mixture of inhibitor and substrate at a fixed saturating concentration (as previously determined by the Michaelis–Menten plot). The increasing inhibitor concentration results in reduction of the peak area related to product formation. Inhibition curves are obtained by plotting the percent inhibition against the apparent inhibitor concentration, namely the concentration of the inhibitor into the IMER (Figure 2.2). The most common problems that can emerge in testing inhibitors are nonspecific interactions between inhibitors and the matrix, which increase the concentration of inhibitors in the interstices of the IMER, thus

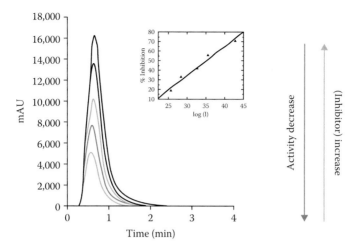

FIGURE 2.2   Overlaid chromatograms of yellow product anion obtained after injection of substrate at saturating concentration and substrate plus increasing concentration of inhibitor (edrophonium).

reducing the enzyme activity recovery. The use of a blank column (a column without any target protein immobilized) represent a successful tool to avoid "false positives" related to nonspecific interaction between the matrix and the ligand. To avoid these drawbacks, the choice of short chromatographic supports characterized by relatively large pores (macropores, 2000–3000 nm) is helpful. The determination of enzyme inhibitors' affinity can be performed either by the determination of the IC50 (concentration which reduces by the half the product peak obtained at saturating conditions) and/or by determining the K_i, which is the dissociation constant for the enzyme–inhibitor complex, by the Lineweaver and Burk plot. However, a correlation curve is required to compare online values with published data obtained by classical in-solution assays. For all tested ChE known inhibitors (i.e., tacrine, donepezil, edrophonium, and ambenonium), data were in agreement with in-solution results obtained by Ellman's method (Bartolini et al., 2004, 2009).

hrBACE-1-micro-IMER was applied to study BACE-1 inhibition by employing reference inhibitors donepezil, pepstatin A, (–)gallocatechin gallate, statine derivative, OM99–2, and inhibitor IV (Mancini et al., 2007; Mancini and Andrisano, 2010), for the screening of new molecule library, for the kinetic characterization of selected hit compound (De Simone et al., 2013), and characterization of natural compound (De Simone et al., 2014). High degree of correlation between data obtained with IMERs and free enzyme was obtained. Several advantages were found over traditional in-solution methods such as FRET assay, for testing BACE-1 activity and its inhibition. In fact, in-solution method presents a lot of drawback such nonspecific inhibition, aggregation-forming inhibitors, and consequently false-positive results generation. These problems are solved using BACE-1-IMER because nonspecific inhibitors show a time-dependent behavior promoted during FRET in-solution assay since it requires more than 1 h to be carried out. So, the described HPLC methods are valid alternative to both identify and confirm the specificity of BACE-1 inhibitors avoiding the formation of aggregates of promiscuous inhibitors. Moreover, the characterization of different types of substrate for online studies allow to test a wide range of compound including natural hit (Mancini et al., 2011).

2.3 IDENTIFICATION AND VALIDATION OF AD BIOMARKERS

A biological marker or biomarker is measured and evaluated as an indicator of normal or pathogenic biological process and pharmaceutical responses to therapeutic interventions. Sensitivity, specificity, and easy-of-use are the most important factors that ultimately define the diagnostic utility of a biomarker.

The identification and validation of biomarkers for diagnosing AD and other form of dementia can be pursued by different approaches, including neuroimaging, genetic testing, proteomics, metabolomics, and neurochemical testing for body fluids, such as cerebrospinal fluid (CSF), plasma, serum, urine, and blood cells (Song et al., 2009).

In order for a biomarker to be useful, certain criteria must be assured (Sjögren et al., 2003; Zetterberg et al., 2003; Desai and Grossberg, 2005; Hampel et al., 2010). According to the current consensus criteria proposed by the National Institute on Aging (NIA) (The Ronald and Nancy Reagan Research Institute of the Alzheimer's Association and the National Institute on Aging Working Group, 1998; Frank et al., 2003), ideal biomarkers for AD should meet the characteristics reported in Table 2.2.

Since the neuropathology of AD represents a complex array (inter- and intracellular and molecular alterations, neuritic amyloid plaque deposits, change in tau phosphorylation, NFTs, inflammation, oxidative stress, loss of synapses, and neuronal cell death), it is clear that the consideration of a single biomarker might not be potent enough. To improve diagnostic specificity with respect to the related disorders, such as mild cognitive impairment (MCI), vascular dementia (VaD),

---

TABLE 2.2
Criteria of a Good Biomarker for the Diagnosis of Alzheimer's Disease

- Detect the fundamental CNS pathophysiology of AD as validated in neuropathologically confirmed cases
- Diagnostic sensitivity >85% for detecting AD
- Diagnostic specificity >75% for distinguishing between other dementias
- React upon pharmacological intervention
- Allow reliability and reproducibility in laboratories worldwide
- Measurable in noninvasive, easy-to-perform, inexpensive tests
- Changes from AD patients and controls should be at least two fold to allow differentiation
- Data reproduced by at least two independent qualified researchers and published in peer-reviewed journals

frontotemporal lobe dementia (FTLD), or Lewy body dementia (LBD), it is essential to develop methods to measure several biomarkers together in a single analysis, in order to create an accurate diagnostic profile.

Only three biomarkers have been well-established and internationally validated to diagnose AD in CSF with ELISA measurement: $A\beta$ 1–42, total tau, and phosphor-tau-181 (Sjögren et al., 2003; Zetterberg et al., 2003; Blennow, 2004, 2005). It is now accepted that only the combination of these three CSF biomarkers increase the diagnostic validity for sporadic AD, with a combined sensitivity of >95% and a specificity of 85%. However, it is a challenge to search for novel biomarkers in biological fluids by using modern potent methods, such as liquid chromatography-ESI-mass spectrometry (LC–MS), surface laser plasmon desorption/ionization (SELDI), matrix-assisted laser desorption/ionization (MALDI), BioChips, GeneChips, and microarrays.

In the present review, we will focus on the approaches to detect established or new potential biomarkers in biological fluids (CSF, plasma, etc.) by chromatography coupled to mass spectrometry in proteomic- or metabolomic-based approaches (Table 2.3). These methods may be used to search for novel biomarkers of disease, underlining the differences in protein metabolism that contribute to the disease and to evaluate how pharmacological treatments in terms of drug pharmacodynamic affect the level of these biomarkers.

2.3.1 Biomarkers Related to $A\beta$ Peptides

$A\beta$ peptide aggregation with the formation of amyloid plaques occurs in the brain under pathological conditions. Analysis (especially with ELISA) of $A\beta$ 1–42 in CSF showed a significant reduction in AD patients compared to controls, this can be caused by reduced clearance of $A\beta$ from brain to CSF/blood, and to enhanced aggregation in the brain (Humpel et al., 2011). As an alternative to the most used ELISA methods, the determination of amyloid peptides in CSF can be performed by chromatographic techniques. Due to the low concentration and complex matrices, the classical approach requires selective protein extraction that involves immunoaffinity precipitation with specific antibodies prior to the chromatography. Afterward, the enzymatic digested proteins are analyzed by reversed-phase liquid chromatography (RP–LC) coupled to mass spectrometry (RP–LC–MS) using a gradient elution. Oe et al. (2006) determined the CSF $A\beta$ 1–42 and 1–40 levels by immunoaffinity purification and RP–LC–MS with negative detection, in multiple reaction monitoring (MRM) mode. MRM is a highly sensitive and selective method for the targeted quantitation of molecules of interest like proteins and peptides. This approach allows for great speed and quantitation of an analyte of interest by monitoring a specific mass transition from parent ion to a selected fragment ion (the most specific or the most abundant one). In Oe et al. (2006) the transition involving 3 water molecule loss was followed for $A\beta$ 1–42 and 1–40. In order to overcome the $A\beta$ losses during sample collection and manipulation, the authors developed a stable isotope dilution method, obtaining good specificity, linearity, and correlation, when compared with the results obtained with ELISA assay. A reduction of $A\beta$ 1–42 and unchanged level of $A\beta$ 1–40 in AD patients were found with respect to controls, according

TABLE 2.3

Examples of Chromatographic Methods to Discover Biomarkers for AD

Biomarker	Sample	Analytical Method	Findings	References
Aβ 1–42, 1–40, Aβ glyco-peptides	CSF	Immunoaffinity purification, RP–LC–MS (MRM)	Isotope dilution, good specificity, linearity, and correlation. Reduction of Aβ 1–42 and unchanged level of Aβ 1–40 in AD patients. Administering a stable isotope-labeled amino acid (^{13}C6-leucine) to monitor synthesis and clearance of Aβ species. Increase of the dominating Tyr glycosylated versus unglycosylated Aβ1-X peptides (X ± 40 and 42) in AD compared to non-AD patients.	Oe et al. (2006), Bateman et al. (2006), Halim et al. (2011)
Aβ 1–42, Aβ 1–40, Aβ 1–38	CSF	SPE, isotope dilution LC, or UHPLC-MS/MS	Decrease level of Aβ 1–42 expressed as Aβ 1–42/Aβ 1–40 ratio. High variability in the content of Aβ peptides, avoiding the differentiation between AD and controls.	Leinenbach et al. (2014), Pannee et al. (2013), Lame et al. (2011)
Aβ digested peptides	CSF	High pH RP–LC for fractionation, LC-MS/MS	Analysis of Aβ 1–5, Aβ 6–16, Aβ 17–28, and Aβ 29–40(42), fragment of either Aβ 1–40 or Aβ 1–42, selection of Aβ 17–28 as a surrogate, representing Aβ.	Kim et al. (2014)
APL1β28	CSF	Immunoprecipitation, RP–LC–MS/MS	Amyloid precursor-like protein-1-derived Aβ-like peptide 28 (APL1β28) was used as a non-amyloidogenic marker for brain Aβ42 production in Presenilin 1 (PS1) mutations associated with familial Alzheimer's disease (FAD).	Tagami et al. (2014), Yanagida et al. (2009)
Tau, phospho-tau	CSF	RP–LC for sample preparation	Together with specific ELISA assays, this approach demonstrated that the ability of CSF tau and phospho-tau to differentiate AD from control.	Meredith et al. (2013)

(Continued)

TABLE 2.3 (*Continued*)
Examples of Chromatographic Methods to Discover Biomarkers for AD

Biomarker	Sample	Analytical Method	Findings	References
Tau, phospho-tau	CSF	Immunopurification, nano UPLC, and LC–MS/MS	Identification of fragments specific to different isoforms, which may hold diagnostic information. By adding exogenous tau increases Aβ levels.	Portelius et al. (2008), McAvoy et al. (2014), Bright et al. (2015)
YKL-40, AD-associated proteins	CSF, nondepleted plasma	2D-PAGE, nano LC–MS/MS	Identification of new biomarkers (like YKL-40) or to follow the AD-associated protein profile changes in different bio-fluids.	Craig-Schapiro et al. (2010), Hu et al. (2005, 2007), Hye et al. (2006)
AD-associated proteins	CSF	2D-PAGE, Cation-exchange LC fractionation, LC–MS/MS	CSF protein changes consistent with inflammation and synaptic loss early in FAD, potential new presymptomatic biomarkers.	Ringman et al. (2012)
AD-associated proteins	Depleted plasma	Labeling with fluorescent dyes, anion exchange and RP LC fractionation, 2D-PAGE, LC–MS/MS	Differentiation of expressed proteins from seven well-characterized AD patients and seven non-AD patients	Henkel et al. (2012)
AD-associated proteins	CSF	SELDI-TOF, purification by anion exchange, metal affinity, reversed-phase LC, 2D-PAGE, LC–MS/MS	Individuation of 30 potential biomarkers of AD, 15 of them were purified and positively identified.	Simonsen et al. (2008)
Phospholipids, sterols	Plasma	LC–MS/MS, GC–MS	Decreasing levels of desmosterol (precursor to cholesterol) in AD patients versus controls.	Sato et al. (2010, 2012)
Untargeted metabolites	CSF	GC–TOF, LC–ECA	Potential of metabolite variables to provide discrimination of the same magnitude as the CSF amyloid and tau proteins.	Motsinger-Reif et al. (2013)
Untargeted metabolites	Plasma	UHPLC–MS/MS	Differentiation of plasma levels of lysophosphatidylcholines, tryptophan, dihydrosphingosine, phytosphingosine, and hexadecasphinganine from controls and AD patients.	Li et al. (2010)

to literature data (Blennow and Humpel, 2003; Humpel et al., 2011). Bateman et al. (2006) described a method to determine the production and clearance rates of proteins within the human CNS. In particular, they developed a method for quantifying the fractional synthesis rate (FSR) and fractional clearance rate (FCR) of Aβ *in vivo* in the human CNS and plasma by administering a stable isotope-labeled amino acid ($^{13}C6$-leucine); sampling CSF; and using immunoprecipitation, trypsin digestion, and liquid chromatography–tandem mass spectrometry (LC–MS/MS) to quantify labeled Aβ. In this study, Bateman et al. (2006) determined the synthesis and clearance of all Aβ species, since Aβ was immunoprecipitated with a specific antibody directed to the central domain of the molecule. In order to measure Aβ species biogenesis and clearance, such as Aβ 1–40 or Aβ 1–42, an immunoprecipitation with C-terminal–specific antibodies was reported. This method may be used to follow the pathophysiology of AD, as a diagnostic method or to monitor new therapeutics (e.g., β- or γ-secretase inhibitors) by measuring the pharmacodynamic effect of the therapy on Aβ level increase or clearance in humans.

Until now, most quantitative measurements of Aβ in plasma or CSF have required antibody-based immunoassays. Since quantification of Aβ by immunoassays is highly dependent on the extent of epitope exposure due to aggregation or plasma protein binding, it is difficult to accurately measure the actual concentration of Aβ, hence new candidate analytical procedure based on antibody-independent methods are proposed. Leinenbach et al. (2014) developed a solid-phase extraction (SPE) and isotope-dilution LC–MS/MS validated method to determine Aβ 1–42 in CSF. This method was found useful to harmonize Aβ 1–42 assays and to facilitate the introduction of a general cutoff concentration for CSF Aβ 1–42 in clinical trials (Leinenbach et al., 2014). A similar method based on SPE and isotope-dilution liquid chromatography–selected reaction monitoring–mass spectrometry (LC–SRM–MS) was applied to determine Aβ 1–42, Aβ 1–40, and Aβ 1–38 in CSF (Pannee et al., 2013), by using a reversed-phase monolithic column. The validated method showed a lower limit of quantification (LLOQ) of 62.5 pg/mL for Aβ 1–42 than ELISA and coefficients of variations below 10%. It was applied to a pilot study on AD patients and controls, verifying disease association with decreased levels of Aβ 1–42 similar to that obtained by ELISA. An even better correlation was obtained using the Aβ 1–42/Aβ 1–40 ratio. This method was found sensitive and not affected by matrix effects, enabling absolute quantification of Aβ 1–42, Aβ 1–40, and Aβ 1–38 in CSF, retaining the ability to distinguish AD patients from controls. Lame et al. (2011) analyzed Aβ 1–38, 1–40, and 1–42 in CSF by ultrahigh performance liquid chromatography, UHPLC-MS (MRM positive polarity), using isotope dilution method with a previous sample purification by SPE. By using a 1.7 μm particle size column, Aβ peptides were eluted using a linear gradient (40°C) over 5.5 min at 0.2 mL/min, with a very short analysis time. The authors reported a high variability in the content of Aβ peptides in CSF. Despite the variability of this method (CV 15%) is lower than ELISA (≥25%), it shows high heterogeneity in pooled CSF lots, the differentiation between AD and controls results are unclear. As an alternative to immunoprecipitation, Kim et al. (2014) detected and quantified Aβ in human plasma by an antibody-free method, in particular, human plasma proteins as a whole were digested by trypsin. High pH

RP–LC was used to fractionate the tryptic digests and to collect peptides, Aβ 1–5, Aβ 6–16, Aβ 17–28, and Aβ 29–40(42) derived from either Aβ 1–40 or Aβ 1–42 (Kim et al., 2014). A C18 column was used with gradient elution, by using a mobile phase at pH 10. Afterward, LC–SRM–MS was used for the determination of the different fractions. Among the above-mentioned peptides, Aβ 17–28 was selected as a representative Aβ surrogate, used to measure the total Aβ level. Detection and quantification of plasma Aβ by LC–MS is a challenge due to its high mass (>4 kDa), hydrophobicity, and low abundance in human plasma. Since plasma proteins such as ApoE, ApoJ, and a2M have high binding affinity toward Aβ, Kim et al. (2014) showed that protease-aided SRM can quantify more Aβ in plasma, compared to free Aβ antibody-based determination. In SRM mode, the 15 most intense ions were automatically selected and matched to theoretically calculate m/z values of product ions by using a designed software. This software allowed also for quantification of the SRM measurements from extracted ion chromatograms (XIC). Moreover, without human plasma digests fractionation, the surrogate peptide was undetectable in LC–SRM–MS. To detect this low-abundance target, high pH RP–LC fractionation for Aβ quantification was necessary. However, the authors were still unable to detect the C-terminal peptide that distinguishes Aβ 1–40, GAIIGLMVGGVV and Aβ 1–42, and GAIIGLMVGGVVIA. A reason could be the very low abundance of this peptide since it is unique only for Aβ 1–40 and Aβ 1–42, unlike the surrogate peptide Aβ 17–28 better representing the Aβ isoforms. Another surrogate of Aβ in CSF, amyloid precursor-like protein-1-derived Aβ-like peptide 28 (APL1β28), was used as a nonamyloidogenic marker for brain Aβ42 production (Yanagida et al., 2009; Tagami et al., 2014). By means of RP–LC–MS/MS, APL1β28 was identified and quantified. This technique was further used to determine the relative ratio of CSF APL1β28 to total APL1β, representative of relative ratio of Aβ 1–42 to total Aβ (the Aβ42 ratio), as biomarker of Presenilin 1 (PS1) mutations associated with familial AD (FAD). It was found that a higher relative ratio of the CSF Aβ42 surrogate in PS1-FAD patients is not due to APL1β28 increase in CSF but to the other surrogates (APL1β25 and APL1β27) decrease, suggesting that Aβ42 accumulation in the PS1-FAD brain occurs without an apparent increase in Aβ42 secretion.

The proteolytic processing of human APP into shorter prone to aggregation amyloid β-peptides, for example, Aβ 1–42, is considered a critical step in the pathogenesis of AD. Halim et al. (2011) studied the possible correlation between glycosylated APP/Aβ peptides and AD. They used the 6E10 antibody to immunopurify Aβ peptides and glycopeptides from CSF samples and then liquid chromatography–tandem mass spectrometry to identify and characterize 64 unique glycopeptides and their glycosylation site. In particular, nanoflow liquid chromatography (C4 nanoscale column) coupled to electrospray ionization (ESI) Fourier transform ion cyclotron resonance (LC–ESI–FTICR) was employed. In this preliminary study and on a small series of CSF samples, Halim et al. (2011) found an increase of the dominating Tyr glycosylated versus unglycosylated Aβ1-X peptides (X ± 40 and 42) in AD compared to non-AD patients, but the number of samples was limited. To establish Aβ glycopeptides as clinically useful biomarkers, the glycopeptide profiles and concentrations need to be confirmed in much larger prospective studies.

2.3.2 BIOMARKERS RELATED TO TAU PROTEIN AND TAU PROTEIN PHOSPHORYLATION

CSF microtubule binding tau protein is a common biomarker for AD. The concentrations of CSF tau are thought to be related to the extent of axonal injury, cell death, and NFT morphology. The combination of CSF Aβ, total tau, and phospho(Thr181)-tau is currently the most-used biomarker test for AD (Kang et al., 2013). Tau is also a very difficult protein to be determined. There are 6 isoforms of tau of varying lengths (352, 381, 383, 410, 412, and 441 amino acids [AAs]) as well as an unknown number of potential degradation products circulating in CSF, many containing several modifications (McAvoy et al., 2014). Thus, analytical specificity is of particular importance for the quantification of tau in CSF. In the case of CSF total tau, multiple immunoassays exist that rely on different combinations of capture and detection antibodies with uncertain analytical specificities. There is currently no standardized reference material or reference method identified that can be used to reconcile the differences between various assay results (Kang et al., 2013). Immunoassays have inherent liabilities in analytical specificity, driven mainly by the specificity of the antibody pair used for capture and detection. In contrast to immunoassays, quantitative LC–MS/MS methods directly measure the target peptide with analytical specificity based on the peptide's retention time, m/z, and fragmentation profile. As a result, LC–MS-based assays have been proposed as alternatives to immunoassays when the immunoassay performance is insufficient or to confirm the analytical specificity of immunoassays.

Tau and phospho-tau exist in many forms. A study performed by Meredith et al. (2013) demonstrates that tau is present in CSF as a series of N-terminal and mid-domain fragments, by using reverse-phase high-performance liquid chromatography and a sample preparation step to enrich and concentrate tau prior to Western blot analysis. Together with specific ELISA assays, this approach demonstrated that the ability of CSF tau and phospho-tau to differentiate AD from control is dependent on the tau species measured. The characterization of the different tau forms were also obtained with an immunoprecipitation-nano flow LC–MS/MS method that allowed the identification of three fragments specific of three different isoforms, which may hold diagnostic information (Portelius et al., 2008). The digested tau samples, acidified with 0.2% TFA (v/v), were first loaded onto a trapping column (C18, 0.3 mm × 5 mm). After 5 min, the trapping column was connected to the nano-flow system, and the analyte molecules were transferred to the analytical separation column (C18, 75 μm × 150 mm). The LC effluent was fractionated onto preformed microcrystalline layers of matrix prepared on pre-structured MALDI sample supports. Ninety-four fractions were collected over a period of 40 min. Mass spectrometric analysis was performed on a MALDI–TOF/TOF. Results were confirmed by using nanoflow LC–ESI combined with a hybrid linear quadrupole ion trap-Fourier transform ion cyclotron resonance mass spectrometer, that unequivocally led to protein characterization and determination. This is a typical approach to evaluate tau isoforms in different matrices and involves protein purification by means of immunoaffinity techniques and Western blot analysis, tryptic digestion, and protein identification by LC–MS/MS (Figure 2.3). A combination of an immunoaffinity assay, employing a monoclonal antibody to selectively enrich tau from routinely available volumes of CSF, and nano UPLC–MS/MS analysis has been recently

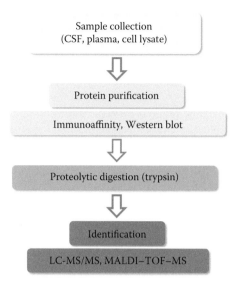

FIGURE 2.3 Scheme of analytical approach for the determination of tau protein isoforms.

applied to measure differences ($p < .0001$) in tau concentrations between healthy controls and patients with AD, the fold difference being 1.7 (McAvoy et al., 2014). In this study, tau, after immunoaffinity purification, underwent tryptic digestion to produce proteotypic peptide fragments that can be identified by reversed-phase nano UPLC–MS/MS, distinguishing the different isoforms. The use of microflow chromatography, combined with a microfluidic device, provides enhanced analytical sensitivity for quantitation of total tau, with the robustness required for clinical assays. Very recently, LC–MS/MS analysis (C8 column) was employed to characterize and quantify immunoaffinity purified secreted tau from AD patient-derived cortical neuron conditioned media (Bright et al., 2015). After immunopurification, Western blot, and tryptic digestion, LC–MS/MS allowed to characterize extracellular species to be composed predominantly of a series of N-terminal fragments of tau, with no evidence of C-terminal tau fragments. In this study, Bright et al. (2015), demonstrated that exogenously added eTau (primary human cortical neurons were treated with eTau, purified secreted Tau, or synthetic Tau, and incubated for 20 days, than the cell lysate was analyzed) increases Aβ levels, suggesting a novel connection between tau and Aβ, with a dynamic mechanism of positive feed forward regulation. Aβ drives the disease pathway through tau, with eTau further increasing Aβ levels, perpetuating a destructive cycle.

2.3.3 RESEARCH OF NEW BIOMARKERS

Although the CSF Aβ 1–42, total tau and phospho-tau-181 are currently considered the most useful biomarkers, they cannot accurately predict conversion from MCI to AD and are not useful for following drug treatment (Sato et al., 2012). In addition, analyzing those markers in CSF, requires unpleasant punctures. Hence, access to less-invasive biomarkers, found in easy-to-acquire fluids such as plasma, would accelerate and reduce the cost of AD diagnosis and offer windows of opportunity for selecting and treating patients with disease-modifying drugs once they are available. On the other hand, CSF offers several advantages as a biological fluid for AD drug discovery research. One of these advantages is that CSF is in direct continuity with the brain and therefore may reflect more specifically CNS chemical changes. CSF has a narrower dynamic range of protein concentration than does blood, with fewer dominant high abundance proteins complicating the analysis. Proteomics and metabolomics on CSF or blood are of particular interest in the discovery of new potential biomarkers for AD.

2.3.3.1 Proteomics

Advancements in liquid chromatography–mass spectrometry-based technology have enabled the identification of proteins differentially expressed in individuals. Such unbiased approaches allow the identification of novel molecular changes and have been used to study various diseases, including AD (Zhang et al., 2004). The proteomic approach usually involves a combination of analytical techniques including preliminary chromatographic fractionation, 2D-gel electrophoresis (2D-PAGE) to resolve proteins. Afterward, enzymatic degradation of isolated protein spots and LC–MS/MS lead to proteolytic peptides identification for new biomarker characterization (Hu et al., 2007; Craig-Schapiro et al., 2010). A scheme of a typical proteomic analysis for biomarker discovery is showed in Figure 2.4. The same approach can be applied to follow the AD-associated protein profile changes in different biofluids (Hu et al., 2005). Proteomic analysis was performed for the first time by 2D-gel electrophoresis and LC–MS/MS on nondepleted peripheral plasma samples (without purifying plasma from the high concentrated proteins) from AD patients and elderly controls by Hye et al. (2006). The use of these combined techniques, validated by immunoblotting, allowed to characterize 15 spots containing, in addition to albumin and immunoglobulin fragments, a total of 11 proteins that differ between patients and controls. The function of almost all these proteins is related to immune regulation suggesting an early inflammatory process in AD. Furthermore, these findings suggest that blood-based biomarkers are a realistic objective in AD. Using almost the same approach, Ringman et al. (2012) identified 56 proteins in CSF, represented by multiple tryptic peptides showing significant differences between individuals with FAD who were mutation carriers (MCs) and related noncarriers (NCs) (46 upregulated and 10 downregulated). Tryptic peptides were first offline fractionated by strong cation-exchange chromatography to be better identified and analyzed by RP–LC–MS/MS with 1 h gradient elution. The authors found more overlap in CSF protein changes between individuals with presymptomatic and symptomatic FAD than those with late-onset AD. The results are consistent with inflammation and synaptic loss early in FAD and suggest new presymptomatic biomarkers of potential usefulness in drug development. A slightly different approach was used by Henkel et al. (2012) to identify and quantify 20 significant differentially expressed proteins from seven well-characterized AD patients and seven non-AD patients. Plasma depleted samples from both diagnostic groups were labeled with different fluorescent dyes, mixed and fractionated by anion exchange chromatography. Elution was carried out, using three discrete salt steps. The ionic strength of the elution

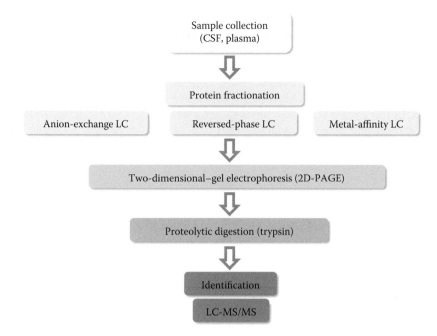

FIGURE 2.4   Scheme of typical proteomic analysis.

buffer was adjusted empirically so that each of the three salt steps released about a third of the column-bound protein content. The 3 fractions were collected by raising the salt concentration stepwise to 180 mM, 260 mM, and 1 M NaCl. The fractions were then subjected to RP-chromatography (fractionation was carried out by three stepwise elevations of the water/acetonitrile ratio) prior to 2D-PAGE and capillary LC–MS/MS characterization of digested peptides. As an alternative to the classical 2D-PAGE, surface enhanced laser desorption/ionization time-of-flight (SELDI-TOF) MS can be used to selectively adsorb proteins to a chemically modified array surface, prior to the addition of an energy-absorbing matrix solution (Conrads et al., 2004). With this technique, a protein profile is generated in which individual proteins are displayed within spectra as unique peaks based on their mass-to-charge ratio (m/z). The identification of the proteins is then performed by the classical approach. With this method, Simonsen et al. (2008) individuated 30 potential biomarkers of AD in CSF and 15 were purified and positively identified. Proteins were purified from pooled CSF by combination of anion exchange chromatography, metal affinity chromatography, reversed-phase chromatography, and cut-off membrane fractionation followed by SDS–PAGE, then tryptic digested gel extracted proteins were analyzed by peptide mapping, using a ProteinChip Reader and by tandem MS, for identification, as described before by Ruetschi et al. (2005).

2.3.3.2   Metabolomics

Metabolomics is a global biochemical approach for biomarker discovery. The metabolome is the collection of small molecules that are found within a system

which basically covers a broad range of small molecules such as glucose, cholesterol, ATP, biogenic amine neurotransmitters, lipid signaling molecules among many
other classes of compounds. The identities, concentrations, and fluxes of metabolites are the final product of interactions between gene expression, protein expression, and the cellular environment. Thus, metabolomic information complements
data obtained from other fields such as genomics and proteomics. Classical analytical platforms for the metabolic profiling are nuclear magnetic resonance (NMR),
LC–MS/MS, and gas chromatography–MS (GC–MS) followed by multivariate statistical analysis. The analytical approach can be untargeted (Greenberg et al., 2009;
Han et al., 2011; Czech et al., 2012; Ibanez et al., 2012), or targeted toward specific
metabolites or metabolite classes (Sato et al., 2010; Kaddurah-Daouk et al., 2011;
Oresic et al., 2011; Sato et al., 2012). These techniques can be applied to metabolomic studies of different samples such as animal models of AD, or AD patients
(*post-mortem* or *in vivo* CSF and blood) (Trushina et al., 2014). Particular interest
is focused on lipidomics because perturbations in lipid metabolism have been noted
since Alois Alzheimer's first reported case of AD (Foley et al., 2010). Several lines
of evidence suggest that the dysregulation of cholesterol, sphingolipid, and fatty
acid metabolism may initiate or accelerate AD pathology and, therefore, be important in the etiopathogenesis of AD and future therapeutic strategies (Mielke and
Lyketsos, 2006). Using RP–LC–MS/MS with gradient elution and GC–MS analysis, Sato et al. (2010) measured phospholipids and sterols in plasma from 10 control,
10 MCI, and 10 AD patients and extended their research to examine group differences in plasma sterols (Sato et al., 2012). After derivatization, silyl sterol ethers
from plasma were analyzed on GC capillary column with programmed temperature
elution and single ion monitoring (SIM) mode. For quantitation, deuterated desmosterol (D6-desmosterol) as internal standard (IS) was used. In an initial test sample
of plasma from 10 AD cases and 10 CN, several peaks were identified that differentiated the groups. Levels of desmosterol, a precursor to cholesterol, were significantly ($p < .009$) decreased in AD patients versus controls. To validate this finding,
the authors examined samples from 42 elderly controls, 26 MCI, and 41 AD patients
and confirmed that levels of desmosterol were sequentially decreased with disease
severity (control > MCI > AD). With the use of two metabolomics platforms, gas
chromatography–time-of-flight mass spectrometry (GC–TOF enables the detection
of over 160 metabolites of intermediary metabolism and provides an overview of
metabolic changes), and LC–electrochemistry array metabolomics platform (LC–
ECA to identify metabolites within key neurotransmitter pathways of tryptophan,
tyrosine, and purine that are implicated in the pathogenesis of AD), Motsinger-
Reif et al. (2013) measured a number of metabolites in CSF from patients with
AD dementia and from cognitively normal controls. The obtained results show the
potential of metabolite variables to provide discrimination potential of the same
magnitude as the CSF amyloid and tau proteins. An untargeted metabolomic method
based on RP–UPLC–MS (C18 column, 30°C, analysis time 16 min) and a multivariate statistical technique has been successfully used to differentiate the plasma levels
of lysophosphatidylcholines, tryptophan, dihydrosphingosine, phytosphingosine,
and hexadecasphinganine from controls and AD patients (Li et al., 2010). These
results indicated the perturbations of lecithin metabolism, amino acid metabolism,

and phospholipids metabolism of AD diseases, which may be important for clinical diagnosis and treatment monitoring.

2.4 CHROMATOGRAPHIC METHODS FOR AD DRUG CANDIDATES PHARMACOKINETICS

Pharmacokinetics (PKs) is the study of the time course of the ADME of a drug, compound, or new chemical entity after its administration to the body (Fan and De Lannoy, 2014). Drug discovery programs incorporate early ADME screens to optimize the selection of successful candidates. While the potency, efficacy, and selectivity of a compound can be facilely determined *in vitro*, to establish its potential as a drug candidate *in vivo* requires an understanding of its PK and ADME properties. Many of the failures of AD drug candidates in development programs are attributed to their undesirable PK properties, such as poor adsorption, the impossibility of reaching the CNS, target organ, too long or too short time life ($t_{1/2}$), and extensive first-pass metabolism (Hughes et al., 2011).

The goal is to assess the overall ADME characteristics before the lead compound can become the drug candidate. The concentration of the drug in the biological matrix changes with time, typically over a broad range, and, therefore, bioanalytical quantification limits are at concentrations much lower than those required for formulated or bulk drugs. Therefore, the determination of the time course of promising molecules in various biological matrices (blood, plasma, urine, etc.) requires sensitive and selective bioanalytical methodologies. Matrix effects and stability issues can also be an obstacle for the accurate analysis of the analyte; these include, among many others, endogenous materials extracted from the biological matrix that may interfere in the analysis, enzymes in the biological fluid that are capable of metabolizing the analyte, plasma proteins to which the analyte can bind and concomitant drugs that might interfere in the analysis. All these factors must be considered when planning a PK analysis (Steinmetz and Spack, 2009).

The molecules that have been shown to have a suitable profile during preclinical trials in animals are promoted to clinical trials on humans. Overall, tandem mass spectrometry (MS/MS) coupled with chromatography (HPLC or UHPLC) is the analytical technique of choice for most assays used during preclinical and clinical studies. SRM along with MRM mass spectrometry determination is a powerful targeted approach for a confident quantitation of molecules in complex biological samples. With this approach great specificity may be achieved by fragmenting the analyte and monitoring both parent and one (SRM) or more product ions (MRM) simultaneously. Important features of LC–MS methodology include shortened analysis time, increased throughput, selectivity, and lower cost of analysis.

Determination of approved and potential drugs for AD treatment in biological matrices for PK studies is usually not an easy issue, thus sample preparation requires special attention to optimize the recovery. Moreover, an appropriate IS with similar extraction recovery and chromatographic retention time needs to be selected. In the case of LC–MS analysis, the use of pure isotope-labeled IS is highly recommended to compensate ion suppression or enhancement due to the matrix composition.

2.4.1 ADME STUDIES FOR AD IN PRECLINICAL TRIALS

2.4.1.1 Absorption

The most important site of compound absorption is the gastrointestinal tract, since oral administration dosing is the most common and preferred route of administration of drugs. Absorption from the gastrointestinal tract, distribution into tissues, and renal excretion are passive processes for most drugs in which the extent and rate are mainly governed by the physicochemical properties of the drug. The evaluation of drug absorption is usually made by analyzing the drug plasma concentration after administration, then determining the PK parameters such as the maximum concentration, half-life $(t_{1/2})$, bioavailability, and so on. The selectivity of chromatographic technique coupled to mass spectrometry allowed in many cases to avoid time-consuming sample preparation like liquid–liquid extraction or SPE, limiting it to plasma protein precipitation.

Many examples of PK studies are available in the scientific literature. We are going to focus this review by examining the studies related to molecules available in nature, because, due to their high *in vitro* activity and low toxicity, some of these compounds could be useful for AD and MCI prevention and treatment.

Among these, (–)-epigallocatechin-3-gallate (EGCG) is the main polyphenolic constituent determined in green tea (Naldi et al., 2014) able to induce α-secretase and prevent Aβ aggregation in animals by directly binding to the unfolded peptide. EGCG also modulates signal transduction pathways, expression of genes regulating cell survival and apoptosis. EGCG is in phase 2–3 of clinical trials in patients with early AD is ongoing (Mangialasche et al., 2010). Lin et al. (2007) published methodologies to measure EGCG in rat plasma after oral administration. Chromatographic separation of EGCG was achieved using a Zorbax Extend-C18 column while MRM detection was used to monitor the transition of the deprotonated molecule [M-H]$^-$, m/z of 457, to the product ion 169 for EGCG. The limit of quantification (LOQ) of EGCG in rat plasma was 5 ng/mL. The absorption of EGCG was determined after its quantitation in the blood of the portal vein and oral bioavailability was about 4.95%.

ZW14, 4-methyl-3-(4-(pyrimidin-2-yl)piperazin-1-yl)-6-(thiophen-3-yl)pyridazine, is a novel anti-neuroinflammation agent with reported *in vivo* efficacy for AD. Its determination in beagle dogs after oral and intravenous administration is reported (Deng et al., 2015). In detail, HPLC–MS/MS quantitative method was developed for the evaluation of ZW14 in dog plasmas which were processed by liquid–liquid extraction with ethyl acetate and separated on a C18 column. The analyte and IS were detected by a MRM scan mode with positive ionization mode. The LLOQ was 0.05 ng/mL. The oral bioavailability of ZW14 resulted to be 26.3% with half-life of 2.6 h.

The investigated compound MH84 is a small molecule, which shows a dual activity concerning two important neurobiological targets of AD: γ-secretase and PPAR-γ (peroxisome proliferator-activated receptor gamma). This molecule has been determined in a validated LC–MS (APCI-qTOF mass analyzer) method (Pellowska et al., 2015). MH84 was administered to mice by oral gavage with a dose of 12 mg/kg. The plasma sample preparation was performed by Extrelut® liquid–liquid separation columns. The required selectivity of the method was achieved by analyzing the [M+H]$^+$ of MH84 and MH4, its structural analogue used as IS ($m/z = 603.2$ and 439.2). The LLOQ in plasma was 1.44 ng/ml. Results show a fast absorption of MH84 into the circulating plasma after a single oral dose reaching its maximum plasma concentrations within 3 h.

The drug absorption after subcutaneous administration of bis(9)-(−)-meptazinol (B9M), a novel AChE inhibitor at nanomolar levels (IC_{50} = 3.9 ± 1.3 nM), was recently reported and validated by Ge et al. (2012). The analytes were separated by a Zorbax Extend-C18 column and monitored by positive ESI in MRM mode; chromatographic separation coupled with mass spectrometry analysis allow the validation of a fast sample preparation process involving a simple dilution with methanol and centrifugation. LLOQ was established at 1 ng/mL. Results confirmed that subcutaneous administration was an optimal route for the administration for B9M, in fact its absolute bioavailability was up to 85.6%.

2.4.1.2   Distribution to the CNS and Other Tissues

Along with good absorption, another key requirement for potential drugs against AD is the ability to pass the blood–brain barrier (BBB) and reach the CNS where they can exert their pharmacological activity. In fact, the BBB is semipermeable and its overcoming is the rate-limiting step for therapeutic drug permeation to the brain. The BBB allows only the entry of therapeutic drug molecules with low molecular weight; however, hydrophobic molecules can easily cross it. Even polar molecules may overcome the BBB, if active transport is activated (Nau et al., 2010). Several bio-analytical methodologies have been validated for the drug determination in the CNS. In these cases, the sample preparation requires to extract the potential drug from the tissue after homogenization.

In vitro studies demonstrated a neuroprotective and antioxidant effect of curcumin (low molecular weight polyphenol) which appeared to be promising for mild AD treatment and prevention and greater than that of tocopherol (Brondino et al., 2014). Moreover, curcumin is described as possessing anti-amyloid properties (Naldi et al., 2012) as well as acting as an anti-inflammatory compound (Baum and Ng, 2004). Ramalingam and Ko (2014) reported a RP–LC-MS/MS validated method for its determination in mouse plasma and brain tissue. The brain homogenate was separated into a supernatant fraction (brain parenchymal fraction) and a pellet fraction (capillary fraction) by centrifugation; the separated brain fractions were then extracted and analyzed for curcumin. Triple quadrupole mass detection with MRM mode was used to monitor the ion transitions, *m/z* of 369 > 285 for curcumin, and *m/z* of 240 > 148 for salbutamol (IS). LLOQ in both matrices were 2.5 ng/mL. The low absorption and CNS distribution of free curcumin resulted from this study and led to the formulation of curcumin-loaded solid lipid nanoparticles administered to mice, which improved the PK profile of the molecule (Ramalingam and Ko, 2014).

Xie et al. (2014) published an LC–MS/MS approach to determine *Mesrine*, a phenylcarbamate of (−)-*S*-meptazinol designed and synthesized as an AChE inhibitor candidate with additional anti-amyloidogenic properties, in mouse plasma, brain, and rat plasma. Reverse-phase chromatography with isocratic elution coupled with MRM detection mode using an ESI in the positive ion mode was performed. Due to the selectivity and sensitivity of the analytical method, a simple protein precipitation was selected as plasma sample preparation method, which could significantly simplify the sample preparation procedure without evaporation and reconstitution. The results showed that meserine is rapidly absorbed with a high subcutaneous absolute bioavailability (>90%) and, mostly, that it could easily cross the BBB to reach the site of drug action (Xie et al., 2014).

Resveratrol (3,5,4-trihydroxy-*trans*-stilbene) is a polyphenol present in a variety of foods (red grapes, blueberries, peanuts, and dark chocolate). Several activities of resveratrol against molecular targets for AD have been discovered. It is able to modulate Aβ neuropathology through reducing both Aβ generation and abnormal Aβ oligomerization and promoting Aβ clearance; it can also modulate tau neuropathology through GSK-3β inhibition, lowering abnormal tau phosphorylation and tau aggregation (Varamini et al., 2013; Pasinetti et al., 2014). Moreover, it has been demonstrated *in vivo* that, resveratrol reduced neurodegeneration and cognitive decline in mice expressing a coactivator of cyclin-dependent kinase 5 and displaying massive forebrain degeneration with AD features (Kim et al., 2007). Investigation of [^{3}H]resveratrol distribution and metabolism after gastric administration was evaluated by Abd El-Mohsen and coworkers (2006) through the use of HPLC with online photodiode array and radioactive detection in order to distinguish between native resveratrol and its metabolites from the different chromatographic retention times. Distribution of resveratrol was evaluated in different rat tissues: kidney, liver, and brain. Results suggest that its glucuronide is the predominant form of resveratrol in plasma while its aglycone represents the main form in the tissues. Moreover, it was found that there are no phenolic degradation products following resveratrol consumption, suggesting that resveratrol is not subjected to degradation by large intestinal bacteria.

Another distribution study involving natural compounds, endowed with interesting activity for AD prevention and treatment, was recently performed by Yang et al. (2014) on ginkgolide B (GB). GB was extracted from ginkgo biloba and was found to be able to inhibit the neurotoxicity of peptide Aβ25–35, a peptide fragment of Aβ, which maintains the same toxicity but a much higher aggregation rate (Naldi et al., 2012). Moreover, clinical studies demonstrated that GB was active in AD patients (Maurer et al., 1998) with mild-to-moderate dementia. However, after PK studies, GB administration was blocked for its drawbacks due to water insolubility, extremely short half-life time ($t_{1/2}$), and low bioavailability. Yang et al. (2014) developed a novel oil-body nano-emulsion (ONE) for GB and performed a PK investigation in rats after its oral administration. LC–MS analyses were performed in negative ion mode and in MRM mode evaluating the GB at [M-H]$^-$, m/z 423.1. Different tissue were analyzed (heart, liver, spleen, lung, kidney, brain, intestine, and spinal cord) and GB showed higher concentration in brain, which showed the brain targeting effect of the new formulation GB-ONE.

2.4.2   PK Studies for Drug Candidate in Clinical Trials for AD

HPLC is the technique of choice for PK studies of the drugs which passed the clinical trials and are currently commercialized for AD treatment. In particular, nowadays, there are five medications approved by the US Food and Drug Administration (FDA) available for the treatment of AD: donepezil (Aricept), rivastigmine (Exelon), galantamine (Razadyne, Reminyl), tacrine (Cognex), and memantine (Namenda). The first four medications are AChE inhibitors while the fifth is a N-methyl-D-aspartate (NMDA) receptor antagonist. Tacrine, due to its poor tolerability and unfavorable pharmacological profile is currently no longer marketed (Bhatt et al., 2007; Han et al., 2009; Arumugam et al., 2011; Pilli et al., 2011, Iordachescu et al., 2012; Konda et al., 2012; Meier-Davis et al., 2012; Park et al., 2012; Steiner et al., 2012; Kumar et al., 2015).

Huperzine A (HupA) is currently tested in clinical trials. It is a natural alkaloid from *Lycopodium* plants with potential clinical application for treating AD. HupA is a highly selective, reversible and potent AChE inhibitor, found in the Chinese medicinal herb *Huperzia serrata* with higher bioavailability and potency when compared to tacrine and donepezil. *In vivo* PKs of HupA in healthy human volunteers after oral administration was investigated by Li and coworkers (2007). The molecule was analyzed by a reverse-phase chromatographic approach using a C18 column and operating in isocratic conditions with ESI–MS/MS detection on a triple quadrupole. In detail, the mass spectrometer was operating in positive ion mode, monitoring the fragment ion at m/z 243/210. The results of this study showed a biphasic profile with rapid distribution followed by a slower elimination rate.

Preclinical evidence suggests that resveratrol, mentioned before, has a potential impact on a variety of human diseases, including AD. Resveratrol is currently in clinical trials for the treatment of AD and MCI (phase 3,4) (Patel et al., 2011).

A number of clinical investigations based on LC–MS analyses have assessed the PKs and metabolism of resveratrol in humans, following oral ingestion of either the single synthetic agent or as a constituent of a particular food or drink (Patel et al., 2011). Results of these studies reported that in humans resveratrol is efficiently absorbed after oral administration; however, rapid phase II metabolism drastically limits its plasma bioavailability. It is extremely encouraging for the future clinical trials of resveratrol that doses of up to 5 g/day, taken for a month, are safe and reasonably well tolerated.

2.5 CONCLUSIONS

Chromatography can play a crucial role in the different AD drug discovery phases, from diagnostic biomarker research and hit compound activity evaluation to PK studies. Due to the complex matrices, such as biological fluids, or brain homogenates and drug physiochemical properties, highly specific analytical methods have to be applied for AD drug discovery studies. To this end, selective chromatographic techniques coupled to sensitive and selective detectors have allowed to solve many of these analytical problems.

Compared to other chromatographic detectors, hyphenation of mass spectrometry (MS) with LC techniques guarantees high sensitivity, high specificity, and the ability to resolve complex mixtures. Overall, tandem mass spectrometry (MS/MS) coupled with HPLC is the analytical technique of choice for most assays used during new drug discovery and bioanalytical methods for the quantification of Alzheimer's biomarkers and drugs in both preclinical and clinical studies. Important features of LC–MS/MS methodology for Alzheimer's drugs include shortened analysis time, increased throughput, selectivity, and lower cost of analysis.

REFERENCES

Abd El-Mohsen, M., Bayele, H., Kuhnle, G. et al. 2006. Distribution of [3H]trans-resveratrol in rat tissues following oral administration. *Br. J. Nut.* 96(1):62–70.

Andrau, D., Dumanchin-Njock, C., Ayral, E. et al. 2003. BACE1- and BACE2-expressing human cells: characterization of beta-amyloid precursor protein-derived catabolites, design of a novel fluorimetric assay, and identification of new in vitro inhibitors. *J. Biol. Chem.* 278:25859–25866.

Andrisano, V., Bartolini, M., Gotti, R., Cavrini, V., Felix, G. 2001. Determination of inhibitors' potency (IC50) by a direct high-performance liquid chromatographic method on an immobilised acetylcholinesterase column. *J. Chromatogr. B Biomed. Sci. Appl.* 753:375–383.

Arumugam, K., Chamallamudi, M.R., Gilibili, R.R. et al. 2011. Development and validation of a HPLC method for quantification of rivastigmine in rat urine and identification of a novel metabolite in urine by LC-MS/MS. *Biomed. Chromatogr.* 25(3):353–361.

Bartolini, M., Cavrini, V., Andrisano, V. 2004. Monolithic micro-immobilized-enzyme reactor with human recombinant acetylcholinesterase for on-line inhibition studies. *J. Chromatogr. A* 1031:27–34.

Bartolini, M., Cavrini, V., Andrisano, V. 2005. Choosing the right chromatographic support in making a new acetylcholinesterase-micro-immobilised enzyme reactor for drug discovery. *J. Chromatogr. A* 1065:135–144.

Bartolini, M., Greig, N.H., Yu, Q.S., Andrisano, V. 2009. Immobilized butyrylcholinesterase in the characterization of new inhibitors that could ease Alzheimer's disease. *J. Chromatogr. A* 1216:2730–2738.

Bartus, R.T., Dean III, R.L., Beer, B., Lippa, A.S. 1982. The cholinergic hypothesis of geriatric memory dysfunction. *Science.* 217:408–414.

Bateman, R.J., Munsell, L.Y., Morris, J.C., Swarm, R., Yarasheski, K.E., Holtzman, D.M. 2006. Human amyloid-β synthesis and clearance rates as measured in cerebrospinal fluid in vivo. *Nature Medicine.* 12:856–861.

Baum, L., Ng, A. 2004. Curcumin interaction with copper and iron suggests one possible mechanism of action in Alzheimer's disease animal models. *J. Alzheimers Dis.* 6(4):367–377.

Bertucci, C., Bartolini, M., Gotti, R., Andrisano, V. 2003. Drug affinity to immobilized target bio-polymers by high-performance liquid chromatography and capillary electrophoresis. *J. Chromatogr. B* 797:111–129.

Besanger, T.R., Hodgson, R.J., Green, J.R., Brennan, J.D. 2006. Immobilized enzyme reactor chromatography: optimization of protein retention and enzyme activity in monolithic silica stationary phases. *Anal. Chim. Acta.* 564:106–115.

Bhatt, J., Subbaiah, G., Kambli, S. et al. 2007. A rapid and sensitive liquid chromatography-tandem mass spectrometry (LC-MS/MS) method for the estimation of rivastigmine in human plasma. *J. Chromatogr. B Analyt. Technol. Biomed. Life Sci.* 852(1–2):115–121.

Blennow, K. 2004. CSF biomarkers for mild cognitive impairment. *J. Intern. Med.* 256:224–234.

Blennow, K. 2005. CSF biomarkers for Alzheimer's disease: use in early diagnosis and evaluation of drug treatment. *Expert Rev. Mol. Diagn.* 5:661–672.

Blennow, K., Hampel, H. 2003. CSF markers for incipient Alzheimer's disease. *Lancet Neurol.* 2:605–613.

Bradford, M.M. 1976. A rapid and sensitive method for the quantitation of microgram quantities of protein utilizing the principle of protein-dye binding. *Anal. Biochem.* 72:248–254.

Bright, J., Hussain, S., Dang, V. et al. 2015. Human secreted tau increases amyloid-beta production. *Neurobiol. Aging* 36:693–709.

Brondino, N., Re, S., Boldrini, A. et al. 2014. Curcumin as a therapeutic agent in dementia: a mini systematic review of human studies. *Sci. World J.* 2014:174282–174286.

Cavalli, A., Bolognesi, M.L., Minarini, A. et al. 2008. The cholinergic hypothesis of geriatric memory dysfunction. *J. Med. Chem.* 51:347–372.

Conrads, T.P., Hood, B.L., Issaq, H.J., Veenstra, T.D. 2004. Proteomic patterns as a diagnostic tool for early-stage cancer: a review of its progress to a clinically relevant tool. *Mol. Diagn.* 8:77–85.

Craig-Schapiro, R., Perrin, R.J., Roe, C.M. et al. 2010. YKL-40: a novel prognostic fluid biomarker for preclinical Alzheimer's disease. *Biol. Psychiatry.* 68:903–912.

Czech, C., Berndt, P., Busch, K. et al. 2012. Metabolite profiling of Alzheimer's disease cerebrospinal fluid. *PLoS One* 7:e31501.

Davies, P., Maloney, A.J. 1976. Selective loss of central cholinergic neurons in Alzheimer's disease. *Lancet.* 2:1403.

De Simone, A., Seidl, C., Santos, C.A., Andrisano, V. 2014. Liquid chromatographic enzymatic studies with on-line Beta-secretase immobilized enzyme reactor and 4-(4-dimethylaminophenylazo) benzoic acid/5-[(2-aminoethyl) amino] naphthalene-1-sulfonic acid peptide as fluorogenic substrate. *J. Chromatogr. B* 953–954:108–114.

De Simone, A., Mancini, F., Cosconati, S. et al. 2013. Human recombinant beta-secretase immobilized enzyme reactor for fast hits' selection and characterization from a virtual screening library. *J. Pharm. Biomed. Anal.* 73:131–134.

Deng, J., Liu, P., Lv, B. et al. 2015. A rapid and sensitive HPLC–MS/MS method for determination of an aminopyridazine derived anti-neuroinflammatory agent (ZW14) in dog plasma: application to a pharmacokinetic study. *J. Pharm. Biomed. Anal.* 111:204–208.

Desai, A. K., Grossberg, G. T. 2005. Diagnosis and treatment of Alzheimer's disease. *Neurology.* 64:S34–S39.

Ellman, G.L., Courtney, K.D., Andres, V. Jr, Featherstone, M. 1961. A new and rapid colorimetric determination of acetylcholinesterase activity. *Biochem. Pharmacol.* 7:88–95.

Fan, J., de Lannoy, I. A. 2014. Pharmacokinetics. *Biochem. Pharmacol.* 87(1):93–120.

Foley, P. 2010. Lipids in Alzheimer's disease: a century-old story. *Biochim. Biophys. Acta* 1801:750–753.

Frank, R.A., Galasko, D., Hampel, H. et al. 2003. Biological markers for therapeutic trials in Alzheimer's disease. Proceedings of the biological markers working group; NIA initiative on neuroimaging in Alzheimer's disease. *Neurobiol. Aging* 24:521–536.

Freije, R., Klein, T., Ooms, B., Kauffman, H.F., Bischoff, R. 2008. An integrated high-performance liquid chromatography-mass spectrometry system for the activity-dependent analysis of matrix metallo proteases. *J. Chromatogr. A* 1189:417–425.

Ge, X.X., Wang, X.L., Jiang, P. et al. 2012. Determination of Bis(9)-(–)-Meptazinol, a bis-ligand for Alzheimer's disease, in rat plasma by liquid chromatography–tandem mass spectrometry: application to pharmacokinetics study. *J. Chromatogr. B Analyt. Technol. Biomed. Life Sci.* 15:881–882.

Girelli, A.M., Mattei, E. 2005. Application of immobilized enzyme reactor in on-line high performance liquid chromatography: a review. *J. Chromatogr. B Analyt. Technol. Biomed. Life Sci.* 819:3–16

Greenberg, N., Grassano, A., Thambisetty, M., Lovestone, S., Legido-Quigley, C. 2009. A proposed metabolic strategy for monitoring disease progression in Alzheimer's disease. *Electrophoresis* 30:1235–1239.

Halim, A., Brinkmalm, G., Rüetschi, U. et al. 2011. Site-specific characterization of threonine, serine, and tyrosine glycosylations of amyloid precursor protein/amyloid β-peptides in human cerebrospinal fluid. *PNAS* 108:11848–11853.

Hampel, H., Frank, R., Broich, K. et al. 2010. Biomarkers for Alzheimer's disease: academic, industry and regulatory perspectives. *Nat. Rev. Drug Discov.* 9:560–574.

Han, J., Jiang, S., Cui, X. et al. 2009. Quantitation of the tacrine analogue octahydroaminoacridine in human plasma by liquid chromatography-tandem mass spectrometry. *J. Pharm. Biomed. Anal.* 50(2):171–174.

Han, X., Rozen, S., Boyle, S.H. et al. 2011. Metabolomics in early Alzheimer's disease: identification of altered plasma sphingolipidome using shotgun lipidomics. *PLoS One* 6:e21643.

Haniu, M., Denis, P., Young, Y.E. et al. 2000. Characterization of Alzheimer's beta -secretase protein BACE. A pepsin family member with unusual properties. *J. Biol. Chem.* 275:21099–21106.

Henkel, A.W., Muller, K., Lewczuk, P. et al. 2012. Multidimensional plasma protein separation technique for identification of potential Alzheimer's disease plasma biomarkers: a pilot study. *J. Neural. Transm.* 119:779–788.

Hodgson, R.J., Besanger, T.R., Brook, M.A., Brennan, J.D. 2005. Inhibitor screening using immobilized enzyme reactor chromatography/mass spectrometry. *Anal. Chem.* 77:7512–7519.

Hu, F.L., Deng, C.H., Zhang, X.M. 2008. Development of high performance liquid chromatography with immobilized enzyme onto magnetic nanospheres for screening enzyme inhibitor. *J. Chromatogr. B Analyt. Technol. Biomed. Life Sci.* 871:67–71.

Hu, Y., Hosseini, A., Kauwe, J. et al. 2007. Identification and validation of novel CSF biomarkers for early stages of Alzheimer's disease. *Proteomics Clin. Appl.* 1:1373–1384.

Hu, Y., Malone, J., Fagan, A., Townsend, R., Holtzman, D. 2005. Comparative proteomic analysis of intra- and interindividual variation in human cerebrospinal fluid. *Mol. Cell. Proteomics* 4:2000–2009.

Hughes, J.P., Rees, S., Kalindjian, S.B., Philpott, K.L. 2011. Principles of early drug discovery. *Br. J. Pharmacol.* 162(6):1239–1249.

Humpel, C. 2011. Identifying and validating biomarkers for Alzheimer's disease. *Trends Biotechnol.* 29:26–32.

Hye, A., Lynham, S., Thambisetty, M. et al. 2006. Proteome-based plasma biomarkers for Alzheimer's disease. *Brain* 129:3042–3050.

Ibanez, C., Simo, C., Martin-Alvarez, P.J. et al. 2012. Toward a predictive model of Alzheimer's disease progression using capillary electrophoresis-mass spectrometry metabolomics. *Anal. Chem.* 84:8532–8540.

Iordachescu, A., Silvestro, L., Tudoroniu, A., Rizea Savu, S., Ciuca, V. 2012. LC-MS-MS method for the simultaneous determination of donepezil enantiomers in plasma. *Chromatographia* 75(15–16):857–866.

Josic, D., Buchacher, A., Jungbauer, A. 2001. Monoliths as stationary phases for separation of proteins and polynucleotides and enzymatic conversion. *J. Chromatogr. B Biomed. Sci. Appl.* 752:191–205.

Kaddurah-Daouk, R., Rozen, S., Matson, W. et al. 2011. Metabolomic changes in autopsy-confirmed Alzheimer's disease. *Alzheimers Dement.* 7:309–317.

Kang, J.H., Korecka, M., Toledo, J.B., Trojanowski, J.Q., Shaw, L.M. 2013. Clinical utility and analytical challenges in measurement of cerebrospinal fluid amyloid-β (1– 42) and tau proteins as Alzheimer disease biomarkers. *Clin. Chem.* 59:903–916.

Kim, D., Nguyen, M.D., Dobbin, M.M. et al. 2007. SIRT1 deacetylase protects against neurodegeneration in models for Alzheimer's disease and amyotrophic lateral sclerosis. *EMBO J.* 26(13):3169–3179.

Kim, J.S., Ahn, H.-S., Cho, S.M., Lee, J.E., Kim, Y.S., Lee, C. 2014. Detection and quantification of plasma amyloid-b by selected reaction monitoring mass spectrometry. *Anal. Chim. Acta* 840:1–9.

Konda, R.K., Challa, B.R., Chandu, B.R., Chandrasekhar, K.B. 2012. Bioanalytical method development and validation of memantine in human plasma by high performance liquid chromatography with tandem mass spectrometry: application to bioequivalence study. *J. Anal. Methods Chem.* 2012:101249–101578.

Kumar, A., Singh, A., Ekavali, E. 2015. A review on Alzheimer's disease pathophysiology and its management: an update. *Pharmacol. Rep.* 67(2):195–203.

Lame, M.E., Chambers, E.E., Blatnik, M. 2011. Quantitation of amyloid beta peptides Aβ 1–38, Aβ 1–40, and Aβ 1–42 in human cerebrospinal fluid by ultra-performance liquid chromatography–tandem mass spectrometry. *Anal. Biochem.* 419:133–139.

Leinenbach, A., Pannee, J., Dülffer, T. et al. 2014. Mass spectrometry-based candidate reference measurement procedure for quantification of amyloid-β in cerebrospinal fluid. *Clin. Chem.* 60:987–994.

Li, N.-J., Liu, W.-T., Li, W. et al. 2010. Plasma metabolic profiling of Alzheimer's disease by liquid chromatography/mass spectrometry. *Clin. Biochem.* 43:992–997.

Li, Y.X., Zhang, R.Q., Li, C.R., Jiang, X.H. 2007. Pharmacokinetics of huperzine a following oral administration to human volunteers. *Eur. J. Drug Metab.* 32(4):183–187.

Lin, L.C., Wang, M.N., Tseng, T.Y., Sung, J.S., Tsai, T.H. 2007. Pharmacokinetics of (-)-epigallocatechin-3-gallate in conscious and freely moving rats and its brain regional distribution. *J. Agric. Food. Chem.* 55(4):1517–1524.

Lowry, O.H., Rosebrough, N.J., Farr, A.L., Randall, R.J. 1951. Protein measurement with the Folin phenol reagent. *J. Biol. Chem.* 193:265–275.

Luckarift, H.R., Johnson, G.R., Spain, J.C. 2006. Silica-immobilized enzyme reactors; application to cholinesterase-inhibition studies. *J. Chromatogr. B Analyt. Technol. Biomed. Life Sci.* 843:310–316.

Mallik, R., Hage, D.S. 2006. Affinity monolith chromatography. *J. Sep. Sci.* 29:1686–1704.

Mancini, F., Andrisano V. 2010. Development of a liquid chromatographic system with fluorescent detection for beta-secretase immobilized enzyme reactor on-line enzymatic studies. *J. Pharm. Biomed. Anal.* 52:355–361.

Mancini, F., De Simone, A., Andrisano, V. 2011. Beta-secretase as a target for Alzheimer's disease drug discovery: an overview of in vitro methods for characterization of inhibitors. *Anal. Bioanal. Chem.* 400:1979–1996.

Mancini, F., Naldi, M., Cavrini, V., Andrisano, V. 2007. Development and characterization of beta-secretase monolithic micro-immobilized enzyme reactor for on-line high-performance liquid chromatography studies. *J. Chromatogr. A.* 1175:217–226.

Mangialasche, F., Solomon, A., Winblad, B., Mecocci, P., Kivipelto, M. 2010. Alzheimer's disease: clinical trials and drug development. *Lancet Neurol.* 9(7):702–716.

Maurer, K., Ihl, R., Dierks, T., Frölich, L. 1998. Clinical efficacy of Ginkgo biloba special extract EGb 761 in dementia of the Alzheimer type. *Phytomedicine* 5(6):417–424.

McAvoy, T., Lassman, M.E., Spellman, D.S. et al. 2014. Quantification of tau in cerebrospinal fluid by immunoaffinity enrichment and tandem mass spectrometry. *Clin. Chem.* 60:683–689.

Meier-Davis, S.R., Meng, M., Yuan, W. 2012. Dried blood spot analysis of donepezil in support of a GLP 3-month dose-range finding study in rats. *Int. J. Toxicol.* 31(4):337–347.

Meredith, J.E., Sankaranarayanan, S., Guss, V. et al. 2013. Characterization of novel CSF tau and ptau biomarkers for Alzheimer's disease. *PLoS One* 8:e76523.

Mielke, M.M., Lyketsos, C.G. 2006. Lipids and the pathogenesis of Alzheimer's disease: is there a link? *Int. Rev. Psychiatry* 18:173–186.

Motsinger-Reif, A.A., Zhu, H., Kling, M.A. et al. 2013. Comparing metabolomic and pathologic biomarkers alone and in combination for discriminating Alzheimer's disease from normal cognitive aging. *Acta Neuropathol. Comm.* 1(28):1–9.

Naldi, M., Fiori, J., Gotti, R. et al. 2014. UHPLC determination of catechins for the quality control of green tea. *J. Pharm. Biomed. Anal.* 88:307–314.

Naldi, M., Fiori, J., Pistolozzi, M. et al. 2012. Amyloid β-peptide 25−35 self-assembly and its inhibition: a model undecapeptide system to gain atomistic and secondary structure details of the Alzheimer's disease process and treatment. *ACS Chem. Neurosci.* 3(11):952–962.

Nau, R., Sörgel, F., Eiffert, H. 2010. Penetration of drugs through the blood-cerebrospinal fluid/blood-brain barrier for treatment of central nervous system infections. *Clin. Microbiol. Rev.* 23(4):858–883.

Oe, T., Ackermann, B.L., Inoue, K. et al. 2006. Quantitative analysis of amyloid beta peptides in cerebrospinal fluid of Alzheimer's disease patients by immunoaffinity purification and stable isotope dilution liquid chromatography/negative electrospray ionization tandem mass spectrometry. *Rapid Comm. Mass Spectrom.* 20:3723–3735.

Oresic, M., Hyotylainen, T., Herukka, S. K. et al. 2011. Metabolome in progression to Alzheimer's disease. *Transl. Psychiatry* 1:e57.

Pannee, J., Portelius, E., Oppermann, M. et al. 2013. A selected reaction monitoring (SRM)-based method for absolute quantification of Aβ 38, Aβ 40, and Aβ 42 in cerebrospinal fluid of Alzheimer's disease patients and healthy controls. *J. Alzheimer Dis.* 33:1021–1032.

Park, Y.S., Kim, S.H., Kim, S.Y. et al. 2012. Quantification of galantamine in human plasma by validated liquid chromatography–tandem mass spectrometry using glimepride as an internal standard: application to bioavailability studies in 32 healthy Korean subjects. *J. Chromatogr. Sci.* 50(9):803–809.

Pasinetti, G.M., Wang, J., Ho, L., Zhao, W., Dubner, L. 2014. Roles of resveratrol and other grape-derived polyphenols in Alzheimer's disease prevention and treatment. *Biochim. Biophys. Acta* 1852(6):1802–1808.

Patel, K.R., Scott, E., Brown, V.A., Gescher, A.J., Steward, W.P., Brown, K. 2011. Clinical trials of resveratrol. *Ann. N Y Acad. Sci.* 1215(1):161–169.

Pellowska, M., Stein, C., Pohland, M. et al. 2015. Pharmacokinetic properties of MH84, a Γ-secretase modulator with PPARγ agonistic activity. *J. Pharm. Biomed. Anal.* 102:417–424.

Pilli, N.R., Inamadugu, J.K., Kondreddy, N., Karra, V.K., Damaramadugu, R., Rao, J.V. 2011. A rapid and sensitive LC-MS/MS method for quantification of donepezil and its active metabolite, 6-o-desmethyl donepezil in human plasma and its pharmacokinetic application. *Biomed. Chromatogr.* 25(8):943–951.

Portelius, E., Hansson, S.F., Tran, A.J. et al. 2008. Characterization of tau in cerebrospinal fluid using mass spectrometry. *J. Proteome Res.* 7:2114–2120.

Prati, F., De Simone, A., Bisignano, P. et al. 2015. Multitarget drug discovery for Alzheimer's disease: triazinones as BACE-1 and GSK-3β inhibitors. *Angewandte.* 54:1578–1582.

Ramalingam, P., Ko, Y.T. 2014. A validated LC-MS/MS method for quantitative analysis of curcumin in mouse plasma and brain tissue and its application in pharmacokinetic and brain distribution studies. *J Chrom. B* 969:101–108.

Ringman, J.M., Schulman, H., Becker, C. et al. 2012. Proteomic changes in cerebrospinal fluid of presymptomatic and affected persons carrying familial Alzheimer disease mutations. *Arch. Neurol.* 69:96–104.

Ruetschi, U., Zetterberg, H., Podust, V.N. et al. 2005. Identification of CSF biomarkers for frontotemporal dementia using SELDI-TOF. *Exp. Neurol.* 196:273–281.

Sato, Y., Nakamura, T., Aoshima, K., Oda, Y. 2010. Quantitative and wide-ranging profiling of phospholipids in human plasma by two-dimensional liquid chromatography/mass spectrometry. *Anal. Chem.* 82:9858–9864.

Sato, Y., Suzuki, I., Nakamura, T., Bernier, F., Aoshima, K., Oda, Y. 2012. Identification of a new plasma biomarker of Alzheimer's disease using metabolomics technology. *J. Lipid Res.* 53:567–576.

Simonsen, A.H., McGuire, J., Podust, V.N. et al. 2008. Identification of a novel panel of cerebrospinal fluid biomarkers for Alzheimer's disease. *Neurobiol. Aging* 29:961–968.

Sjögren, M., Andreasen, N., Blennow, K. 2003. Advances in the detection of Alzheimer's diseaseuse of cerebrospinal fluid biomarkers. *Clin. Chim. Acta* 332:1–10.

Song, F., Poljak, A., Smythe, G.A., Sachdev, P. 2009. Plasma biomarkers for mild cognitive impairment and Alzheimer's disease. *Brain Res. Rev.* 61:69–80.

Steiner, W.E., Pikalov, I.A., Williams, P.T. et al. 2012. An extraction assay analysis for galanthamine in Guinea pig plasma and its application to nerve agent countermeasures. *J. Anal. Bioanal. Tech.* 3:1–5.

Steinmetz, K.L., Spack, E.G. 2009. The basics of preclinical drug development for neurode-generative disease indications. *BMC Neurol.* 9(Suppl 1):S2.

Tagami, S., Okochi, M., Yanagida, K. et al. 2014. Relative ratio and level of Amyloid-β 42 surrogate in cerebrospinal fluid of familial Alzheimer disease patients with presenilin 1 mutations. *Neurodegener. Dis.* 13:166–170.

The Ronald and Nancy Reagan Research Institute of the Alzheimer's Association and the National Institute on Aging Working Group, 1998. *Neurobiol. Aging* 19(2):109–116.

Trushina, E., Mielke, M.M. 2014. Recent advances in the application of metabolomics to Alzheimer's disease. *Biochim. Biophys. Acta* 1842:1232–1239.

Varamini, B., Sikalidis, A.K., Bradford, K.L. 2013. Resveratrol increases cerebral glycogen synthase kinase phosphorylation as well as protein levels of drebrin and transthyretin in mice: an exploratory study. *Int. J. Food Sci. Nutr.* 7486(1):1–8.

Vilanova, E., Manjon, A., Iborra, J.L. 1984. Tyrosine hydroxylase activity of immobilized tyrosinase on enzacryl-AA and CPG-AA supports: stabilization and properties. *Biotechnol. Bioeng.* 26:1306–1312.

Wheatley, J.B., Schmidt, D.E. Jr. 1999. Salt-induced immobilization of affinity ligands onto epoxide-activated supports. *J. Chromatogr. A* 849:1–12.

Wu, S., Sun, L., Ma, J. et al. 2011. High throughput tryptic digestion via poly (acrylamide-co-methylenebisacrylamide) monolith based immobilized enzyme reactor. *Talanta.* 83:1748–1753.

Xie, Y., Jiang, P., Ge, X. 2014. Determination of a novel carbamate AChE inhibitor meserine in mouse plasma, brain and rat plasma by LC–MS/MS: application to pharmacokinetic study after intravenous and subcutaneous administration. *J. Pharm. Biomed. Anal.* 96:156–161.

Yanagida, K., Okochi, M., Tagami, S. et al. 2009. The 28-amino acid form of an APLP1-derived Aβ-like peptide is a surrogate marker for Aβ42 production in the central nervous system. *EMBO Mol. Med.* 1:223–235.

Yang, P., Cai, X., Zhou, K., Lu, C., Chen, W. 2014. A novel oil-body nanoemulsion formulation of ginkgolide B: pharmacokinetics study and in vivo pharmacodynamics evaluations. *J. Pharm. Sci.* 103(4):1075–1084.

Zetterberg, H., Wahlund, L.-O., Blennow, K. 2003. Cerebrospinal fluid markers for prediction of Alzheimer's disease. *Neurosci. Lett.* 352:67–69.

Zhang, R., Barker, L., Pinchev, D. et al. 2004. Mining biomarkers in human sera using proteomic tools. *Proteomics* 4:244–256.

3 Characterization of the Kinetic Performance of Silica Monolithic Columns for Reversed-Phase Chromatography Separations

Gert Desmet, Sander Deridder, and Deirdre Cabooter

CONTENTS

3.1 INTRODUCTION

3.1.1 EMERGENCE OF SILICA MONOLITHIC COLUMNS

The first generation of silica monoliths for chromatographic separations was introduced almost 20 years ago by Nakanishi, Soga, and Tanaka [1–3] and was commercialized by Merck a few years later [4]. At that time, the search for faster and more efficient separations had resulted in efforts to reduce the diffusion path length by decreasing the particle size. The faster mass transfer obtained by the smaller particle size led to a dramatic increase in column back pressure making high operating pressures or high temperatures mandatory to obtain acceptable separations. As a possible solution for this problem, attempts were made to reduce the diffusion path length, while maintaining the size of the convection path. This resulted in the introduction of continuous silica gel monoliths with a broad range of well-defined, controllable pores in the micrometer range (throughpores) as well as nanometer range mesopores. In contrast with particle-packed columns, the size of the monolith silica skeleton could be controlled independently from the size of the interstitial voids within certain limitations. This led to an initial introduction of silica monoliths with skeleton sizes of ~1 μm, throughpores of ~2 μm, and mesopores of 10–25 nm in the silica skeleton. The ratio of throughpore-to-skeleton size larger than unity resulted in porosities that were more than 20% higher than those of particle-packed columns. This gave rise to large permeability values allowing monolithic columns to be used at high linear velocities or in long column lengths leading to unprecedented efficiencies [5]. The small size of the silica skeleton branches on the other hand resulted in efficiencies that were comparable or even better than those obtained on 5 μm particle-packed columns, especially when the columns were operated at high flow rates. For high molecular weight solutes, the silica monoliths performed better than their 5 μm particle-packed counterparts due to the shorter diffusion path lengths of the silica monoliths.

Despite these promising features, monolithic silica columns were not able to compete with the novel generation of small (sub-2 μm) particle-packed columns that were commercialized only a few years later in combination with ultrahigh pressure chromatographic equipment [6–8]. This lack in performance of the silica monoliths was attributed to their fabrication process, which failed to deliver radially homogeneous columns, resulting in an excessively large contribution of the eddy diffusion term in their van Deemter equation [9].

More recently, efforts have been undertaken to improve the performance of silica monoliths by improving their radial homogeneity and altering the size of their throughpores and skeleton branches. This has led to the introduction of the so-called second generation of silica monoliths [10–12].

This chapter discusses the evolution that silica monoliths have undergone over the past 20 years. Kinetic performance of silica monoliths is discussed in terms of efficiency and permeability. It is demonstrated how these parameters can be determined experimentally and theoretically based on correlations that have been obtained through numerical simulations. Finally, all relevant separation parameters are combined using the so-called kinetic plots to give a comprehensive overview of the merits of silica monoliths in comparison with some other state-of-the-art chromatographic supports.

For the sake of simplicity, the discussion in this chapter mainly focuses on the fundamental evaluation of analytical bore columns with inner diameters (I.D.) of 2.0–4.6 mm.

3.1.2 PRODUCTION PROCESS OF SILICA MONOLITHIC COLUMNS

Silica-based monolithic columns are produced via a sol–gel process involving sequential hydrolysis and polycondensation of alkoxysilanes, such as tetraethyl orthosilicate or tetramethyl orthosilicate (TMOS) in the presence of a water-soluble organic polymer or porogen [13,14]. The two components (alkoxysilane and polymer) are dissolved under slightly acidic conditions and the resulting solution is poured into a gelation tube and heated at moderate temperature [11]. Successive condensation reactions lead to the growth of siloxane oligomers which link together to form a gel network. Concurrently with the sol–gel transition, spinodal decomposition occurs and phase separation takes place between the silica-rich and solvent-rich phase, representing the future silica skeletons and throughpores, respectively. To manipulate phase separation and in this way control the pore size of the gel, polyethylene glycol (PEG) (or polyethylene oxide [PEO]) is typically used as porogen [1,2,5,15]. Varying the concentration of the porogen, the size of the throughpores can be controlled. The monolithic rod shrinks noticeably after phase separation has taken place. Aging in a siloxane solution increases the stiffness and strength of the gel by adding new monomers to the silica skeleton and improving the degree of siloxane cross-linking. Adding ammonium to the aging solution, mesopores are formed. Finally, the gel is dried which causes the monolith to shrink further [16,17].

The shrinkage of the rod can result in detachment from the inner surface of the gelation tube or mold wherein it is prepared. The silica rod therefore needs to be cladded with polyether ether ketone (PEEK) to obtain a silica monolith suitable for chromatographic separation. The PEEK plastic cover is shrink-wrapped around the silica rod to ensure no void spaces remain between the monolith and the PEEK material [16]. The obtained silica rods can be derivatized with alkyl chains or any of the other chemical groups typically used to prepare reversed-phase stationary phases using essentially the same methods as for conventional silica particles. The commercially available first generation of monolithic silica rods is not retained by frits. Flow distributors made of a polymer disk with six equidistant openings at about half the column inner diameter are used to introduce the sample at the column inlet and collect it at the column outlet [18].

3.1.3 EVOLUTION OF SILICA MONOLITHS

To enhance the performance and homogeneity of the first generation monoliths, several improvements to their production process have recently been proposed. This has resulted in the introduction of the second generation monoliths.

Cabrera et al. reported the use of a constant amount of TMOS and acetic acid in the initial synthesis mixture, while increasing the amount of PEO. It was demonstrated that an increase in PEO content led to a decrease of the macropore diameter and domain size, and the formation of monoliths with narrow macropore size distributions, while maintaining the macropore volume [19]. The PEO content could, however, not be increased indefinitely since this would lead to a larger degree of inhomogeneity of the silica structure and a concomitant loss in efficiency, while also the backpressure increased dramatically. Minakuchi et al. made similar observations and reported that an increased PEO content in the starting mixture delays the phase separation, leading to structures with smaller skeleton and macropore sizes [2]. The increased viscosity of the reaction mixture caused by the ongoing polymerization moreover results in structures that do not deform easily during phase separation, leading to an improved structural homogeneity.

Nakanishi et al. reported the use of poly(acrylic acid) (PAA) as an alternative to PEG to induce phase separation [12]. When phase separation occurs, the water-soluble polymer is distributed to either the solvent-rich or the silica-rich phase. PEG is distributed to the silica-rich phase, resulting in the formation of a hydrophobic layer at the surface of the gelling phase due to the specific adsorption of PEG chains onto surface silanol groups of silica oligomers. This process is more pronounced when a hydrophobic mold is used. In this case a dense layer, called the skin layer, is formed on the outermost part of the gelled silica rods. The formation of this layer results in a deformation of the gel framework before actual gelation. Deformation of the gelling skeleton beneath the skin layer may also occur, resulting in structural inhomogeneities in the outermost part of the column. Both phenomena have a negative effect on the performance of the column. PAA, on the other hand, is distributed to the solvent phase upon phase separation. This results in less formation of skin layer and less deformation of the skeleton in the vicinity of the mold wall, hence resulting in structures with an improved radial homogeneity. Monolithic columns with small domain sizes can easily be produced using PAA, resulting in improved column efficiencies. The cladding process of this second generation monoliths is also different, as it uses a partially molten glass tube [12]. Additionally, monolithic silica frits are used at the ends of the columns that possibly allow a more uniform flow distribution of the eluent at both the column inlet and outlet [18].

Since 2012, manufacturers such as Merck (Darmstadt, Germany) (increased PEO amount approach) and Kyoto Monotech (Kyoto, Japan) (PAA approach) have released prototypes of these second generation silica monoliths which have since then been evaluated by several researchers in terms of performance and morphology [20–25].

3.2 EFFICIENCY OF SILICA MONOLITH COLUMNS

3.2.1 Column Efficiency Evaluation Using Plate-Height Models

Column efficiency is often derived from the width of a peak when it passes through the detector. Assuming a Gaussian peak profile and assuming peak broadening takes place only in the column, the plate count N can be calculated as

$$N = f \left(\frac{t_R}{w} \right)^2 \tag{3.1}$$

In this equation, t_R represents the retention time of the peak and w the width of the peak. The factor f depends on how the peak width is measured. For example, $f = 16$ when the peak width is measured at the base of the peak (i.e., the distance on the baseline between the intersections of the tangents with the baseline) and $f = 5.545$ when the peak width is measured at half the peak height. Note that N is a measure for the quality of a separation based on a single peak.

The height equivalent to a theoretical plate (HETP or H) can be derived from the plate count N and the length of the column L:

$$H = \frac{L}{N} \tag{3.2}$$

Observations made from the peak width measured at a fractional height of the peak are only valid when the peak is symmetrical. For asymmetrical peaks, the column HETP should be calculated from the first moment and the second central moment, σ^2_t, of the peak:

$$H = L \cdot \frac{\sigma^2_t}{t_R^2} \tag{3.3}$$

In Equation 3.3, t_R is the first moment, which gives the most accurate estimate of the retention time of a non-Gaussian peak, whereas the second central moment represents the variance of the peak.

Most chromatographic instruments nowadays are equipped with software that allows calculating the first and second moment automatically. However, the results obtained via moment analysis strongly depend on the definition of the start and the end of the peak, which can result in a large scatter on the obtained data, especially for peaks with a low signal-to-noise ratio. Therefore, despite the fact that moment analysis should lead to the most accurate assessment of the true HETP of a peak, many analysts still resort to the peak width determined at half the peak height for the assessment of plate counts and plate heights.

Alternatively, when dealing with asymmetrical (tailing or fronting) peaks that can be described by the exponentially modified Gaussian model, the column plate count can be determined using the Foley–Dorsey method [26]:

$$N = \frac{41.7(t_R/w_{0.1})^2}{(b/a)+1.25} \tag{3.4}$$

where:
   $w_{0.1}$ is the peak width at 10% of the peak height
   b/a is the asymmetry factor at 10% of the peak height (a is the width of the front
      and b the width of the tail of the peak at 10% of the peak height)

Note that Equation 3.4 is applicable only when the asymmetry factor b/a is greater than unity, corresponding to a tailing peak. For fronting peaks, the asymmetry factor needs to be inverted (hence the ratio of a/b is used). Using the Foley–Dorsey method, one should also be aware that it is based on a mathematical model which for some peak shapes might not always be the most correct one. However, in many cases, the Foley–Dorsey method gives a good approximation of the plate count for asymmetrical peaks.

Experimental data should be corrected for the system variance (σ^2_{ext}) and dead time (t_{ext}) that can be measured experimentally by removing the column from the system and replacing it with a zero-dead volume connection piece. System peaks are ideally analyzed using the method of moments to account for peak asymmetry. By injecting a sufficiently large concentration of analyte, the signal-to-noise ratio can be kept high to minimize the influence of the baseline noise on the calculated variances.

Plate heights can be measured as a function of the flow rate under isocratic conditions, plotted versus the mobile phase velocity and fitted to a plate-height equation such as the van Deemter equation or the Knox equation [27,28]. Although these equations are not satisfying from a fundamental point of view, they can be valuable to compare data obtained with different columns or different compounds. Since mass transfer depends on the retention factor k, the composition of the mobile phase should be adjusted in such a way that different columns are compared for the same or similar compounds yielding the same retention factor.

Many examples of plate-height data can be found in literature for both first and second generation silica monoliths.

In a study by Minakuchi et al. [1], the plate-height values obtained for amylbenzene were compared on first generation silica monoliths with skeleton sizes of 1.0–1.7 μm and throughpores of 1.5–1.8 μm and conventional packed columns with a particle size of 5 μm. It was demonstrated that both column types displayed similar values of minimal plate height ($H_{min} = 10$–15 μm) and hence yielded similar efficiencies at their optimal velocity. The efficiency of the silica monoliths was, however, largely maintained at higher velocities, whereas the HETP of the particle-packed column rapidly decreased. These results suggested that the silica monoliths are characterized by a large A-term and small C-term in comparison with the particle-packed columns. The large A-term of the silica monoliths was ascribed to the relatively large and straight throughpores that hinder efficient mixing to average the contribution of eddy diffusion, while the small C-term was attributed to the short diffusion path length of the small skeleton size of the silica rods [1]. For a large molecule (insulin), the performance of the silica monoliths was much better than that of the silica particles (mesopore size ~30 nm), especially at high flow rates. Silica rods with large mesopores (~25 nm) provided the best results, but also silica rods with mesopores of 12 nm performed better than the wide-pore silica particles with a particle size of 5 μm. This was attributed to the short diffusion path length of the small skeletons. Looking at the effect of the silica skeleton size on column performance, it was demonstrated for both amylbenzene and insulin that smaller skeleton sizes result in smaller slopes of the van Deemter curve in the C-term region. This was attributed to the dependence of the HETP at high mobile phase velocity of a term proportional to the square of the skeleton size.

For silica monoliths with domain sizes varying between 2.3 and 5.9 μm, but with a constant ratio of throughpore diameter to silica skeleton diameter, it was shown that a smaller domain size resulted in a better performance [2]. The performance of the silica rods with the smallest domain sizes was, however, not as high as expected based on their pore and skeleton sizes. This was evident from plots of the reduced plate height (h) against the reduced velocity (ν), which indicated that higher h values were observed for decreasing domain sizes. The plots indicated that the A-term increased and the C-term decreased when the domain size decreased, suggesting that the rods with the smallest domains had less regularity than their larger domain size counterparts, and improvements in the preparation conditions were in order to achieve a higher regularity of the structure to reduce the A-term contribution.

Reduced values of the plate height ($h = H/d_p$) and velocity ($\nu = u \cdot d_p/D_{mol}$) were obtained using the domain size d_{dom} as the characteristic length of the structure, instead of the particle diameter d_p that is typically used for packed particle columns. The domain size d_{dom} was taken as the sum of the throughpore size (d_{tp}) and the size of the silica skeletons at narrow (saddle) portions (d_{skel}), since this constitutes the repeating unit size of domains in the preparation process of a silica rod involving phase separation [1]

$$d_{dom} = d_{tp} + d_{skel} \tag{3.5}$$

Correlations that relate the domain size to the throughpore and skeleton size for monolithic structures with different degrees of clustering and different external porosities are presented in Deridder et al. [29]. These correlations can be used to estimate the different characteristic sizes of a monolith by measuring ε_e (see also Section 2.2.1) and using scanning electron microscopy (SEM) pictures to measure one characteristic length (e.g., the skeleton size).

McCalley compared the performance of 3.0–3.5 μm and 5 μm particle columns with that of a commercially available first generation silica monolith ($d_{skel} \sim 1.0$ μm, $d_{tp} \sim 2.0$ μm) for neutral and basic compounds [30]. For benzene, similar observations to Minakuchi et al. [1] were made: the silica monolith resulted in similar H_{min} values as the 5 μm particle-packed phase ($H_{min} = 10$–12.5 μm) and largely maintained this value at high flow rates. At 5 mL/min, this resulted in plate-height values that were almost 3.5 times higher for the 5 μm particle-packed columns, compared to the silica monoliths, emphasizing the clear advantage of the monoliths at high flow rates. For weak bases, and for strong bases analyzed under acidic conditions at room temperature, the monolithic column still showed advantages over the 5 μm particles at high flow rate, despite the fact that these compounds gave somewhat greater peak tailing on the monolithic column and the Foley–Dorsey method was used to calculate plate heights.

Guiochon and Kele studied the batch-to-batch reproducibility of the chromatographic results obtained on six commercially available first generation silica monoliths ($d_{skel} \sim 1.0$ μm, $d_{tp} \sim 2.0$ μm) and found a high degree of reproducibility for the absolute retention times, retention factors, and separation factors. For peak efficiency, the relative standard deviation (RSD) values were larger, especially for the polar compounds, although all reproducibility data were below 15%. In any case, the reproducibility figures obtained on the silica monoliths were reported to be better than or closely match the values obtained on five brands of particle-packed columns [31]. Despite the obvious advantages of the monolithic columns compared with particle-packed columns originating from the high interstitial volume, some drawbacks were observed such as tailing peaks under all conditions.

More fundamental studies revealed that the lower-than-expected performance and significant peak tailing of the first generation monoliths was mainly attributed to structural inhomogeneities resulting from their fabrication process. Relative trans-column velocity differences of some 3%–4% were demonstrated in a 10 mm I.D. silica monolith [32], while differences of some 2% were observed in 4.6 mm I.D. silica rods [33]. These radial velocity differences lead to radial concentration gradients which are the major cause of eddy diffusion. Normally these radial concentration gradients are relaxed by diffusion. The low radial dispersion coefficient of silica monoliths, resulting from their high external porosity and low tortuosity, however, leads to slow radial mixing, making it more difficult for radial concentration gradients to be relaxed [9]. Gritti and Guiochon demonstrated that it is mainly the trans-column eddy diffusion term that is responsible for the relatively poor performance of the silica monoliths. The authors stressed that the macroscopic structure of the silica monoliths should be made more radially homogeneous to improve the performance of the monolithic silica columns and that column inlet and outlet should be optimized to minimize the influence of the residual structure heterogeneity. Finally, they suggested that the domain size could be reduced to about 2 μm [9].

An improved structural homogeneity has been demonstrated for several second generation monoliths with reduced domain sizes. Hormann et al. [20] compared the morphology and separation efficiency of first and second generation Chromolith monoliths (Merck) with an I.D. of 4.6 mm and reported minimum plate heights of 6.7 μm and smaller slopes at higher flow velocities for the second generation monoliths, be it that the slope of silica monoliths is generally shallow compared to particle-packed columns. They also reported a reduced peak asymmetry for the second generation monoliths, indicating an improved radial homogeneity. This was also reflected in the reduction of the A-term, which was 61%–76.5% smaller than that of the first generation, depending on the method of analysis of the plate-height data (statistical moments vs. width at half the peak height). A more homogeneous macropore space of the second generation monoliths was also demonstrated by chord length distributions. Scanning electron microscopy images revealed a denser structure for the second generation monoliths, with finer skeletons and a larger number of smaller macropores. It was shown that the average macropore size decreased from 1.98 to 1.33 μm compared to the first generation of monoliths, while the average skeleton size decreased from 1.2 to 0.9 μm.

Similar observations were made by Gritti and Guiochon, who evaluated the performance of four second generation Chromolith columns with identical dimensions of 4.6 × 100 mm [21] and reported average throughpore and skeleton sizes of 1.2 and 0.9 μm, respectively. Minimum HETPs of 6.5–7.0 μm were obtained for a retained compound (naphthalene). On the contrary, the minimum HETP of a nonretained compound such as uracil was much larger (9.9–10.2 μm) indicating that some radial heterogeneity still remained in the column since it is easier for retained compounds to relax radial concentration gradients caused by the radial structure heterogeneity due to their larger residence times and faster diffusion across the stationary phase [22]. In any case, the minimum plate heights obtained on the second generation monoliths for moderately retained compounds were shown to be much lower than those obtained on the first generation silica rods and comparable to those on 3.5 μm conventional fully porous particle columns.

Although eddy diffusion largely decreased for these second generation silica monoliths compared to the first generation, the residual eddy diffusion term of uracil was about twice as large as that of naphthalene, confirming that the structure of the new monolithic columns largely improved, but is not yet perfectly radially homogeneous [21]. This improved radial homogeneity was also reflected in the improved peak shapes obtained on the second generation monoliths. The second generation of monolithic columns hence benefits from the significant reduction of the domain size and an improvement of the radial heterogeneity of the rod.

For prototype second generation monoliths from Kyoto Monotech, available in 2.3 and 3.2 mm I.D., Gritti and Guiochon derived average throughpore size values of 1.1 μm and skeleton sizes of 0.8 μm, resulting in a domain size of 1.9 μm [23]. These values were confirmed by Cabooter et al., who measured skeleton sizes of 0.8 μm on SEM pictures taken from the same type of columns [24]. Minimum plate heights of 4–5 μm were measured for a nonretained compound (uracil), while the HETPs for a retained compound (naphthalene) were some 10% larger [23]. The performance of the 3.2 mm I.D. column was reported to be slightly better than that of the 2.3 mm I.D. columns. The fact that the nonretained compound showed a somewhat better performance than the retained compound, suggested that the eddy diffusion of retained and nonretained compounds proceeded at similar velocities. This was confirmed by measuring the actual eddy diffusion term, confirming the radial uniformity of these new monolithic rods. Gritti and Guiochon attributed this in part to the narrower inner diameter of these columns (2.3–3.2 mm I.D. vs. 4.6 mm I.D.), to the improved cladding procedure and to the use of frits instead of polymer disks with six equidistant apertures as flow distributors [23].

The improved radial homogeneity of the second generation monoliths from Kyoto Monotech was also evidenced from the improved peak shape observed for these columns. Using the Foley–Dorsey method to analyze the peak profiles, Cabooter et al. measured slightly larger minimum HETP values (~7 μm for 3.2 mm I.D. columns) for these columns than Gritti and Guiochon [24].

3.2.2   MODELING CHROMATOGRAPHIC BAND BROADENING IN SILICA MONOLITHS

To improve the performance of silica monoliths further, models that accurately describe band broadening in silica monoliths as a function of their geometrical characteristics and the physicochemical properties of the mobile phase and the analytes (diffusion coefficients and retention factors) are an important asset. The most commonly employed band broadening models in liquid chromatography are expressed using A-, B-, and C-terms that are usually considered as black box variables. The ultimate band broadening model, however, should provide mathematical expressions for A-, B-, and C-terms, based on the geometrical characteristics of the packing and the diffusion coefficients and retention factors of the analytes.

Geometrical characteristics of a silica monolith can be hard to quantify considering the high degree of geometrical freedom compared to a packed bed column. A limited number of (semi)empirical correlations expressing the relation between the morphology of the silica monolith and the A, B, and C parameters are nevertheless available [1,2,15,34–40]. Those correlations involve the use of geometrical concepts such as random capillary networks, throughpore and domain sizes, equivalent sphere diameters, and cylindrical skeleton equivalents.

Correlations that describe the relation between the diffusion coefficients and retention factors of the analytes and the A-, B-, and C-terms can be obtained by solving the general diffusion and convection mass balance. This has been done by a number of renowned researchers such as Lapidus and Amundson [41], van Deemter et al. [28], and Giddings [42,43]. Specific equations for silica monoliths have been developed by Miyabe and Guiochon [36,37,44], approximating the geometry of the monolith as a cylindrical skeleton concentrically surrounded by a cylindrical pore.

All these plate-height expressions can in essence be brought back to a single expression: *the general plate-height model*:

$$H = H_{ax} + H_{Cm} + H_{Cs} \tag{3.6}$$

where:

H_{ax} is the band broadening originating from the eddy dispersion and longitudinal diffusion

H_{Cm} and H_{Cs} are the contributions of the mass transfer resistance in the through-pore region and skeleton zone, respectively, given by

$$H_{Cm} = 2 \cdot \frac{k''^2}{(1+k'')^2} \cdot u_i \cdot \frac{V_{tp}}{\Omega} \cdot \frac{1}{k_{f,tp}} \tag{3.7}$$

$$H_{Cs} = 2 \cdot \frac{k''}{(1+k'')^2} \cdot u_i \cdot \frac{V_{tp}}{\Omega} \cdot \frac{1}{k_{f,skel} \cdot K_P} \tag{3.8}$$

In the following sections, each of the parameters in Equations 3.6 through 3.8 will be explained in more detail and experimental procedures to determine their values will be discussed.

3.2.2.1 Denotation of k″, u_i, and K_P

Equations 6.7 and 6.8 are expressed in terms of the zone retention factor k'' which is based on the interstitial velocity u_i:

$$k'' = \frac{u_i \cdot t_R}{L} - 1 \tag{3.9}$$

$$u_i = \frac{F}{S \cdot \varepsilon_e} \tag{3.10}$$

where:

F is the flow rate

S the cross-sectional area of the column

ε_e the external porosity of the column

The interstitial velocity should not be confused with the linear velocity u_0:

$$u_0 = \frac{L}{t_0} \tag{3.11}$$

that can be determined from the elution time of an unretained marker (t_0).

The zone retention factor k'' is a measure for the time an analyte spends within the mesopores of the monolith skeleton, independent of the fact whether this time is spent in the stagnant mobile phase of the mesopores or the stationary phase covering the mesopore surface, compared to the time spent in the throughpores.

The much more frequently used phase retention factor k or k' is based on the linear velocity u_0 (or on the elution time of an unretained marker) and describes the retention of an analyte in the stationary phase covering the mesopores, compared to the time spent in the mobile phase (either free flowing in the throughpores of the monolith or stagnant in the mesopores):

$$k' = \frac{u_0 \cdot t_R}{L} - 1 = \frac{t_R}{t_0} - 1 \tag{3.12}$$

Since band broadening essentially only depends on how long the analyte spends in the mesopores (regardless of the fact whether this is in the stagnant mobile phase or the stationary phase), the general plate-height model can conveniently be expressed in terms of the zone retention factor k''.

The zone retention factor k'' can be calculated from the phase retention factor k' as follows:

$$k'' = \left(1 + k'\right) \cdot \frac{\varepsilon_T}{\varepsilon_e} - 1 \tag{3.13}$$

where ε_T is the total porosity of the column, calculated as

$$\varepsilon_T = \frac{V_0}{V_G} = \frac{t_0 \cdot F}{\pi \cdot r^2 \cdot L} \tag{3.14}$$

The total porosity ε_T is a measure for the volumetric fraction of the column that is accessible to the mobile phase (both inside and outside the particles). It is calculated as the ratio of the void volume V_0 to the geometrical volume V_G of the column. The void volume can, for example, be determined from the elution time of an unretained marker or by pycnometry measurements wherein the mass of the column is measured when it is sequentially filled with solvents (x and y) of different densities. V_0 can then be deduced from the difference in mass (m_x and m_y) and the difference in density (ρ_x and ρ_y) [45]:

$$V_0 = \frac{m_x - m_y}{\rho_x - \rho_y} \tag{3.15}$$

Typical solvents used in pycnometry measurements are tetrahydrofuran (THF) ($\rho_{THF} = 0.886$ g/cm³) and dichloromethane ($\rho_{CH2Cl2} = 1.322$ g/cm³) [46].

The external porosity ε_e is a measure for the volumetric fraction outside the mesopores accessible to the mobile phase—the interstitial or external volume V_e—and hence relates the volume of the throughpores to the geometrical volume of the column:

$$\varepsilon_e = \frac{V_e}{V_G} \tag{3.16}$$

The interstitial volume V_e of a column can be determined under chromatographic conditions via inverse-size exclusion chromatography [47–49]. This technique implies injecting a number of polystyrene standards, spanning a large range of molecular weights (typically between 500 and 2 million Da), into the column using THF as mobile phase to avoid any adsorption of the probe compounds onto the stationary phase. When the obtained retention volumes of the polystyrene standards are subsequently plotted against the cubic root of their molecular weight ($MW^{1/3}$), the data points will be grouped in two zones. The smaller polystyrene standards can enter and explore the mesopore volume and will hence elute with a higher retention volume than the larger polystyrene standards that are excluded from the mesopore volume. Extrapolating the straight line that connects the retention volumes of the excluded polystyrene standards to $MW^{1/3} = 0$, allows deriving the interstitial volume of the column.

Alternatively, the interstitial volume of a reversed-phase column can be determined using the so-called total pore blocking (TPB) method [50,51]. This method involves blocking the mesopores of a porous packing (particle-packed or monolithic columns) by flushing the column with a hydrophobic solvent such as octane or decane. The hydrophobic solvent is then pushed out of the interstitial volume by flushing the column with an aqueous buffer that is immiscible with the hydrophobic liquid occupying the pores of the porous support. During this flushing step, the hydrophobic solvent will remain inside the pores, due to the strong hydrophobic contact with the nonpolar coating inside the pores, and because it is immiscible with the hydrophilic buffer flowing outside the particles. After this flushing step, the originally porous support will behave as a completely nonporous, impermeable support. Subsequently injecting a small MW, nonretained (i.e., polar) tracer molecule such as potassium iodide (KI), the interstitial volume can be deduced from the measured elution time provided the hydrophobic liquid can be removed from every corner of the interstitial space. An advantage of the TPB method is that very small molecules can be used to measure the void volume, so that even the smallest corner of it can be accessed and sampled.

The interstitial volume can also be determined via electrostatic (Donnan) exclusion chromatography [52]. When a mobile phase containing an electrolyte is used at a sufficiently high pH (pH > 8), the residual surface silanol groups will be dissociated and form an electric double layer at the silica surface. The thickness of this double layer can be controlled by varying the ionic strength of the mobile phase. In this way, the extent to which an ionic analyte (such as NO_3^-) is excluded from the pore space (Donnan exclusion) can be controlled. At high ionic strength, a thin electrostatic double layer is formed allowing the ionic tracer to experience the total pore space. At low ionic strength, the thickness of the double layer will increase to a point where it overlaps completely excluding the ionic tracer from the mesopore space. At this point, the elution volume of the ionic tracer will reflect the interstitial volume, from which the external porosity can be deduced.

The skeleton-based equilibrium constant K_P, appearing in Equation 3.8, is defined as

$$K_P = \frac{m_{eq,skel}}{V_{skel} \cdot C_{m,eq}} \tag{3.17}$$

where:

V_{skel} is the total monolithic skeleton volume of the column (hence everything out-side of the flow-through pores)

$m_{eq,skel}$ is the total mass of analyte occupying this volume in equilibrium with the mobile phase

This parameter can be rewritten in terms of k'' as follows [53]:

$$K_P = k'' \frac{\varepsilon_e}{1 - \varepsilon_e} \tag{3.18}$$

3.2.2.2   Determination of H_{ax}

H_{ax} represents the band broadening caused by the differences in path length of the analyte experiences when flowing through the column (eddy dispersion) and the lon-gitudinal diffusion, which is often represented by the B-term constant. The charac-teristic length for eddy dispersion is the domain size d_{dom}, since this dispersion is related to the distance between adjacent parallel flow paths and the characteristic length should therefore include the skeleton size d_{skel} and the throughpore size d_{tp}.

Many different expressions have been suggested in literature to describe eddy dispersion. In combination with longitudinal diffusion, the following expressions are among the most popular:

$$H_{ax} = d_{dom} \cdot \left[A + \frac{B}{\nu_{i,dom}} \right] \tag{3.19}$$

$$H_{ax} = d_{dom} \cdot \left[A \cdot \nu_{i,dom}^{1/3} + \frac{B}{\nu_{i,dom}} \right] \tag{3.20}$$

$$H_{ax} = d_{dom} \cdot \left[\sum_j \left(\frac{1}{A_j} + \frac{1}{D_j \cdot \nu_{i,dom}} \right)^{-1} + \frac{B}{\nu_{i,dom}} \right] \tag{3.21}$$

In these equations, $\nu_{i,dom}$ is the reduced interstitial velocity:

$$\nu_{i,dom} = \frac{u_i \cdot d_{dom}}{D_{mol}} \tag{3.22}$$

and B is defined as [54]:

$$B = 2 \cdot \frac{D_{eff}}{D_{mol}} (1 + k'') \tag{3.23}$$

D_{eff} represents the effective mobile phase diffusion coefficient, which is a measure for the effective longitudinal diffusion experienced by an analyte in the monolithic column. D_{eff} is hence a combination of the diffusion rate in the mobile phase in the throughpores of the monolith (determined by the bulk molecular diffusion coeffi-cient D_{mol}) and the diffusion in the mesopores of the monolith D_{skel}.

Very recently, Deridder et al. demonstrated that the B-term band broadening in first and second generation silica monoliths can accurately be predicted over a wide range of zone retention factors using the second-order accurate effective medium theory expression given by [25,55]:

$$\frac{D_{eff}}{D_{mol}} = \frac{1}{\left(1+k''\right)\varepsilon_e} \cdot \frac{1+2\beta_1\left(1-\varepsilon_e\right)-2\zeta_2\beta_1^2\varepsilon_e}{1-\beta_1\left(1-\varepsilon_e\right)-2\zeta_2\beta_1^2\varepsilon_e} \tag{3.24}$$

ζ_2 is the so-called three point parameter which was determined to be $\sim$0.49 for monolithic columns with an external porosity $\varepsilon_e = 0.60 - 0.65$ via numerical calculations [25].

β_1 is the polarizability constant, equal to

$$\beta_1 = \frac{\alpha_{skel}-1}{\alpha_{skel}+2} \tag{3.25}$$

and α_{skel} is the relative porous zone permeability, equal to

$$\alpha_{skel} = \frac{\varepsilon_e k''}{1-\varepsilon_e} \cdot \frac{D_{skel}}{D_{mol}} \tag{3.26}$$

With D_{skel} the diffusion coefficient in the stationary zone. The value of D_{skel}/D_{mol} was determined via curve-fitting to be of the order of 0.4, which is relatively large compared to D_{skel}/D_{mol} values typically found in particle-based columns. This large value was ascribed to a higher surface diffusion coefficient and/or better connected pore wall structures in the monolithic columns [25].

Alternatively, D_{eff} can be determined experimentally via peak-parking measurements [56,57]. Peak-parking measurements imply injecting an analyte onto a column at a certain flow rate and stopping the mobile phase flow when the analyte peak reaches the middle of the column. During this time, the analyte will diffuse freely in the column under the same experimental conditions. After a specific time t_{park}, the flow is resumed and the analyte peak is eluted via the detector. The variance of the peak σ_t^2 can be calculated from the total retention time of the peak t_{tot} (which includes the elution time and the parking time) and the peak efficiency N:

$$\sigma_t^2 = \frac{t_{tot}^2}{N} \tag{3.27}$$

and subsequently transformed into spatial coordinates as follows:

$$\sigma_x^2 = \sigma_t^2 \cdot u_R^2 \tag{3.28}$$

where u_R is the linear velocity of the retained compound, calculated as

$$u_R = \frac{L}{t_R} \tag{3.29}$$

t_R is the time needed to pass through the column with length L when the mobile phase is actually flowing, and can hence be determined as $t_R = t_{tot} - t_{park}$.

When this experiment is repeated for a number of parking times (typically ranging between 0 and 90 min, in intervals of 15 min) the obtained peak variances σ_x^2 can be plotted against the applied parking time t_{park} to obtain a plot with a straight line. The effective diffusion coefficient can be obtained from the slope of this line since

$$\sigma_x^2 = 2 \cdot D_{eff} \cdot t_{park} \tag{3.30}$$

Apart from modeling and predicting the B-term constant of the van Deemter curve, the expressions established in Equations 3.24–3.26 also allow making the reverse calculation, that is, determining the value of D_{skel} from a measurement of D_{eff} [58].

3.2.2.3 Assessment of D_{mol}

To model band-broadening in silica monoliths, an accurate knowledge of the molecular diffusion coefficient D_{mol} is a prerequisite. Several correlations can be used to calculate D_{mol}, such as the Wilke–Chang equation [59], the Scheibel correlation [60], and Lusis–Rattcliff correlation [61]. However, a certain error is always involved in using these equations, especially when diffusion coefficients in large percentages of organic solvent are considered [62]. Therefore, a better option is to determine D_{mol} experimentally. This can be done using the Taylor–Aris method [63] wherein the variance of a peak is measured after elution through a long, coiled capillary at low flow rate. The radius and length of the capillary should be chosen such that radial equilibration of the sample concentration is effective along the tubing length. The flow rate should be low enough to avoid secondary flow in the coiled capillary. Finally, the system variance should be negligible compared to the total variance of the peak. Assuming longitudinal diffusion of the solute is small, the molecular diffusion coefficient can be deduced from the peak variance σ_t^2, the elution time of the peak t, and the capillary diameter d_t as follows:

$$D_{mol} = \frac{d_t^2 \cdot t}{96 \cdot \sigma_t^2} \tag{3.31}$$

3.2.2.4 Expressions for $k_{f,tp}$ and $k_{f,skel}$

The mass-transfer coefficients $k_{f,tp}$ and $k_{f,skel}$ given in Equations 3.7 and 3.8 are generally expressed in dimensionless terms using the so-called Sherwood number (Sh). The definition of the Sherwood number requires the use of a characteristic length d_{ref} and a diffusion coefficient D:

$$Sh = \frac{k_f \cdot d_{ref}}{D} \tag{3.32}$$

For the mass transfer inside the mesopores of a silica monolith, where convection is zero, the Sherwood number is a constant geometry-dependent shape factor. The characteristic length for the intra-skeleton mass transfer is the skeleton diameter d_{skel}, while the diffusion coefficient is D_{skel}. This results in the following expression for $k_{f,skel}$:

$$k_{f,skel} = Sh_{skel} \frac{D_{skel}}{d_{skel}} \tag{3.33}$$

The value of Sh_{skel} has been demonstrated to range between 7.39 and 7.65 for an ordered monolithic skeleton structure (tetrahedral skeleton model or TSM) with external porosities ranging between $\varepsilon_e = 0.6$ and $\varepsilon_e = 0.9$, respectively.

For the mass transfer inside the throughpores of the silica monolith ($k_{f,tp}$), both convective and diffusive effects need to be taken into account. This typically results in an expression for the Sherwood number that consists of two terms, one related to the diffusion and one related to the convection via the dimensionless velocity:

$$Sh_{tp} = \gamma_d + \gamma_c \cdot v_i^n \tag{3.34}$$

Giving the following expression for $k_{f,tp}$:

$$k_{f,tp} = Sh_{tp} \cdot \frac{D_{mol}}{d_{tp}} = (\gamma_d + \gamma_c \cdot v_{i,tp}^n) \cdot \frac{D_{mol}}{d_{tp}} \tag{3.35}$$

Note that the relevant characteristic length for the mass transfer is the size of the throughpores. The γ_d and γ_c parameters appearing in Equations 3.34 and 3.35 are geometrical shape factors. Their values and an expression for Equation 3.34 have been determined via numerical calculations on a TSM with an external porosity of $\varepsilon_e = 0.60$ in Deridder and Desmet [87]:

$$Sh_{tp} = 4,5 + 0,20 \ v_i^{0.56} \quad 0.5 < n_i < 250 \tag{3.36}$$

This equation leads to values of the mobile zone mass transfer coefficient that deviate from the values obtained with more traditional correlations such as those of Wilson and Geankoplis and Kataoka et al. [36,64]. This is mainly because the latter neglect the effect of axial dispersion, which leads to values of the mass transfer term that tend to zero when the velocity becomes very small ($v_i \to 0$).

3.2.2.5 Calculating the V_{tp}/Ω Factor

The V_{tp}/Ω factor appearing in Equations 3.7 and 3.8 can most conveniently be calculated by introducing the volume of the stationary zone V_{skel}:

$$\frac{V_{tp}}{\Omega} = \frac{V_{tp}}{V_{skel}} \frac{V_{skel}}{\Omega} = \frac{\varepsilon_e}{1-\varepsilon_e} \frac{d_{skel}}{\alpha} \tag{3.37}$$

from which we can also estimate the throughpore size using the definition of the hydraulic pore diameter [65]:

$$d_{tp} = 4\frac{V_{tp}}{\Omega} = 4\frac{V_{tp}}{V_{skel}} \frac{V_{skel}}{\Omega} = 4\frac{\varepsilon_e}{1-\varepsilon_e} \frac{d_{skel}}{\alpha} \tag{3.38}$$

In this equation, α is a shape factor equaling 4 for infinitely long cylinders. For silica monolith-like structures α varies almost linearly with the external porosity between values of 3.18 ($\varepsilon_e = 0.60$) and 3.70 ($\varepsilon_e = 0.90$). These values have been obtained via numerical simulations based on a TSM.

3.2.2.6   Complete Plate-Height Equation

With appropriate expressions for the mass transfer coefficients, Equation 3.6 can now be written in full as a function of the interstitial velocity u_i and the different characteristic mass-transfer distances:

$$H = H_{ax} + 2\frac{k''^2}{(1+k'')^2}\frac{\varepsilon_e}{1-\varepsilon_e}\frac{1}{\alpha}d_{skel}\cfrac{1}{\left(\gamma_d + \gamma_c\left(\dfrac{u_i d_{tp}}{D_{mol}}\right)^n\right)}\frac{d_{tp}}{D_{mol}}u_i$$

$$+2\frac{k''}{(1+k'')^2}\frac{1}{\alpha}\frac{1}{Sh_{skel}}\frac{d_{skel}^2}{D_{skel}}u_i \tag{3.39}$$

Using Equation 3.38, this can be rewritten as

$$H = H_{ax} + \frac{1}{2}\frac{k''^2}{(1+k'')^2}\cfrac{1}{\left(\gamma_d + \gamma_c\left(\dfrac{u_i d_{tp}}{D_{mol}}\right)^n\right)}\frac{d_{tp}^2}{D_{mol}}u_i + 2\frac{k''}{(1+k'')^2}\frac{1}{\alpha}\frac{1}{Sh_{skel}}\frac{d_{skel}^2}{D_{skel}}u_i \tag{3.40}$$

Equation 3.40 directly shows how H_{Cm} and H_{Cs} depend on the square of the through-pore and skeleton size, respectively.

Equations 3.39 and 3.40 can be reduced to dimensionless coordinates by choosing a characteristic size and writing all terms of Equations 3.39 and 3.40 in terms of the same reference length. In principle, both characteristic lengths d_{tp} and d_{skel} appearing in Equations 3.39 and 3.40 can be used for this purpose. Using for example the skeleton size d_{skel} as characteristic length, this results in:

$$h_{skel} = \frac{H}{d_{skel}} = h_{ax} + \frac{1}{2}\frac{k''^2}{(1+k'')^2}\cfrac{1}{\left(\gamma_d + \gamma_c\left(\dfrac{d_{tp}}{d_{skel}}\right)^n v_{i,skel}^n\right)}\left(\frac{d_{tp}}{d_{skel}}\right)^2 v_{i,skel}$$

$$+2\frac{k''}{(1+k'')^2}\frac{1}{\alpha}\frac{1}{Sh_{skel}}\frac{D_{mol}}{D_{skel}}v_{i,skel} \tag{3.41}$$

Equations 3.39 through 3.41 allow predicting the influence of retention (k''), through-pore size, and skeleton size on the contribution of mass transfer to band broadening and could potentially be used to determine how throughpore and skeleton sizes need to evolve to further optimize the performance of silica monoliths.

3.3   RELATION BETWEEN MONOLITHIC COLUMN PERMEABILITY AND POROSITY VALUES

So far, only band broadening and plate-height data have been discussed as a measure for column performance. Performance, however, also depends to a large extend on permeability (K_v) since this parameter determines how long the column can be made (to maximize efficiency) or how high the linear velocity can be raised (to maximize the speed of analysis). Column permeability can be measured experimentally by recording the column pressure obtained at a certain velocity u and using the following correlation [66]:

$$K_v = \frac{u \cdot \eta \cdot L}{\Delta P_{total} - \Delta P_{ext}} \qquad (3.42)$$

In this equation, the column pressure is obtained by subtracting the system pressure (ΔP_{ext}) from the total pressure (ΔP_{total}) and η is the mobile phase viscosity that can be deduced from correlations [62,67]. Note that Equation 3.42 is written here in very general terms, since u can be defined in many different ways.

Column permeability can be measured for the linear velocity u_0 defined in Equation 3.11 and resulting in the linear velocity-based permeability K_{v0}; for the interstitial velocity u_i defined in Equation 3.10 and resulting in the interstitial velocity-based permeability K_{vi}; and for the superficial velocity u_s:

$$u_s = \frac{F}{\pi r^2} \qquad (3.43)$$

resulting in the superficial velocity-based permeability K_{vs}.

Tanaka and coworkers measured linear velocity-based permeability values for first generation monoliths ranging between $K_{v0} = 5.6 \times 10^{-14}$ and 4.0×10^{-13} m², depending on the throughpore size of the monolitihic rod [2,15]. For a commercially available first generation monolith with a throughpore size of $d_{tp} \sim 2$ μm (Onyx, Phenomenex), Desmet et al. measured a value of $K_{v0} = 8.06 \times 10^{-14}$ m² [68]. This value is in excellent agreement with values that were obtained on six different Chromolith first generation monoliths with dimensions of 4.6×100 mm ($K_{v0} = 7.8 - 8.3 \times 10^{-14}$ m²) [31]. Cabooter et al. measured values of $K_{v0} \sim 8.3 \times 10^{-14}$ m² and $K_{v0} \sim 4.7 \times 10^{-14}$ m² on first generation Chromoliths with I.D. of 3.0 mm and 2.0 mm, respectively. The low permeability values obtained for the 2.0 mm I.D. rods were attributed to their smaller skeleton sizes [24].

Superficial velocity-based permeability values ranging between $K_{vs} = 5.3 \times 10^{-14}$ and 7.3×10^{-14} m² have been reported by several authors for both prototype and commercially available first generation monoliths [9,69,70]. Note that the permeability values obtained using the superficial velocity are somewhat lower than the values obtained using the linear velocity, since the superficial velocity is lower than the linear velocity at a specific flow rate ($u_s = u_0 \varepsilon_T$) and hence also $K_{vs} = K_{v0} \varepsilon_T$.

Anyhow, the permeability values measured in monoliths are several times higher than those typically observed for 5 μm particle-packed columns. In fact, Leinweber et al. determined via dimension analysis that the permeability of a first generation silica monolith is equivalent to that of a fixed bed of 11.0 μm spheres [40].

For the second generation of silica monoliths, superficial velocity-based permeability values in the range of $K_{vs} = 1.3 - 3.3 \times 10^{-14}$ m² have been reported [18,20,21,69], while Cabooter et al. measured permeability values based on the linear velocity of $K_{v0} = 1.8 - 2.0 \times 10^{-14}$ m² [24], which is on an average some four times smaller than that of the first generation as a direct consequence of the smaller domain sizes of the former ($d_{dom} \sim 2.0$ μm).

Although the permeability values of the second generation monoliths are significantly smaller than those of the first generation, they are still well above those of particle-packed columns obtained with sub-2 μm fully porous or sub-3 μm core-shell particles and are not within reach of the backpressure limit of the instrumentation. These large permeability values of the silica monoliths compared to those of particle-packed columns can be attributed to the large diameter of the channels open for percolation of the mobile phase through the bed of stationary phase and the high external porosity of the silica rods.

External porosity values of first and second generation silica monoliths have been measured via inverse-size exclusion, TPB, and Donnan exclusion experiments by numerous research groups and reported to be in the range of $\varepsilon_e = 0.61\%-0.70\%$ for both generation monoliths [1–4,15,18,20,21,24,70,71]. These values are significantly higher than the external porosity values that are typically measured for packed particle columns ($\varepsilon_e \sim 0.40$) [46,48,71,72].

Total porosity values of first and second generation silica monoliths have been determined from the elution time of an unretained marker or via pycnometry measurements and are typically $\varepsilon_T = 0.74\%-0.87\%$ [1–4,9,15,20,21,24,70].

These values indicate that the first and second generation monoliths are very similar in terms of porosities. The large difference in permeability must hence be attributed to their different domain sizes.

For particle-packed columns, the superficial velocity-based permeability can be related to the particle size and external porosity via the Kozeny–Carman equation [65]:

$$K_{vs} = \frac{d_p^2}{180} \frac{\varepsilon_e^3}{(1-\varepsilon_e)^2} \quad \left(\text{packed bed of spheres}\right) \tag{3.44}$$

From this correlation, it can be deduced that the permeability of a packed column will increase with increasing particle size and external porosity.

Until recently, clear-cut and ready-to-use expressions such as Equation 3.44 did not exist for monolithic columns, which was attributed to the fact that the geometry of a packed bed of spheres is always more or less the same when a sufficient packing pressure is applied, whereas a wealth in packing morphologies exist for monolithic columns, depending on their manufacturing process.

In any case, the permeability of any porous medium will always depend on its external porosity and on the size and the shape of its throughpores and structural elements. For monolithic columns, these structural features are usually estimated by visual determination from SEM pictures. Characteristic lengths that are typically employed for monoliths are the throughpore size (d_{tp}), the skeleton or globule size (d_{skel}), and the domain size (d_{dom}). As was already stated in Section 3.2.1, it is generally adopted that $d_{dom} = d_{tp} + d_{skel}$.

In Deridder et al. [29], general correlations for the permeability of silica monoliths have been derived using computational fluid dynamics simulations on simplified mimics of the typical geometry of silica monoliths. These correlations link the permeability of silica monoliths to their degree of globule clustering, their external porosity and the size of their skeletons.

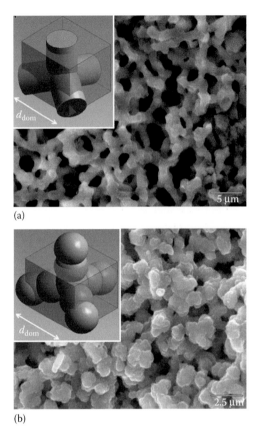

(a)

(b)

FIGURE 3.1 SEM pictures obtained for (a) a first generation monolith with an I.D. of 3.0 mm and (b) a second generation monolith with an I.D. of 2.0 mm. The insets show graphical representations of the building blocks from which the different ordered monolithic structures were built to derive general correlations for the permeability: (a) cylindrical TSM and (b) single globule chain-type TSM. (Adapted from Cabooter, D. et al., *J. Chromatogr. A*, 1325, 72–82, 2014; Deridder, S. et al., *J. Sep. Sci.*, 34, 2038–2046, 2011.)

For a perfectly ordered TSM with purely cylindrical branches (Figure 3.1a), the following relation between column permeability K_{vs}, external porosity ε_e, and skeleton size d_{skel} was derived:

$$K_{vs} = \frac{1}{51} \varepsilon_e \cdot d^2_{skel} \left(\frac{\varepsilon_e}{1-\varepsilon_e} \right)^{1.59} \tag{3.45}$$

For monoliths with branches that rather resemble a single string of interconnected globules (Figure 3.1b), the so-called single globule chain-type TSM, this correlation reduced to:

$$K_{vs} = \frac{1}{51} \varepsilon_e \cdot d^2_{skel} \left(\frac{\varepsilon_e}{1-\varepsilon_e} \right)^{1.43} \tag{3.46}$$

Although these expressions were derived for perfectly ordered structures, Cabooter et al. [24] demonstrated that these correlations can be used to predict the average skeleton size of first and second generation monoliths from experimentally measured permeability and porosity values. SEM pictures taken from the monolithic columns studied in Cabooter et al. [24] (see also Figure 3.1) revealed that the branches of the second generation monoliths resembled a single string of interconnected globules and were hence rather of the single globule chain TSM-type (cf. Equation 3.46), whereas the first generation monoliths were more of the cylindrical TSM-type (cf. Equation 3.45). The discrepancy between skeleton sizes derived using Equations 3.45–3.46 and d_{skel} values that were assessed from SEM pictures, was generally less than 10% for all columns evaluated and hence considered very small. Considering that the external porosity of both generation monoliths evaluated in Cabooter et al. [24] was largely the same, the decreased permeability of the second generation monoliths (that were obtained from Kyoto Monotech) was largely attributed to the reduction of the skeleton sizes. The single globule chain-type structure of the second generation monoliths, moreover lead to a further reduction in permeability of some 10% compared to the purely cylindrical branches of the first generation.

3.4   KINETIC PLOT METHOD

3.4.1   COMPARING CHROMATOGRAPHIC TECHNIQUES

The data cited so far have demonstrated that the smaller domain sizes and the improved radial homogeneity of the second generation monoliths have resulted in a much improved efficiency compared to the first generation monoliths. This has, however, been achieved at the cost of an increased backpressure. To account for the effect of the improved efficiency and the decreased permeability on the resulting kinetic performance of the rods and to relate their performance to that of some typical packed bed columns, several authors have proposed the use of the separation impedance [73]:

$$E_0 = \frac{t_0\,\Delta P}{N^2\eta} = \frac{H^2}{K_{v0}} = h^2\phi_0 \tag{3.47}$$

In this equation, ϕ_0 is a reduced measure for the column permeability, called the flow resistance and defined as

$$\phi_0 = \frac{d^2}{K_{v0}} \tag{3.48}$$

Wherein d can be any preferred characteristic length, provided the same value of d is used for the calculation of the reduced plate height h.

The separation impedance is a measure for column performance that primarily depends on the reduced plate height and the flow resistance and is generally considered as the ultimate figure of merit in LC supports. It circumvents the need to define a common characteristic length when columns with different morphologies are to be compared. Generally, the lower the value of the separation impedance is, the better the performance of the column is considered to be. Typical values of E_0 for packed columns are between 2000 and 5000, while Tanaka et al. reported E_0 values for first generation silica rods of some 300–900, owing to the large throughpore size/skeleton size ratio and smaller skeleton size in a silica monolith [1,2,15]. The E_0 number, however, does not give any real information on the speed of analysis since it is a dimensionless number. It can therefore not be used to compare different chromatographic supports for a specific separation problem (i.e., the time required to obtain a specific number of plates) [74]. To meet with this shortcoming, the kinetic plot method has been developed as an alternative plate height representation method that automatically produces the relevant performance characteristics for two important chromatographic optimization problems: obtaining the maximum number of plates in a given analysis time and minimizing the analysis time to obtain a certain number of plates. The kinetic plot method allows comparing the separation performance of different packing structures in a uniform and standardized way by expressing the performance of a chromatographic system in a universal and practically relevant unit: the time required to obtain a certain efficiency, resolution, or peak capacity.

3.4.2 CONSTRUCTING KINETIC PLOTS

Kinetic plots are as easy to establish as the more commonly used van Deemter curves and are based on a simple transformation of the different equations that determine the performance of a chromatographic system [74,75]. These are the pressure drop Equation 3.42 and the basic equation that relates the plate height and plate number with the column length:

$$L = N \cdot H \tag{3.49}$$

Equations 3.44 and 3.49 can be combined to obtain the following expression:

$$N = \left(\frac{\Delta P}{\eta} \right) \left[\frac{K_{v0}}{u_0 H} \right] \tag{3.50}$$

Inserting Equation 3.50 into the equation that determines the dead time of a chromatographic separation:

$$t_0 = \frac{L}{u_0} \tag{3.51}$$

it is found that:

$$t_0 = \left(\frac{\Delta P}{\eta} \right) \left[\frac{K_{v0}}{u_0^2} \right] \tag{3.52}$$

For a certain value of ΔP and η, Equations 3.50 and 3.52 allow transforming any experimental data-point of u_0 versus H, obtained in a column with permeability K_{v0}, into a data-point of N versus t_0. This (N, t_0)-data-point represents the efficiency N that can be obtained in a certain time t_0 would the support under consideration be used in a column that is exactly long enough to generate the pressure ΔP at the given velocity u_0. The calculations needed to obtain these kinetic plot data-points can easily be implemented into a spreadsheet calculator such as MS Excel [75]. Once t_0 has been calculated, it can be transformed into the corresponding retention time t_R of the compound under consideration, by incorporating the retention factor of the compound:

$$t_R = t_0(1+k) \tag{3.53}$$

To obtain the ultimate performance limits of the columns or supports under consideration, the plots should be constructed for the largest possible ΔP_{max} value (either the maximum pressure that the instrument can deliver or the maximum allowable column pressure). Kinetic plots can also be used to investigate the effect of different pressures by recalculating the same set of plate-height data for a series of different ΔP_{max} values. Care should, however, be taken when extrapolating low pressure data into the range of pressures above 400 bar. In this case effects such as the occurrence of radial temperature gradients and changes in viscosity that take place under ultra-high pressure conditions could become influential [76,77]. Downscaling the pressure is never a problem as it can only lead to an underestimation of the performance [78].

The viscosity η in Equations 3.50 and 3.52 can be treated as a normalization variable, using the same reference value for all the data series represented in the same kinetic plot. The effect of viscosity differences in experiments conducted with a different mobile phase will in this way be eliminated. The actual viscosity of the mobile phase used in each experiment can, however, also be used. This approach allows to account for the fact that more open-porous reversed-phase systems need mobile phases with a higher percentage of water and hence a higher viscosity to obtain a sufficiently large retention factor, which is particularly relevant when studying monolithic columns.

In the experiments discussed below, the mobile phase composition was adapted in such a way that the same retention factor was obtained for the analyte under consideration on each column. The viscosity of the actual mobile phase was used to construct the plots, to account for the fact that monolithic columns require a more retentive mobile phase to obtain the same retention factor as the packed particle columns.

As an example, Figure 3.2 shows the graphical representation of the data rearrangement determined by Equations 3.50 and 3.52 for a sub-2 μm particle column and a commercial first generation monolith, considered at a maximum pressure of 400 bar [86]. As can be noted by following the numbered data points, each (u_0, H)-data point in Figure 3.2a is transformed into a unique (t_0, N)-data-point in Figure 3.2b. As low velocities allow the use of long columns at the maximum pressure, the data points corresponding to the B-term dominated region of the van Deemter curve transform into the long analysis time and high efficiency end of the kinetic plot curve (upper-right side of Figure 3.2b). Inversely, the data-points originating from the C-term dominated region transform into the short analysis time and low efficiency end of the plot (lower-left side of Figure 3.2b). Looking at the van Deemter curves and permeability data in Figure 3.2a it is hard to say which column is the best one. The sub-2 μm column has the lowest plate heights and hence the highest efficiency, whereas the monolithic column has the highest permeability. By transforming the

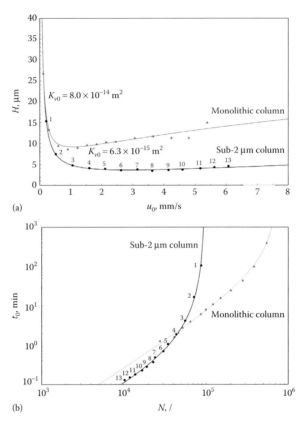

(a)

(b)

FIGURE 3.2 Transformation obtained using Equations 3.50 and 3.52 of (a) experimental van Deemter data into (b) kinetic plot data for a fully porous sub-2 μm particle column (black data) and a commercial silica monolith (gray data). Both columns were compared at a pressure of 400 bar. Experimental conditions: $T = 30°C$, test compound: 10 ppm methylparaben ($k \sim 2$), mobile phase: 35%/65% (v/v) ACN/H$_2$O for the sub-2 μm column (Agilent Zorbax SB 2.1 × 50 mm, $d_p = 1.8$ μm) and 27%/73% (v/v) ACN/H$_2$O for the monolith (Phenomenex Onyx 2.1 × 100 mm, $d_{dom} = 3$ μm). Injection volume: 2 μL, detection at 254 nm. (Adapted from Broeckhoven, K. et al., *J. Chromatogr. A*, 1228, 20–30, 2012.)

van Deemter curves into the kinetic plots in Figure 3.2b it can, however, easily be assessed that the monolithic column should be preferred for separations requiring large efficiencies (more than 50,000 plates), while the sub-2 μm column will deliver the fastest separations when small efficiencies (less than 50,000 plates) are needed. This difference in behavior is caused by the much more open structure and accompanying high permeability of the monolithic support, allowing to use very long columns at a sufficiently large velocity on the one hand, and due to the smaller diffusion distances and the better packing homogeneity of the packed bed column, leading to high separation efficiencies in relatively short columns, on the other hand. It must be noted that Figure 3.2 compares the two support types for the same maximum pressure of 400 bar and in this way provides a view on the intrinsic differences between both support types that are related to their different geometry. This can be

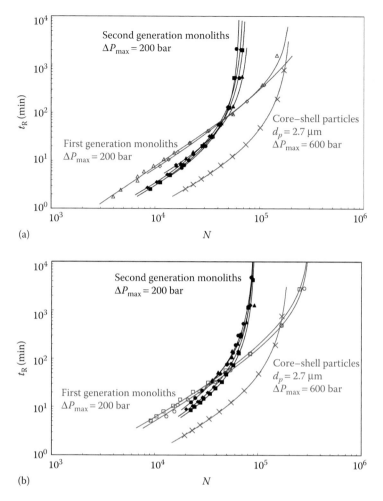

(a)

(b)

FIGURE 3.3   Kinetic plots of analysis time (t_R) versus plate count (N) for benzophenone and for different monolithic columns evaluated at 200 bar: (a) monolithic columns with an I.D. of 2.0 mm, first generation monoliths (◇ and △) and second generation monoliths (■ ◆ ● ▲) and (b) I.D. of 3.0–3.2 mm, first generation monoliths (○ and □) and second generation monoliths (■ ◆ ● ▲). Open, blue symbols refer to first generation monoliths, closed, black symbols to second generation monoliths. The red curves (×) are obtained for a core–shell column (2.0 × 100 mm, d_p = 2.7 μm) operated at a maximum pressure of 600 bar. The mobile phase was adapted on all columns to obtain k = 8.7 for benzophenone. (Adapted from Cabooter, D. et al., *J. Chromatogr. A*, 1325, 72–82, 2014.)

interesting when one wants to compare the kinetic quality of different supports. To know which column performs best in a given range of desired efficiencies, however, it is more relevant to compare columns at their proper pressure limit.

This has been done in Figure 3.3 wherein the performance of first and second generation monoliths is evaluated at a pressure of 200 bar and compared with the performance of core–shell particles at a maximum pressure of 600 bar for a compound eluting with a retention factor $k \sim 9$. Combining the data on efficiency and permeability, it is evident from Figure 3.3 that the second generation monoliths perform better than the first generation for separations requiring $N = 10,000–45,000$ plates. The plots in Figure 3.3 moreover indicate that a gain in analysis time of some 1.5–2.5 times can be obtained in this practical range of separations when switching from a first generation to a second generation monolith. For separations requiring more than 50,000 plates, the first generation monoliths clearly perform better due to their large permeability, which allows them to be used in long columns without compromising the speed of the separation too much [24].

Comparing the performance of the first and second generation monoliths to that of a core–shell column, evaluated at its own maximum pressure of 600 bar, the plots in Figure 3.3 show that the second generation monoliths are not able to outperform the core–shell particles in the range of $N = 10,000–45,000$ when each column is considered at its proper pressure limit, despite their much improved efficiency. To assess the kinetic performance of the monolithic columns at higher operating pressures, kinetic plots were also constructed for first and second generation monoliths at pressures of 400 and 600 bar. These plots were obtained by selecting $\Delta P = 400$ and $\Delta P = 600$ bar in Equations 3.50 and 3.52. As was mentioned before, this extrapolation merely gives a qualitative prediction of the performances that are maximally achievable under (ultra-)high pressure conditions since they do not take any additional band broadening into account that could originate from trans-column temperature gradients that can develop at high pressures.

The resulting plots are shown in Figure 3.4 (dashed and dotted lines) and demonstrate that the largest gain in plate count at a fixed analysis time is obtained for performances that are situated and thus attained at velocities in the B-term dominated range of the kinetic plots. This large increase in separation performance is attributed to the fact that high pressures allow to bring the velocities at which these performances are obtained closer to the optimum velocity, leading to a strong plate height decrease and hence a higher plate count. Performances obtained in the C-term dominated range of the kinetic plot benefit much less from this pressure increase, since they are already attained at velocities in excess of what is needed to operate the support at its optimum. An additional increase in inlet pressure will therefore generate a much weaker effect in the C-term dominated range. Since the performances in the C-term region correspond to plate counts in the practical range of efficiencies ($N = 10,000–50,000$ plates), Figure 3.4 suggests that even if the second generation silica monoliths evaluated in this study would be able to operate at high inlet pressures (~ 600 bar), they would still not be able to compete with state-of-the art core–shell particle columns. Further structural improvements of the silica monoliths hence seem to be in order to make them competitive with state-of-the-art small particle columns.

The approach of displaying time against efficiency to compare different chromatographic supports is certainly not new and was actually already used in the very beginning of modern chromatographic history. Giddings presented this type of plot in 1965 when he compared the performance limits of LC with those of GC on a system-independent basis [79]. Later, Knox [80] and Guiochon [81] used the same approach to compare the performance of packed bed columns with open-tubular columns.

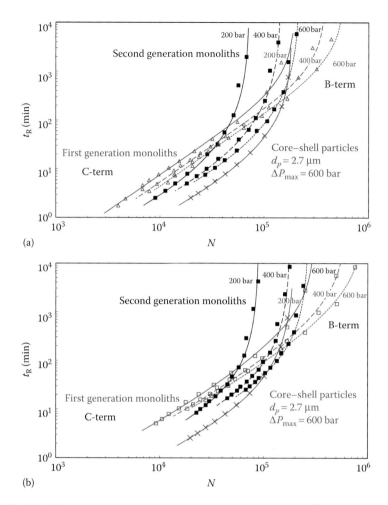

FIGURE 3.4   Kinetic plots of analysis time (t_R) versus plate count (N) for benzophenone obtained at different pressures for representative columns of each type (generation and diameter): (a) monolithic columns with an I.D. of 2.0 mm and (b) I.D. of 3.0–3.2 mm. Symbols are the same as in Figure 3.3. Full lines: 200 bar, dashed lines: 400 bar, dotted lines: 600 bar. The red curves (×) are obtained for a core–shell column (2.0 × 100 mm, d_p = 2.7 μm) operated at a maximum pressure of 600 bar. Conditions same as in Figure 3.3. (Adapted from Cabooter, D. et al., *J. Chromatogr. A*, 1325, 72–82, 2014.)

In 1997, Hans Poppe proposed to plot t_0/N versus N instead of t_0 versus N, creating the so-called *Poppe plots* [82]. Tanaka et al. have used Poppe plots in a logarithmic scale to compare the performance of particle-packed columns to that of prototype silica monoliths with different dimensions (normal bore and microbore) at 400 bar [5,83]. They demonstrated the advantage of monolithic columns for high column efficiencies owing to their large permeability which allows them to be used in long column lengths at reasonably low column pressures.

To obtain a more zoomed-in view on the kinetic plots and Poppe plots described above, the t_0 expression in Equation 3.52 can be divided by the square of the N expression in Equation 3.50, yielding:

$$\frac{t_0}{N^2} = \frac{\eta\,H^2}{\Delta P\,K_{v0}} \tag{3.54}$$

Equation 3.54 shows that the ratio t_0/N^2 is, except for the proportionality constant $(\Delta P/\eta)$, equal to the separation impedance (E_0) (see Equation 3.47).

Hara et al. used plots of log (t_0/N^2) versus log N to compare the performance of microbore second generation monoliths (200 μm I.D.) with that of particle-packed columns at 400 bar. The second generation monoliths were prepared at a constant TMOS concentration and different PEG concentrations, which resulted in columns with larger or smaller domain sizes (varying between 3.0 and 2.2 μm). It was demonstrated that the curves for the monolithic silica columns with smaller domain sizes were below the curves representing the performance of 2 μm particle-packed columns in a region where the plate number is larger than 30,000, indicating that these second generation silica monoliths can compete with 2 μm particle columns for plate counts above 30,000 plates when both column types are evaluated at 400 bar. At 1000 bar, however, the monolithic columns would only become more favorable than the particle-packed columns when plate counts above 80,000 plates would be required [10].

Morisato et al. used plots of log (t_0/N^2) versus log N to compare the performance of prototype second generation monoliths, prepared using PAA as the porogen, with that of a first generation monolith at 400 bar. The second generation monoliths revealed performances that were similar to those of 3 μm particle-packed columns, rather than those of the first generation monolith [12].

To investigate the pure performance of columns in more detail, the use of reduced kinetic plots of separation impedance (E_0) versus plate number ratio (N_{opt}/N) has been proposed [84,85]. These plots allow assessing the pure quality of a packing or support structure without having to specify a reference length (such as a domain or skeleton size) and without taking the pressure into consideration. The separation impedance is calculated according to Equation 3.47, while the plate number ratio can be derived from Equation 3.50:

$$\frac{N_{opt}}{N} = \frac{u_{opt}H_{min}}{u\,H} = \frac{v_{opt}h_{min}}{v\,h} \tag{3.55}$$

The plate number ratio quantifies the number of times a given plate number N is smaller than the optimal plate number N_{opt} of the investigated support at the employed physicochemical conditions. Equations 3.47 and 3.55 demonstrate that a plot of E_0 versus N_{opt}/N will only depend on the reduced variables h, v, and φ_0. As a consequence, packings with similar packing quality, intra-particle diffusion characteristics, and retention factors, but a different characteristic length, will have coinciding curves. The lower this curve, the better the packing quality of the column will be. A plot of E_0 versus N_{opt}/N can hence be considered as the kinetic plot equivalent of the reduced van Deemter curve, with the advantage that it also contains information on the column permeability.

Cabooter et al. constructed reduced kinetic plots for first generation and second generation monoliths and compared them to plots obtained for a core–shell column [24]. These plots showed a slightly better structural quality for the second generation monoliths compared to that of the first generation, owing to the large permeability values obtained for the first generation monoliths. The minimum impedance values of the core–shell column were below those obtained for the second generation monolithic columns, suggesting a better bed quality of the former.

3.5   CONCLUSIONS

The first generation of silica monoliths that was introduced almost 20 years ago displayed efficiencies comparable to those of 5 μm particle-packed columns and large permeability values. At that time, they held great promise to revolutionize liquid chromatographic separations. Despite this great potential, it gradually became clear that the efficiency obtained on these silica rods—typically minimum plate heights of some 10–15 μm were obtained—was not as high as could be expected based on their throughpore and skeleton sizes. This was attributed to their production process, which failed to deliver radially homogeneous rods resulting in an excessively large A-term contribution in their van Deemter equation.

Efforts made to improve this production process have recently resulted in the introduction of a second generation of silica monoliths. These second generation silica monoliths display a larger radial homogeneity and have smaller domain sizes (domain sizes of 2.4–2.9 μm for the first generation vs. 1.9–2.0 μm for the second generation), which has led to efficiencies that are about 1.5 times higher than those of the first generation.

The decreased domain size of the second generation monoliths, however, has also resulted in reduced permeability values that are some four times lower than those of the first generation, implying higher pressures are required to operate these columns. Despite this decrease in permeability values, they are still well above those obtained in columns packed with small particles (sub-2 μm) and are well out of reach of the backpressure limitations of conventional chromatography instruments.

Combining the effect of column permeability and efficiency, kinetic plots demonstrate that the second generation performs better than the first generation for separations requiring $N = 10,000–45,000$ plates for a compound eluting with a retention factor $k \sim 9$. The first generation, however, maintains its advantage over the second generation for separations requiring higher efficiencies, due to their large permeability. This indicates that silica monoliths of the first generation can still be useful for challenging separation problems. The large macropore size and rigid bed structure of the first generation makes them also particularly suitable for the direct analysis of *dirty samples* that could easily block the stationary bed of small particle columns.

In comparison with the performance of the state-of-the-art small particle columns (sub-2 μm and core–shell particles) that can be operated at much higher inlet pressures, the second generation is, however, still not able to compete with these particles in the practical range of separations ($N \leq 30,000–45,000$), when all column types are evaluated at their own maximum pressure and despite the improved performance of the second generation in this range.

It is expected that the structural features of the silica-based monoliths will be further improved by adapting and optimizing their sol–gel production process to enhance their performance. It will be particularly challenging to reduce the domain size further while maintaining or further improving the homogeneity of the porous silica structure. Monolithic silica columns clad with higher pressure-resistant material are currently also under development and will allow investigating the effect of a higher inlet pressure on the obtained performance in a practical manner.

A number of correlations are presented in this chapter that relate the performance of first and second generation monoliths to their structural features and the physicochemical properties of the analytes and mobile phases used to evaluate them. These expressions could potentially be used to guide future research on the further improvement of silica monoliths.

REFERENCES

1. H. Minakuchi, K. Nakanishi, N. Soga, N. Ishizuka, N. Tanaka, Effect of skeleton size on the performance of octadecylsilylated continuous porous silica columns in reversed-phase liquid chromatography, *J. Chromatogr. A* 762 (1997) 135–146.
2. H. Minakuchi, K. Nakanishi, N. Soga, N. Ishizuka, N. Tanaka, Effect of domain size on the performance of octadecylsilylated continuous porous silica columns in reversed-phase liquid chromatography, *J. Chromatogr. A* 797 (1998) 121–131.
3. H. Minakuchi, K. Nakanishi, N. Soga, N. Ishizuka, N. Tanaka, Octadecylsilylated porous silica rods as separation media for reversed-phase liquid chromatography, *Anal. Chem.* 68 (1996) 3496–3501.
4. K. Cabrera, G. Wieland, D. Lubda, K. Nakanishi, N. Soga, H. Minakuchi, K. Unger, SilicaROD, a new challenge in fast high-performance liquid chromatography separations, *TrAC* 17 (1998) 50–53.
5. N. Kobayashi, H. Minakuchi, K. Nakanishi, K. Hirao, K. Hosoya, T. Ikegami, N. Tanaka, Monolithic silica columns for high-efficiency separations by high-performance liquid chromatography, *J. Chromatogr. A* 960 (2002) 85–96.
6. S.A. Wren, Peak capacity in gradient ultra performance liquid chromatography (UPLC), *J. Pharm. Biom. Anal.* 38 (2005) 337–343.
7. J.S. Mellors, J.W. Jorgenson, Use of 1.5-μm porous ethyl-bridged hybrid particles as a stationary-phase support for reversed-phase ultrahigh-pressure liquid chromatography, *Anal. Chem.* 76 (2004) 5441–5450.
8. R.E. Majors, Fast and ultrafast HPLC on sub-2μm porous particles—where do we go from here? *LC–GC N. Am.* 23 (2005) 1248–1255.
9. F. Gritti, G. Guiochon, Measurement of the eddy diffusion term in chromatographic columns. I. Application to the first generation of 4.6 mm I.D. monolithic columns, *J. Chromatogr. A* 1218 (2011) 5216–5227.
10. T. Hara, H. Kobayashi, T. Ikegami, K. Nakanishi, N. Tanaka, Performance of monolithic silica capillary columns with increased phase ratios and small-sized domains, *Anal. Chem.* 78 (2006) 7632–7642.
11. K. Cabrera, A new generation of silica-based monolithic HPLC columns with improved performance, *LC–GC N. Am.* 4 (2012) 56–60.
12. K. Morisato, S. Miyazaki, M. Ohira, M. Furuno, M. Nyudo, H. Terashima, K. Nakanishi, Semi-micro-monolithic columns using macroporous silica rods with improved performance, *J. Chromatogr. A* 1216 (2009) 7384–7387.

13. K. Nakanishi, N. Soga, Phase separation in silica sol-gel system containing polyacrylic acid. I. Gel formation behavior and effect of solvent composition, *J. Non-Cryst. Sol.* 139 (1992) 1–13.

14. K. Nakanishi, N. Soga, Phase separation in silica sol-gel system containing polyacrylic acid. II. Effects of molecular weight and temperature, *J. Non-Cryst. Sol.* 139 (1992) 14–24.

15. N. Tanaka, H. Kobayashi, K. Nakanishi, H. Minakuchi, N. Ishizuka, A new type of chromatographic support could lead to higher separation efficiencies, *Anal. Chem.* 73 (2001) 420A–429A.

16. A.-M. Siouffi, Silica gel-based monoliths prepared by the sol-gel method: Facts and figures, *J. Chromatogr. A* 1000 (2003) 801–818.

17. G. Guiochon, Monolithic columns in high-performance liquid chromatography, *J. Chromatogr. A* 1168 (2007) 101–168.

18. F. Gritti, G. Guiochon, Measurement of the eddy dispersion term in chromatographic columns. II. Application to new prototypes of 2.3 and 3.2 mm I.D. monolithic silica columns, *J. Chromatogr. A* 1227 (2012) 82–95.

19. S. Altmaier, K. Cabrera, Structure and performance of silica-based monolithic HPLC columns, *J. Sep. Sci.* 31 (2008) 2551–2559.

20. K. Hormann, T. Müllner, S. Bruns, A. Höltzel, U. Tallarek, Morphology and separation efficiency of a new generation of analytical silica monoliths, *J. Chromatogr. A* 1222 (2012) 46–58.

21. F. Gritti, G. Guiochon, Measurement of the eddy dispersion term in chromatographic columns: III. Application to new prototypes of 4.6 mm I.D. monolithic columns, *J. Chromatogr. A* 1225 (2012) 79–90.

22. F. Gritti, G. Guiochon, Impact of retention on trans-column velocity biases in packed columns, *AIChE J.* 56 (2010) 1495–1509.

23. F. Gritti, G. Guiochon, Measurement of the eddy dispersion term in chromatographic columns. II. Application to new prototypes of 2.3 and 3.2 mm I.D. monolithic silica columns, *J. Chromatogr. A* 1227 (2012) 82–95.

24. D. Cabooter, K. Broeckhoven, R. Sterken, A. Vanmessen, I. Vandendael, K. Nakanishi, S. Deridder, G. Desmet, Detailed characterization of the kinetic performance of first and second generation silica monolithic columns for reversed-phase chromatography separations, *J. Chromatogr. A* 1325 (2014) 72–82.

25. S. Deridder, A. Vanmessen, K. Nakanishi, G. Desmet, D. Cabooter, Experimental and numerical validation of the effective medium theory for the B-term band broadening in 1st and 2nd generation monolithic silica columns, *J. Chromatogr. A* 1351 (2014) 46–55.

26. J.P. Foley, J.G. Dorsey, Equations for calculation of chromatographic figures of merit for ideal and skewed peaks, *Anal. Chem.* 55 (1983) 730–737.

27. E. Grushka, L.R. Snyder, J. H. Knox, Advances in band spreading theories, *J. Chromatogr. Sci.* 13 (1975) 25–37.

28. J.J. van Deemter, F.J. Zuiderweg, A. Klinkenberg, Longitudinal diffusion and resistance to mass transfer as causes of non-ideality in chromatography, *Chem. Eng. Sci.* 5 (1956) 271–289.

29. S. Deridder, S. Eeltink, G. Desmet, Computational study of the relationship between the flow resistance and the microscopic structure of polymer monoliths, *J. Sep. Sci.* 34 (2011) 2038–2046.

30. D.V. McCalley, Comparison of conventional microparticulate and a monolithic reversed-phase column for high-efficiency fast liquid chromatography of basic compounds, *J. Chromatogr. A* 965 (2002) 51–64.

31. M. Kele, G. Guiochon, Repeatability and reproducibility of retention data and band profiles on six batches of monolithic column, *J. Chromatogr. A* 960 (2002) 19–49.

32. K.S. Mriziq, J.A. Abia, Y. Lee, G. Guiochon, Structural radial heterogeneity of a silica-based wide-bore monolithic column, *J. Chromatogr. A* 1193 (2008) 97–103.

33. J. Abia, K. Mriziq, G. Guiochon, Radial heterogeneity of some analytical columns used in high-performance liquid chromatography, *J. Chromatogr. A* 1216 (2009) 3185–3191.

34. J.J. Meyers, A.I. Liapis, Network modeling of the convective flow and diffusion of molecules adsorbing in monoliths and in porous particles packed in a chromatographic column, *J. Chromatogr. A* 852 (1999) 3–23.

35. N. Tanaka, H. Kobayashi, N. Ishizuka, N. Minakuchi, K. Nakanishi, K. Hosoya, T. Ikegami, Monolithic silica columns for high-efficiency chromatographic separations, *J. Chromatogr. A* 965 (2002) 35–49.

36. K. Miyabe, G. Guiochon, The moments equations of chromatography for monolithic stationary phases, *J. Phys. Chem. B* 106 (2002) 8898–8909.

37. K. Miyabe, A. Cavazzini, F. Gritti, M. Kele, G. Guiochon, Moment analysis of mass-transfer kinetics in C18 silica monolithic columns, *Anal. Chem.* 75 (2003) 6975–6986.

38. U. Tallarek, F.C. Leinweber, A. Seidel-Morgenstern, Fluid dynamics in monolithic adsorbents: Phenomenological approach to equivalent particle dimensions, *Chem. Eng. Technol.* 25 (2002) 1177–1182.

39. F.C. Leinweber, D. Lubda, K. Cabrera, U. Tallarek, Characterization of silica-based monoliths with bimodal pore size distribution, *Anal. Chem.* 74 (2002) 2470–2477.

40. F.C. Leinweber, U. Tallarek, Chromatographic performance of monolithic and particulate stationary phases. Hydrodynamics and adsorption capacity, *J. Chromatogr. A* 1006 (2003) 207–228.

41. L. Lapidus, N.R. Amundson, Mathematics of adsorption in beds. VI. The effect of longitudinal diffusion in ion exchange and chromatographic columns, *J. Phys. Chem.* 56 (1952) 984–988.

42. J.C. Giddings, Role of lateral diffusion as a rate-controlling mechanism in chromatography, *J. Chromatogr. A* 5 (1961) 46–60.

43. J.C. Giddings, *Dynamics of Chromatography Part 1*, Marcel Dekker, New York, 1965.

44. K. Miyabe, G. Guiochon, A kinetic study of mass transfer in reversed-phase liquid chromatography on a C18 silica gel, *Anal. Chem.* 72 (2000) 5162–5171.

45. R.M. McCormick, B.L. Karger, Distribution phenomena of mobile-phase components and determination of dead volume in reversed-phase liquid chromatography, *Anal. Chem.* 52 (1980) 2249–2257.

46. H. Song, E. Adams, G. Desmet, D. Cabooter, Evaluation and comparison of the kinetic performance of ultra-high performance liquid chromatography and high-performance liquid chromatography columns in hydrophilic interaction and reversed-phase liquid chromatography conditions, *J. Chromatogr. A* 1369 (2014) 83–91.

47. D. Lubda, W. Lindner, M. Quaglia, C. Du Fresne von Hohenesche, K.K. Unger, Comprehensive pore structure characterization of silica monoliths with controlled mesopore size and macropore size by nitrogen sorption, mercury porosimetry, transmission electron microscopy and inverse size exclusion chromatography, *J. Chromatogr. A* 1083 (2005) 14–22.

48. H. Guan, G. Guiochon, Study of physico-chemical properties of some packing materials: I. Measurements of the external porosity of packed columns by inverse size-exclusion chromatography, *J. Chromatogr. A* 731 (1996) 27–40.

49. Y. Yao, A.M. Lenhoff, Pore size distributions of ion exchangers and relation to protein binding capacity, *J. Chromatogr. A* 1126 (2006) 107–119.

50. D. Cabooter, J. Billen, H. Terryn, F. Lynen, P. Sandra, G. Desmet, Detailed characterisation of the flow resistance of commercial sub-2 micrometer reversed-phase columns, *J. Chromatogr. A* 1178 (2008) 108–17.

51. D. Cabooter, F. Lynen, P. Sandra, G. Desmet, Total pore blocking as an alternative method for the on-column determination of the external porosity of packed and monolithic reversed-phase columns, *J. Chromatogr. A* 1157 (2007) 131–41.
52. S. Jung, S. Ehlert, M. Pattky, U. Tallarek, Determination of the interparticle void volume in packed beds via intraparticle Donnan exclusion, *J. Chromatogr. A* 1217 (2010) 696–704.
53. G. Desmet, K. Broeckhoven, Equivalence of the different Cm- and Cs-term expressions used in liquid chromatography and a geometrical model uniting them, *Anal. Chem.* 80 (2008) 8076–8088.
54. G. Desmet, K. Broeckhoven, J. De Smet, S. Deridder, G. Baron, P. Gzil, Errors involved in the existing B-term expressions for the longitudinal diffusion in fully porous chromatographic media–part I: Computational data in ordered pillar arrays and effective medium theory, *J. Chromatogr. A* 1188 (2008) 171–188.
55. G. Desmet, S. Deridder, Effective medium theory expressions for the effective diffusion in chromatographic beds filled with porous, non-porous and porous-shell particles and cylinders. Part II: Numerical verification and quantitative effect of solid core on expected B-term band broadening, *J. Chromatogr. A* 1218 (2011) 46–56.
56. J.H. Knox, Band dispersion in chromatography – a new view of A-term dispersion, *J. Chromatogr. A* 831 (1999) 3–15.
57. J.H. Knox, L. McLaren, A new gas chromatographic method for measuring gaseous diffusion coefficients and obstructive factors, *Anal. Chem.* 36 (1964) 1477–1482.
58. G. Desmet, S. Deridder, Effective medium theory expressions for the effective diffusion in chromatographic beds filled with porous, non-porous and porous-shell particles and cylinders. Part I: Theory, *J. Chromatogr. A* 1218 (2011) 32–45.
59. C.R. Wilke, P. Chang, Correlation of diffusion coefficients in dilute solutions, *AICHE 1* (1955) 264–270.
60. E.G. Scheibel, Liquid diffusivities, *Ind. Eng. Chem.* 46 (1954) 2007–2008.
61. M.A. Lusis, C.A. Ratcliff, Diffusion in binary liquid mixtures at infinite dilution, *Can. J. Chem. Eng.* 46 (1968) 385–387.
62. J. Li, P.W. Carr, Accuracy of empirical correlations for estimating diffusion coefficients in aqueous organic mixtures, *Anal. Chem.* 69 (1997) 2530–2536.
63. J.G. Atwood, J. Goldstein, Measurements of diffusion coefficients in liquids at atmospheric and elevated pressure by the chromatographic broadening technique, *J. Phys. Chem.* 88 (1984) 1875–1885.
64. E.J. Wilson, C.J. Geankoplis, Liquid mass transfer at very low reynolds numbers in packed beds, *Ind. Eng. Chem. Fundam.* 5 (1966) 9–14.
65. J.F. Richardson, J.H. Harker, J.R. Backhurst, *Coulson & Richardson's Chemical Engineering*, vol. 2, 5th edn., Elsevier Butterworth-Heinemann, Oxford, 2002.
66. J. Bear, *Dynamics of Fluids in Porous Media*, Dover Publications, New York, 1988.
67. D. Guillarme, S. Heinisch, J.-L. Rocca, Effect of temperature in reversed phase liquid chromatography, *J. Chromatogr. A* 1052 (2004) 39–51.
68. G. Desmet, D. Cabooter, P. Gzil, H. Verelst, D. Mangelings, Y. Vander Heyden, D. Clicq, Future of high pressure liquid chromatography: Do we need porosity or do we need pressure? *J. Chromatogr. A* 1130 (2006) 158–166.
69. D. Hlushkou, K. Hormann, A. Höltzel, S. Khirevich, A. Seidel-Morgenstern, U. Tallarek, Comparison of first and second generation analytical silica monoliths by pore-scale simulations of eddy dispersion in the bulk region, *J. Chromatogr. A* 1303 (2013) 28–38.
70. B. Bildlingmaier, K.K. Unger, N. von Doehren, Comparative study on the column performance of microparticulate 5-μm C18 bonded and monolithic C18-bonded reversed-phase columns in high-performance liquid chromatography, *J. Chromatogr. A* 832 (1999) 11–16.

71. D. Cabooter, F. Lynen, P. Sandra, G. Desmet, Total pore blocking as an alternative method for the on-column determination of the external porosity of packed and monolithic reversed-phase columns, *J. Chromatogr. A* 1157 (2007) 131–41.

72. D. Cabooter, J. Billen, H. Terryn, F. Lynen, P. Sandra, G. Desmet, Detailed characterisation of the flow resistance of commercial sub-2 micrometer reversed-phase columns, *J. Chromatogr. A* 1178 (2008) 108–17.

73. P.A. Bristow, J.H. Knox, Standardization of test conditions for high performance liquid chromatography columns, *Chromatographia* 10 (1977) 279–289.

74. G. Desmet, D. Clicq, P. Gzil, Geometry-independent plate height representation methods for the direct comparison of the kinetic performance of LC supports with a different size or morphology, *Anal. Chem.* 77 (2005) 4058–4070.

75. G. Desmet, P. Gzil, D. Clicq, Kinetic plots to directly compare the performance of LC supports, *LC-GC Eur.* 18 (2005) 403–408.

76. M. Martin, G. Guiochon, Effects of high pressure in liquid chromatography, *J. Chromatogr. A* 1090 (2005) 16–38.

77. C. Horvath, H.-J. Lin, Band spreading in liquid chromatography: General plate height equation and a method for the evaluation of the individual plate height contributions, *J. Chromatogr.* 149 (1978) 43–70.

78. A. Jerkovich, J. Mellors, J. Jorgenson, Recent developments in LC column technology, *LC-GC Eur.* 16 (2003) 20–23.

79. J.C. Giddings, Comparison of theoretical limit of separating speed in gas and liquid chromatography, *Anal. Chem.* 37 (1965) 60–63.

80. J.H. Knox, M. Saleem, Kinetic conditions for optimum speed and resolution in column chromatography, *J. Chromatogr. Sc.* 7 (1969) 614–622.

81. G. Guiochon, Conventional packed columns vs. packed or open tubular microcolumns in liquid chromatography, *Anal. Chem.* 53 (1981) 1318–1325.

82. H. Poppe, Some reflections on speed and efficiency of modern chromatographic methods, *J. Chromatogr. A* 778 (1997) 3–21.

83. N. Tanaka, H. Nagayama, H. Kobayashi, T. Ikegami, K. Hosoya, N. Ishizuka, H. Minakuchi, K. Nakanishi, K. Cabrera, D. Lubda, Monolithic silica columns for HPLC, Micro-HPLC and CEC, *J. High Resol. Chromatogr.* 23 (2000) 111–116.

84. D. Cabooter, S. Heinisch, J.L. Rocca, D. Clicq, G. Desmet, Use of the kinetic plot method to analyze commercial high-temperature liquid chromatography systems. I: Intrinsic performance comparison, *J. Chromatog. A* 1143 (2007) 121–33.

85. D. Cabooter, J. Billen, H. Terryn, F. Lynen, P. Sandra, G. Desmet, Detailed characterisation of the flow resistance of commercial sub-2 micrometer reversed-phase columns, *J. Chromatogr. A* 1178 (2008) 108–17.

86. K. Broeckhoven, D. Cabooter, S. Eeltink, G. Desmet, Kinetic plot based comparison of the efficiency and peak capacity of high-performance liquid chromatography columns: Theoretical background and selected examples, *J. Chromatogr. A* 1228 (2012) 20–30.

87. S. Deridder, G. Desmet, New insights in the velocity dependency of the external mass transfer coefficient in 2D and 3D porous media for liquid chromatography, *J. Chromatogr. A* 1227 (2012) 194–202.

4 Recent Advances in the Characterization and Analysis of Therapeutic Oligonucleotides by Analytical Separation Methods Coupling with Mass Spectrometry

Su Pan and Yueer Shi

CONTENTS

4.1 INTRODUCTION

Oligonucleotide therapeutics (OGNTs) are a class of short-chain oligonucleotides (OGNs) consisting typically of 15–50 nucleotide units in length. OGNTs are designed to target the gene directly or at the mRNA-expression stage, to disrupt the production of disease-related proteins at the cell level.[1] With a more complete understanding of how OGNs are used to regulate cellular functions and defense mechanisms in nature, therapeutic applications of this class of molecules has seen increased interest, resulting in OGNTs being more thoroughly studied and applied in preclinical models and clinical applications.[2] OGNTs have been explored as drug candidates for a wide range of diseases, including cancer, AIDS, Alzheimer's, and cardiovascular disorders.[1] Numerous clinical trials have been conducted; however, to date only three OGNs have received FDA approval. Fomivirsen (Vitravene), an 18-mer phosphothioate antisense OGN to treat cytomegalovirus retinitis in AIDS patients, was the first FDA-approved OGNT drug in 1998. Pegaptanib (Macugen) was approved in 2004 as an anti-angiogenic 28-mer aptamer RNA for the treatment of neovascular (wet) age-related macular degeneration (AMD). Finally, Kynamro is an antisense OGN approved to reduce cholesterol by targeting the mRNA for apolipoprotein B.

OGNs are usually several kDa in size and possess high degrees of negative charges. Because their physicochemical properties are very different from conventional drugs, OGNTs present new analytical challenges for drug development and manufacturing. For example, since OGNTs are prevailing regulators of gene expression inside cells, they are typically present with concentrations in the nanomolar range. To measure these low levels of OGNs and related metabolites requires highly sensitive quantitative methods to determine the pharmacodynamic (PD) and pharmacokinetic (PK) properties, toxicological profiles, and terminal phase elimination for therapeutic use. Another analytical challenge is, under certain circumstances, the measurement of both individual strands and duplex where the mechanism of action of the OGNTs requires formation of a sense–antisense duplex. Sample preparation from biological matrices is also required, for example, in plasma, tissues, and cells. As a result, matrix-specific compounds like salts, lipids, and proteins can potentially interfere with the characterization and quantitation of OGNs. Therefore, sample preparation is a critical step to separate the target OGNTs and several techniques such as protein precipitation (PP), liquid–liquid extraction (LLE), solid-phase extraction (SPE), and combination are widely used in OGN purification.[3]

In addition to the hydrophobicity of the OGN phosphate backbone that arises from its high degree of negative charge, OGNT drug delivery is also challenged by the universal presence of nucleases in biological fluids which result in rapid hydrolysis *in vivo*.[4] Therapeutic DNA and RNA are routinely modified to enhance their stability in biological matrices using non-natural chemistries.[2b] Chemical modification to the backbone or sugar residues of OGNs is currently the most effective strategy to improve their resistance to

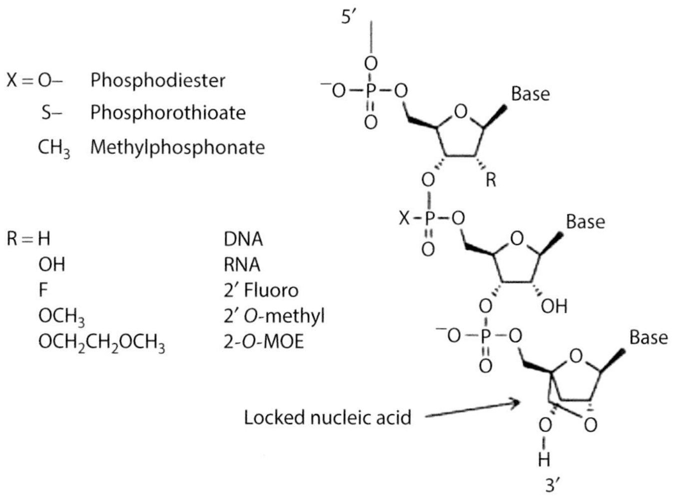

X = O−   Phosphodiester
   S−   Phosphorothioate
   CH$_3$ Methylphosphonate

R = H   DNA
    OH   RNA
    F   2′ Fluoro
    OCH$_3$ 2′ O-methyl
    OCH$_2$CH$_2$OCH$_3$ 2-O-MOE

Locked nucleic acid

FIGURE 4.1   Common chemical modifications made to oligonucleotides to improve resistance to nuclease degradation. (From Beverly, M. B., *Mass. Spectrom. Rev.* 30, 979–998, 2011.)

nucleases. In addition, chemical modifications to OGNs offer additional advantageous benefits such as lowered toxicity levels and an overall increase in potency, which is a result of the binding specificity being amplified along with the volume of distribution being increased. The structures of the most common modifications are shown in Figure 4.1.

4.2   MECHANISMS OF OGNTs

For more than two decades, OGNTs have presented a new paradigm for drug discovery. Mechanisms and strategies for OGNTs have been focused on either correcting the genetic defect or silencing gene expression.

The first category of OGNTs are those that hybridize to RNA and mediate gene silencing.[2c,4] The most common RNA-based therapeutic OGNs are antisense,[5] small interfering RNA (siRNA),[6] and microRNA (miRNA).[7] These OGNTs downregulate gene expression by inducing enzyme-dependent RNA degradation. Antisense OGNs are small single-stranded DNA or RNA which hybridizes to specific sequence of the mRNA to inhibit or decrease expression of the target protein[8] (Figure 4.2b). Antisense DNA or RNA (asDNA or asRNA) bind with the complimentary mRNA to form an antisense/mRNA duplex which activates the nuclear enzyme RNaseH which then degrades the mRNA strand. The most common antisense modification is substituting phosphonate for the phosphorothioate backbone which maintains the OGN-like configuration making it an excellent substrate for enzyme RNaseH. These modified antisense OGNs have a different mechanism which no longer recognizes RNaseH but directly interacts with the mRNA substrate to alter the protein translation. Fomivirsen (brand name Vitravene) is an example of asRNA. It is an approved 18-mer phosphothioate OGN developed by ISIS to treat cytomegalovirus retinitis in AIDS patients.[5]

In contrast to antisense, therapeutical siRNA is based on a naturally occurring gene-silencing mechanism (Figure 4.2a).[6b,9] They are small double-stranded RNA

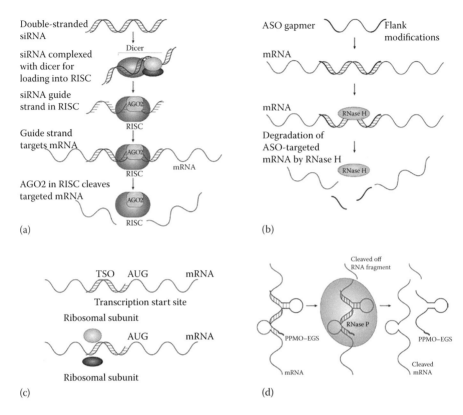

FIGURE 4.2 Mechanisms of oligonucleotide-induced downregulation of gene expression. (a) Small interfering RNA (siRNA). Synthetic double-stranded siRNA is complexed with components of the RNA interference pathway, dicer, argonaute 2 (AGO2), and other proteins to form an RNA-induced silencing complex (RISC). The RISC binds to a targeted mRNA via the unwound guide strand of siRNA, allowing AGO2 to degrade the RNA. The RISC-bound siRNA can also bind with mismatches to unintended mRNAs, leading to significant off-target effects (see main text). (b) Antisense gapmer oligonucleotides. These commonly have a phosphorothioate backbone with flanks that are additionally modified with 2'-O-methoxyethyl (2'-MOE) or 2'-O-methyl (2'-OMe) residues (highlighted in bold in figure). Flank modifications increase the resistance of the antisense oligonucleotide (ASO) to degradation and enhance binding to targeted mRNA. The unmodified "gap" in a gapmer–mRNA duplex is recognized by ribonuclease H (RNase H), a ribonuclease that degrades duplexed mRNA. (c) Translation-suppressing oligonucleotides (TSOs). Phosphorodiamidate morpholino oligomers (PMOs) and their derivatives, or oligonucleotides fully substituted with 2'-MOE or 2'-OMe residues, are not recognized by RISC or RNase H and therefore do not lead to RNA degradation. Nevertheless, they lead to downregulation of gene expression via steric blockade of ribosome access to mRNA and suppression of protein translation. (d) External guide sequences (EGSs) and RNase P. A peptide-conjugated PMO (PPMO) is designed to hybridize to targeted bacterial mRNA and form stem–loop structures such that the resulting duplex resembles tRNA. In bacteria, a tRNA-processing ribozyme—RNase P—recognizes this structure and cleaves mRNA. (From Kole, R. et al., *Nat. Rev. Drug Discov.*, 11, 125–140, 2012.)

which can knock down gene expression in the RNA interference (RNAi) pathway. When long, double-stranded RNA molecules are introduced, they are chopped by the endonuclease Dicer into 21- to 23-nucleotide small siRNA followed by incorporating into the multicomponent RNA-induced silencing complex (RISC), which unwinds the duplex and degrades homologous mRNAs. Synthetic siRNA are directly incorporated in the RISC and are routinely modified at the 2′ ribose position because they do not interfere with the configuration that is necessary for activity. Substitutions of the 2′ hydroxyl group (2′-OH) enhance the binding specificity of the guide strand in the RISC and increase affinity for the target mRNA for cleavage. These include 2′-O-methyl (2′-O-Me), O-ethyl (2′-O-ethyl), O-methoxyethyl (2′-MOE), fluorine (2′-F) substitutions, and locked nucleic acids (LNAs). Synthetic siRNA have been demonstrated to target genes *in vivo* for multiple diseases, including bone cancer, liver cirrhosis, hepatitis B, and human papillomavirus.[3a,6b]

miRNA is a small noncoding RNA molecule (containing about 22 nucleotides), which can play important regulatory roles in animals and plants by targeting mRNAs for cleavage or translational repression.[10] The function of miRNAs in cell to cell communications has been recognized only in recent years. miRNAs resemble siRNAs of the RNAi pathway, except they derive at different locations regions of RNA transcripts. miRNAs are able to exist both in cells and in the extracellular space of tissues after the transportation, thereby shuttling regulatory signals from the source to the recipient cells. The studies of miRNAs in diagnostic and cardiovascular therapeutic areas have been reported and are currently one of the most rapidly expanding areas in OGNTs.[11]

Enzymatic nucleic acid molecules such as ribozyme (RNAzymes) and DNAzymes (Figure 4.2c, d) are OGNs that are capable of catalyzing specific biochemical reactions to cleave a specific location in a target RNA. Unlike antisense or siRNA, these catalytically functioned OGNs do not depend on cellular nucleases for activity. Ribozymes and DNAzymes have become important tools for blocking gene expression and antiviral agents. In recent years, these OGNTs have been evaluated in pre-clinical studies for the treatment of viral infections, cancer, genetic disorders, as well as nervous system diseases.[12]

A second category of OGTs are those which hybridize to DNA to either modulate gene expression at its earliest stage or to correct harmful sequences arising from mutations. Rather than trying to separate the two DNA strands, either intra- or intermolecular triple helical DNA (triplex-forming OGNs or TFOs) structures are formed to take up residence in the major groove of the double helix and form Hoogsteen or reverse Hoogsteen base pairs with one of the strands. TFOs have been shown to inhibit elongation of the growing RNA and, may stimulate transcription if they bind upstream of a promoter. In addition, TFOs are reported to induce site-specific recombination and gene correction in mammalian cells.[13]

The third category of OGNTs are those that bind targets at certain proteins or small molecules to form complex secondary and tertiary structures of analogous antibodies. Aptamers are nonbiological OGNs that can bind to protein-like antibodies. *In vitro* selection of functional OGNs for their binding or catalytic activities (called SELEX) is carried out by screening massive DNA or RNA residues of 20–100 units in length.[14] In the past two decades, some high-affinity aptamers have been identified in a broad cross-section of protein families including cytokines, proteases, kinases, cell-surface receptors, and cell-adhesion molecules.[15] So far however, only one aptamer has been approved by FDA. Macugen, approved in 2004, is an anti-angiogenic 28-mer RNA for the treatment of neovascular (wet) AMD. An L-ribonucleic acid aptamer (trade name Spiegelmer) is an RNA-like molecule built from L-ribose units.[16] RNA Spiegelmers are structured OGNs of typically 30–60 units in length. Due to their L-mirror configuration, they are highly resistant to degradation by nucleases.[17] Spiegelmers are considered potential drugs and are currently being tested in clinical trials.

Immunomodulatory OGNs are synthetic cytosine phosphate-guanosine (CpG) oligodeoxynucleotides (ODNs). CpG DNAs are being used as therapeutic vaccines in various animal models of infectious diseases, tumors, allergic diseases, and autoimmune diseases.[18] A core unmethylated hexameric sequence motif consisting of 5-purine-purine-cytosine-guanine-pyrimidine-pyrimidine-3 is the key to activate the innate impurity in infectious diseases. Currently, CpG DNA and CpG ODN are being used as therapeutic vaccines in various animal models of infectious diseases, tumors, allergic diseases, and autoimmune diseases.[19] Immunomodulatory regimens offer an attractive approach as they often have fewer side effects than existing drugs, including less potential for creating resistance to microbial diseases.

Recently, ODN-based approaches termed decoy ODN have used synthetic ODN containing an enhancer element that can penetrate cells to bind to sequence-specific DNA-binding proteins and interfere with transcription *in vitro* and *in vivo*. Decoy ODNs are used to attract proteins away from their binding sites in DNA and a variety of therapeutic applications have been suggested in tumor and cancer treatment.[20]

4.3 SAMPLE PREPARATION FOR OGNT BIOANALYSIS

Due to the complexity of biological matrices (e.g., blood, plasma, urine, and tissue), sample preparation/purification is a critical element of all bioanalytical methods for OGNs. The salt, small organic and inorganic compounds, proteins, lipids, and non-protein macromolecules in the biological matrices can potentially interfere with the chromatographic separation and ionization efficiency for MS detection. PP, LLE, SPE, and combinations are commonly applied in OGNT purification from biological matrices.[3a] Although PP using ammonium acetate, methanol, or acetonitrile is the most direct method for sample preparation, the strong protein binding of OGNs typically results in significant losses of the target analyte with low recovery (<10%),[21] therefore this approach is rarely used for OGNT bioanalysis.

LLE using phenol/chloroform solvent system is a widely used extraction method for OGNs with high recoveries.[21–23] Turnpenny et al. developed a simple LLE procedure with high extraction recovery of 89 ± 2% for all eight antisense OGNs in plasma and 92 ± 5% in liver and kidney homogenate.[24] Although the LLE step helps to extract the OGNs from the biological matrices, it cannot fully remove matrix-related organic impurities. Therefore, the combination of LLE step followed by either a one- or two-step SPE is typically applied to provide further clean-up.[21,25]

Several different SPE phases have been used for OGNT bioanalysis: (1) strongly hydrophilic, water-wettable polymer with a unique hydrophilic–lipophilic balance such as Waters Oasis HLB material; (2) reversed-phase–LC stationary phase such as regular C18 and phenyl; and (3) a mixed-mode, reversed-phase/anion-exchange water-wettable polymer materials such as Waters Oasis MAX/AMCX/WCX/WAX and Phenomenex Clarity Biosolutions. Zhang et al. reported an extraction of OGNT from rat plasma samples using a phenol/chloroform LLE followed by Oasis HLB SPE with the absolute recovery of 70%–80% and less than 6% ionization suppression.[21] Phenomenex Clarity Biosolutions SPE column was reported with a promising 80% recovery for the extraction of siRNA from rat plasma.[26]

4.4 ANALYTICAL SEPARATION METHODS FOR OGNs

Increasing drug development demands reliable and high-throughput bioanalytical methods for OGNs and their metabolites in biological fluids. Among traditional analytical techniques including radiolabel tracers, LC-UV or fluorescence detection, immunoaffinity, quantitative polymerase chain reaction (PCR) assays, LC-MS provides major advantages of selectivity and sensitivity and has become the method of choice for bioanalysis of OGNTs. Separation techniques used for the analysis of OGNT include capillary electrophoresis (CE), anion-exchange high-performance LC, ion-pairing reversed-phase LC and mixed-mode LC.

4.4.1 CAPILLARY ELECTROPHORESIS

CE provides several advantages over HPLC such as higher separation power and speed, smaller sample amount requirements, and lower cost. Based on its superior resolving power, capillary gel electrophoresis (CGE, a type of CE), has been used for analysis of OGN and OGN-based therapeutics since the 1990s.[27] Carmody et al. have shown separation of a metabolite from its OGN-based parent drug, and resolution of single base pairs has been demonstrated on the replaceable gels developed in recent years.[27f] A review by Chen and Bartlett provides a good overview of CGE development and covers gel types, separation conditions, sample introduction modes, and extraction procedures.[28]

Ultraviolet (UV) detection is the most common detection technique employed with CE. In recent years, advances in CE detection coupling mass spectrometry as a detector, has provided molecular mass and structural information to be obtained, and also offers unique selectivity. Several types of interfaces to couple the end of the CE capillary to mass spectrometer for sample transfer have been developed for CE-MS, including the sheathless interface, the liquid junction, and the sheath–liquid interfaces (Figure 4.3).[29,30] Despite the advantages provided by MS detection, CE-MS often suffers from poor concentration sensitivity. To overcome this, transient on-column capillary isotachophoresis, sample stacking, and addition of preconcentration membranes or solid-phase packing at the CZE column have been developed to preconcentrate sample and subsequently enhance sensitivity. Strong cation adducts are formed between the sodium or potassium electrolyte buffers used in CE and the negative charges at sugar–phosphate backbone of OGNs during CE-MS analysis. This leads to significant analyte ion suppression, which in turn decreases sensitivity. Several methods were developed to reduce this effect. One strategy is to replace sodium or potassium ions with an excess of ammonium ion, which forms a weaker bond with the phosphodiester groups than sodium or potassium ion and can dissociate during the electrospray process, the MH$^+$ ion of the OGN.

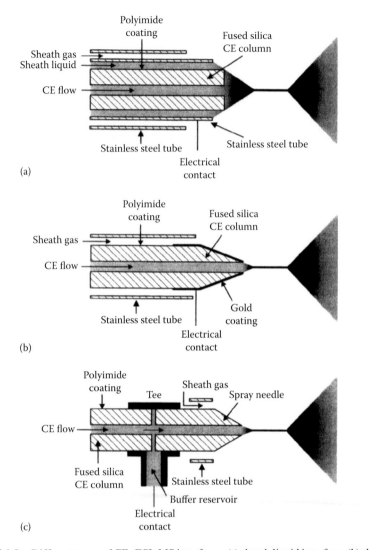

FIGURE 4.3 Different types of CE–ESI–MS interfaces: (a) sheath liquid interface, (b) sheathless interface, and (c) liquid junction interface. (From Willems, A. V. et al., *Electrophoresis*, 26, 1221–1253, 2005. With permission.)

4.4.2 Ion-Exchange HPLC

Anion-exchange chromatography separates the multiple negatively charged OGNs based on their competition with the positively charged stationary phases and increasing anion strength in the mobile phase. An ion-exchange LC offers excellent selectivity for the separation of OGNs based on length (N-x deletion) via the number of charges on the phosphate backbone. The mobile phase typically contains sodium chloride and sodium perchlorate in Tris or sodium phosphate as ion-exchange

TABLE 4.1
Ion Exchange Can Provide a Solution to the Analysis of Essentially All of These Classes of Oligonucleotides

Separation	Ion Exchange	IPRP
N-X failure sequence	Good	Good
PS to a PO	Good	Possible
Diastereoisomer separation	Possible	No
2′-5′ linkage isomers	Possible	Very difficult
Antisense from sense	Good	Possible
Double stranded from single stranded	Good	Good
ON extraction from tissue	Good	No
Metabolite identification	Possible with MS	Possible with MS

Note: IPRP, Ion-pair reversed phase; ON, oligonucleotide; PO, phosphodiester; PS, phosphorothioate.

reagent. Sodium perchlorate has been shown to offer selectivity for nucleobase composition. Thayer et al. reported using two different methacrylate-based stationary phases, eluent pH and modifiers, and eluent salts (e.g., Cl^- and ClO^-_4) to affect the retention time of 21–25 nucleotide unmodified ssDNA.[31]

Advantages provided by ion-exchange chromatography for analyzing OGNs are reviewed by Cook and Thayer.[32] A summary of the capabilities for OGN analysis between ion-exchange and ion-pair reverse phase is listed in Table 4.1. The most significant advantage ion-exchange chromatography provides is the ability to distinguish phosphorothioate diastereoisomers.[33]

Anion-exchange chromatography for analysis of OGNs is typically coupled with UV or fluorescence detectors. Direct coupling of anion-exchange chromatography with mass spectrometry is not feasible. Typical mobile phases contain high salt content or nonvolatile buffers including sodium chloride and sodium perchlorate in Tris or phosphate buffers, which are not compatible with direct ESI mass spectrometry interfaces unless proper desalting processes are applied prior to ESI interface. Thayer et al. developed a rapid desalting method in conjunction with anion-exchange chromatography for OGNs.[33a] Full-length product and impurities resolved by high-resolution anion exchangers are collected into vials or well plates in a fraction-collecting autosampler (Figure 4.4). The fractions are loaded onto a C18 trap column followed by an extended wash for desalting before they are reinjected for ESI–MS analysis.

4.4.3   Ion-Pair Reversed-Phase LC

Ion-pair RP–LC (IPLC) using ESI–MS compatible organic bases has become the LC method of choice for OGNs. The use of salt modifiers and ionic liquids to enhance strongly acidic or basic analyte retention has been firmly established for many years and has been known as "ion-pair" reversed-phase LC. OGN, as the analyte ion of interest is anionic, a cationic IP reagent can be used to suppress charge and form

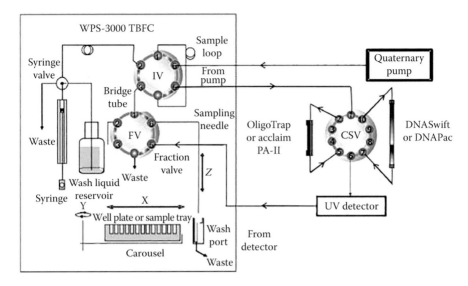

FIGURE 4.4 A titanium-based inert HPLC system with a WPS-3000 BTFC fraction-collecting autosampler configured for program-controlled column selection using anion exchange oligonucleotide purification as the first step followed by reversed-phase desalting. IV, injection valve; FV, fraction collection valve; CSV, column selection valve. (From Thayer, J. R. et al., *Anal. Biochem.*, 399, 110–117, 2010.)

neutral ion pairs by electrostatic attraction. When volatile mobile phase additives are used, this chromatographic approach could be easily interfaced with mass spectrometry to provide information on OGN molecular weight and structure.

Resolution of OGNs is based primarily on their length and ion-pairing induced-hydrophobicity. However, sequence and hydrophobic modifications can also have a chromatographic influence. Common IP agents employed include acetate or bicarbonate salts of triethylammonium (TEA, TEAA, TEAB),[34] hexylammonium acetate (HAA), tripropylammonium, tributylammonium, dimethylbutylammonium, and TEA. While increasing concentrations of IP reagents may enhance the OGN separation efficiency, it may also cause reduction of MS signal. Studies investigating the optimal IP reagent concentration are a topic of several research papers. The introduction of 1,1,1,3,3,3-hexafluoro-2-propanol (HFIP) into IP reagents such as TEA as an ion-pair system is so far the best IP solvent combination for OGN analysis by LCMS.[35] With 400 mM HFIP, only 15 mM TEA is needed to achieve the same ion-pairing effect as 100 mM TEAA alone. In the HFIP/TEA solvent system, TEA serves as an ion-pairing reagent for the negatively charged phosphate group in OGN and HFIP as a dynamic liquid/gas phase pH adjuster.[3a,35] The physicochemical parameters of HFIP such as hydrophobicity, relative volatility, and pK_a are assumed to be the fundamental reasons for this effect. In the presence of HFIP, the hydrophobicity of TEA might be greatly enhanced, thus a much lower concentration of TEA is needed. Furthermore, MS signal intensity of OGNs is increased along with a reduced cation adduction. An explanation for this is that HFIP selectively removes the cation counterion during the droplet evaporation step in ESI process, which alters the pH of the electrospray droplets. Reported in 2014, Gong and McCullagh performed a comprehensive comparison of six ion-pairing reagents to determine their performance as mobile phase modifiers for OGN LC/MS analysis.[36] The study was conducted using a Waters OST column (2.1 mm × 100 mm, 1.7 μm) with four OGNs (10–40 mer) as model compounds. Selected ion-pairing reagents such as triethyl-amine (TEA), tripropylamine (TPA), hexylamine (HA), *N,N*-dimethylbutylamine (DMBA), dibutylamine (DBA), or *N,N*-diisopropylethylamine (DIPEA) were combined with hexafluoro-2-propanol (HFIP). Adduct ion formation, chromatographic separation, and MS signal intensity at different conditions were collected and compared. It was shown that maximum separation for small, medium, and large OGNs was obtained with TPA, DMBA, and DBA as ion-pairing reagents, respectively. The effect of dissolution solvents on separation, MS signal intensity, and adduct ion formation was also examined. It is shown that the combination of 15 mM HA with 50 mM hexafluoro-2-propanol gave the best chromatographic performance and dissolution solvent type can have a significant impact on adduct ion formation with OGNs (resolution values: 14.1 ± 0.34, 11.0 ± 0.17, and 6.4 ± 0.11 for the pairs of OGNs T10 & T15, T15 & T25, and T25 & T40, respectively [3 replicates; see Figure 4.5]).

The Buszewski research group evaluated the influence of various IP reagents and the impact of different stationary phases on the retention behavior of OGNs.[37] In their studies, six different commercial UHPLC columns (Kinetex C18 core–shell with two different dimensions, Kinetex PFP core–shell, Hypersil Gold C18, C8, and phenyl) and three lab-packed stationary phases (octadecyl, cholesterol, and alkylamide) were studied. Their results demonstrated that among the commercial UHPLC columns, the octadecyl and phenyl stationary phases provided the most effective separation of OGNs with one single base difference in the sequence. They also reported for the first time that two novel types of packing materials, cholesterol and alkylamide were more effective at separating OGNs compared with the octadecyl packing, which was achieved under conditions of low temperature (30°C) and neutral pH. Separation of OGNs based on IP-RP–LC interaction using core–shell columns was studied by Biba et al.[38,39] Commercially available columns such as Ascentis Express C18, Ascentis Express Phenyl-Hexyl, Ascentis Express RP-Amide, Poroshell EC-C18, Kinetex XB-C18, and Kinetex C18 were selected for the investigation. Improved peak shape and resolution of closely related impurities comparing to porous C18 column were observed on core–shell-based column (Figure 4.6). Under neutral pH mobile phase and elevated column temperatures (≥60°C) which are required for IP-RP–LC separation, a modest column stability is observed for the core–shell columns.

4.4.4 Hydrophilic Interaction Liquid Chromatography

Hydrophilic interaction liquid chromatography (HILIC) is used extensively for the separation of polar and hydrophilic compounds which are difficult to retain. HILIC uses hydrophilic stationary phases which can be grouped into five categories of neutral polar or ionic surfaces: (1) simple unbounded silica silanol or diol-bonded phases; (2) amino- or anionic-bonded phases; (3) amide-bonded phases; (4) cationic-bonded phases; and (5) zwitterionic-bonded phases. A recent review article highlighted new

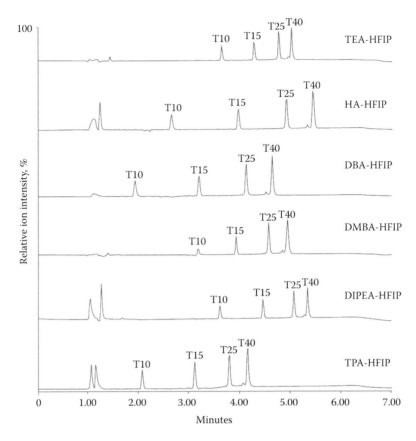

FIGURE 4.5 MS chromatograms of d(pT) using different IP reagents with HFIP at their optimized conditions. MPA: 15 mM TEA–400 mM HFIP; 15 mM HA–50 mM HFIP; 15 mM DBA–25 mM HFIP; 5 mM DMBA–300 mM HFIP; 30 mM DIPEA–200 mM HFIP; 15 mM TPA–50 mM HFIP, MPB: MPA in 80% MeOH, 4% MeOH in 5 min, 0.2 mL/min, injected 2 μL.

trends in column technology applied for HILIC stationary phase for separation of nucleosides, nucleotides, and OGNs.[40]

Mixed-mode columns contain two or more interaction sites and can offer unique selectivity than reversed-phase and ion-exchange separation modes alone. This could bring some value for some challenging OGN separations. Biba and co-workers studied separation of a library of RNA oligomers with N-X deletions and isomeric RNA 21-mers using mixed-mode columns from Imtakt.[41] The columns evaluated consist of C18 and ion-exchange selectors with ion-exchange capacities ranging from low to high. Using ammonium acetate or triethylammonium acetate as buffers on a Scherzo SW-C18 column (weak ion-exchange capacity), the separation of RNA oligomers with N-X deletions was predominantly by IP-RP–LC interaction mode and significant peak tailing were observed. When NaCl or NaBr was added to the mobile phase, different retention patterns were observed with improved peak shape for the N-X deletion mixture (Figure 4.7). Interestingly, RNA 21-mers with base-flip, which co-eluted under

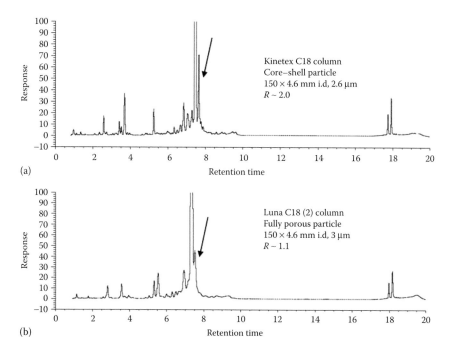

FIGURE 4.6 Comparison of (a) core–shell versus (b) fully porous particles for oligonucleotide analysis by IP-RPLC method. (Data from Biba, M. et al., *J. Pharm. Biomed. Anal.*, 72, 25–32, 2013.)

single IP-RP–LC or ion-exchange condition, were baseline resolved. The authors also noted that higher NaCl or NaBr concentrations were required to elute larger RNA oligomers when a Scherzo SM-C18 column (higher ion-exchange capacity) was used. However, a different retention pattern from the IP-RP–LC or ion-exchange separation was still reproduced. Mixed-mode stationary phases are typically available on 200 Å pore size supports with the lower surface areas compared to typical 100 Å pore size supports, as a result, lower retention times with equal selectivity are achievable.

4.4.5 CAPILLARY HPLC

Application of capillary liquid chromatography (CLC) using monolithic columns for OGNs has been explored in recent years.[42] There are several advantages of CLC when compared to regular HPLC. First, a much smaller sample amount is needed since CLC has better signal height-to-sample mass ratio.[43] Second, CLC coupled with ESI offers significant mass signal increase since it is inversely proportional to flow rate. Third, resolution of liquid chromatography increases with decreasing inner diameter of the HPLC column. CLC using a monolithic column based on continuous, rod-shaped, porous copolymer of styrene and divinylbenzene was reported for separation of single- and double-stranded OGNs.[42b] Mixed-sequence single-stranded OGNs ranging in size from 8 to 32 nucleotide units and differing in length by two nucleotide units were baseline resolved. In their study, at elevated temperatures

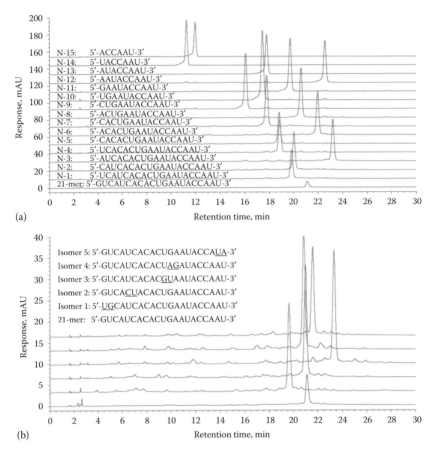

FIGURE 4.7 Chromatographic separation of RNA oligonucleotide samples by mixed-mode chromatography using Scherzo SW-C18 column with NaCl gradient. (a) Separation of N-X deletion series and (b) separation of "base-flip" isomer standards. Conditions: the mobile phase consisted of 100 mM Tris pH 7.4 in water (mobile phase A) and 1 M NaCl in 100 mM Tris pH 7.4 water/acetonitrile (90/10 (v/v%)). It is also important to note that Tris buffer (at pH 7.4) was used because the Tris buffer components are neutral or positively charged and therefore do not compete or interfere with the anion-exchange of the chloride anion. A linear gradient of 35%–65% B over 30 min was used. The flow rate was 1 mL/min, column temperature was 50°C, the injection volume was 10 μL and the UV detection was set at 260 nm. (Data from Biba, M. et al., *J. Chromatogr. A.*, 1304, 69–77, 2013.)

specifically ranging from 55.8°C to 59.8°C, mismatch based on single nucleotide from A to G for a 209 bp amplicon can be reproduced by the capillary poly(styrene-divinylbenzene)-based monolithic column. Furthermore, at 75°C, the wild-type single strands including newly formed mutated single strands were separated from a short 62 bp PCR product. In addition to IP-RP–LC, weak anion exchange-based separation was demonstrated on monolithic poly(glycidyl methacrylate-*co*-divinylbenzene) (GMA-*co*-DVB) column. Bisjak et al.[44] demonstrated efficient separation of a homologous series of phosphorylated oligothymidylic acids [d(pT)12–18]. Further

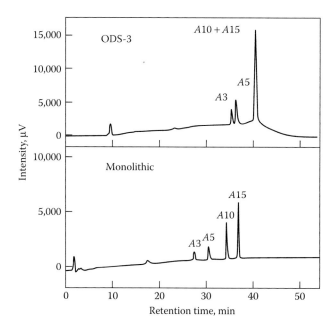

FIGURE 4.8   Separation of 3-mer ($A3$), 5-mer ($A5$), 10-mer ($A10$), and 15-mer ($A15$) oligoad-enylates using an ODS packed column and a silica-based monolithic column. Top: ODS-3 (GL Science), flow rate: 5 μL/min, bottom: monolithic column (Monocap C18 WideBore, GL Science), flow rate: 25 μL/min; column temperature: 60°C; mobile phase: 0.1 M TEAA (pH 7.0) in water (component A), methanol (component B); gradient: initial eluent 1% B, final 30% B in 60 min; detection: 260 nm. (From Kawamura, K. et al., *Chromatographia*, 78, 487–494, 2015.)

modification with installation of quaternary ammonium functionalities enable the GMA-*co*-DVB column to provide strong anion exchange capability.[45]

Kawamura et al. researched separation of short OGNs with silica-based C18 monolithic column based on ion-pairing reverse-phase separation mechanism.[46] Complete baseline resolution was achieved for OGNs shorter than 30-mer with a single nucleotide difference; however, it has not been observed on packed capillary column nor on a semi-μ-HPLC system (Figure 4.8). Separation of single-nucleotide polymorphisms (SNPs) was attempted on the silica-based monolith column. However, SNPs formed as a result of modifications at the 3′-end could not be separated indicating that the SNP column interactions were not strongly affected by the 3′ terminus. Finally, the author also noted that clogging of the column was encountered much less frequently on the monolithic than on the packed capillary columns.

4.5   MASS SPECTROMETRY

The current state-of-the-art MS instrumentation and techniques offer many different commercially available mass analyzers that are suitable for detecting and character-izing structural modifications of OGNs. These different instruments have their own advantages and disadvantages for the analysis of OGNs based on their resolution power, duty cycle, dynamic range, MS/MS function, cost, and ease of use. In this section, we will briefly discuss the main types of MS instruments which have been employed in the analysis of OGNs.

4.5.1 ESI–MS

ESI is a "soft" ionization technique that is widely used for macromolecules because it overcomes the propensity of these molecules to fragment when ionized. In an electrospray ion source, liquid is passed through a narrow orifice with high voltage applied to produce a mist of finely charged aerosol particles. The size of charged droplets gradually decreases within the droplets, resulting in the ejection of gas phase ions under atmospheric pressure conditions.[47] It is the most common ionization technique for the analysis of OGNs due to its direct interface of LC eluent to mass spectrometer. The majority of LC–MS applications for OGNs reported have utilized ion-pairing reversed-phase (RP-IP) chromatography. While ion pairing helps retain OGNs, ion-pairing interactions can also suppress the ESI ionization leading to low MS sensitivity as noted previously in Section 4.3. Several groups have developed methods using low levels of IP reagent in mobile phase along with an acidic solvent (TEA/HFIP) to maximize retention while minimizing ion suppression. Excellent reviews on this subject are given by Huber,[48] Lin,[3b] van Dongen,[3a] and Basiri.[49]

4.5.1.1 ESI-Quadrupole

A single quadrupole mass analyzer can be operated in two modes: full scan which allows all ions entering the quadrupole to be transmitted or single ion mode (SIM) where only selected ions are allowed to pass through the mass filter. In order to obtain structural information of a certain ion via its fragmentation pattern, a triple quadrupole MS systems can be utilized. These systems utilize an rf-only mass quadrupole or other multipole to serve as a collision cell (Q2) positioned between the two quadrupole mass filters (Q1 and Q3). Triple quadrupoles are widely used for the quantitative analysis of OGNs and their metabolites due to their high selectivity and sensitivity, wide dynamic range, excellent precision, relatively low cost, and ease of operation.[50]

Liu et al. reported using a LC-MS/MS method on a triple quadrupole MS to quantify regional DNA methylation as a potential clinical biomarker for cancer classification.[51] The methylated PCR amplicons were subjected to enzymatic hydrolysis, and the 5-methyl-2′-deoxycytidine (5 mdC) moiety was separated on Hypersil Aquasil C18 column and quantitated by mass spectrometry (Sciex API300) using 2′-deoxycytidine (2dC) as an internal standard. This assay was shown to differentiate 5% of promoter methylation level with an intraday precision ranging from 3% to 16% using two tumor suppressor genes.[52] A synthetic 24-mer antisense drug and its metabolites 5′N-1/3′N-1, 5′N-2, and 5′N-3 were separated by IR-RP–LC with a Phenomenex Genimi C18 column.[53] The detection was carried out in multiple reaction monitoring (MRM) mode using negative-ion ESI on an API 4000 triple quadrupole mass spectrometer. The LOQ was as low as 4.0 ng/mL for this OGN and its four metabolites. Similar quantification work has been published for two 15-mer OGNs in human plasma by Hemsley and his coworkers using online SPE and IP-UPLC with

API 5000 triple quadrupole detection. Their method was validated over the range of 10–4000 pM (~50–20,000 pg/ml).

4.5.1.2   ESI-Ion Trap

The quadrupole ion trap mass analyzer is a three-dimensional analog of the quadrupole mass analyzer and consists of a ring electrode and two end cap electrodes. Ions enter in through one end cap electrode and RF voltage is then applied to generate a quadrupole field that can trap ions over a large mass range. Following the trapping period, additional RF voltage is applied, which destabilize the ions inside and eject them out according to their m/z values. MS/MS experiments are carried out by isolating a single m/z parent ion within the ion trap and inducing collisions with a He buffer gas. The resulting fragment ions are then ejected out of the ion trap and analyzed by a detector. This process of ion isolation, collision, and fragment analysis can be repeated several times to allow for the complete acquisition of MS^n spectra ($n \geq 3$), which can provide a powerful tool for OGN analysis and sequencing.[54] The MS^n feature of ion trap has been used for sensitive detection of minor DNA adduct in human tissues.[55] Easter and coworkers studied the separation and identification of phosphorothioate OGNs using HILIC coupled with ESI-ion trap mass spectrometry. Baseline resolution was achieved for 21-mer and 23-mer phosphorothioate at low sample concentration of 50 nM, which is similar to that demonstrated by ion-pairing chromatography.[56] However, ion traps provide for lower duty cycles and dynamic range which tends to limit its use for quantitative analysis.

4.5.1.3   ESI-Q-TOF

As mentioned earlier, MS/MS based on triple quadrupole or ion trap instruments has been widely used in analyzing OGNs. However, the low resolution of these two techniques make it very challenging in OGN sequencing data interpretation, especially for larger OGNs, due to the presence of multiple charged species and adduct ions. The quadrupole time-of-flight (Q-TOF) mass spectrometer provides the MS/MS capability of quadrupole mass analyzers with the high mass resolution of time-of-flight (TOF) mass analyzers to provide unique capabilities for sequencing OGNs. In TOF analysis, ions accelerated to the same kinetic energy travel through a field-free drift tube toward an ion detector. Ions with higher m/z move faster than those with lower m/z, hence ions are separated in space by their velocities. The flight time is inversely proportional to ion velocity, which can be converted to m/z value. Current TOF instrumentation typically has resolving powers of 10,000 or better and typical mass accuracies of 2–5 ppm or even lower. Since its inception, the Q-TOF has been utilized as a powerful tool for structure confirmation/elucidation of OGNs.[57]

Gilar et al. and Gong's lab reported the use of a Waters Micromass LCT (ESI-TOF) coupled with IP-RP–LC or HILIC LC for the characterization and identification of synthetic and chemically modified OGNs achieving picomole level detection limits.[58] In other reports, Waters SYNAPT qTOF mass spectrometers were used for sequence confirmation of truncated OGNs and metabolites based on accurate mass MS/MS and MS^E methods.[59] In MS^E experiment, two scan functions are used for data acquisition with the first scan using low collision energy and second scan the collision energy is ramped from low to high (20–40 eV, for example). This "all-in-one" approach enables the acquisition of intact molecular ion information from the first scan and collection of fragment ion data from the ions in the preceding scan from the second scan. It was demonstrated that MS^E is equivalent to the traditional MS/MS method to produce fragment ions of the truncated OGNs or their metabolites. More importantly, MS^E provides a method for rapid assessment of metabolite or truncated OGN's structures. Turner et al. reported an approach for the sequence determination of Spiegelmers, which are enantiomers of naturally occurring RNA with high resistance to enzymatic degradation using Agilent LCMS system (Q-TOF). This method can be easily applied to the *de novo* sequencing of mirror-image ONGs and be able to generate all desired fragments in one step without MS/MS experiments.[60]

4.5.1.4 ESI-Orbital Trap

Since the introduction of the first orbital trap-based mass spectrometer in 2006, the orbital trap has proven to be an important analytical tool with a wide range of applications.[61] The orbitrap mass analyzer offers high-resolving power (up to 240,000 at m/z 200 for the latest generation) with a mass accuracy error as low as 1 ppm. The orbitrap is an ion trapping device consisting of outer and inner electrodes with a spindle shape. Ions with specific kinetic energy are trapped and oscillate between the inner and outer electrodes following a circular orbit around the z-axis. The m/z value of an ion is related to the frequency of ion oscillation (ω), which is recorded and converted to mass spectrum using a Fourier transformation (FT). The resolving power of orbitrap is a function of the number of axial oscillations. High-field orbitraps have a narrower gap between the inner and the outer electrodes, which increases the oscillation frequencies of the ions and doubles the acquisition rate for a defined resolving power.[62] In addition to the original hybrid ion trap–orbital trap configuration, the Orbitrap technology is now available as a standalone device (Exactive Plus Orbitrap™). The orbitrap instrument is sensitive to space charging effects and only a defined maximum quantity of ions can be properly trapped and analyzed at a time. When the sample matrix is complex, the capacity of the trap may be reached quickly due to a considerable background. The quadrupole-orbital trap hybrid instrument which was introduced in 2011 (Q Exactive™ Hybrid Quadrupole-Orbitrap Mass Spectrometer) was designed to overcome this issue. In this configuration, the quadrupole mass filter serves as a low-resolution mass filter which only allows defined precursors to be collected and analyzed at high-resolution within the trap in a mode called SIM. The Quadrupole-Orbitrap is less prone to space charging because of the selectivity provided by quadrupole.[63] Tretyakova et al. successfully implemented orbital trap technology in analyzing DNA in human samples. They observed excellent sensitivity and greatly improved signal-to-noise ratios for complex samples containing trace amounts of DNA adducts in the presence of a large excess of normal nucleosides.[57] Smith demonstrated the use of ESI(−) with a combination of a Scientific Orbitrap mass spectrometer for the characterization of RNAi-related 20-mer OGNs with the intention of supporting an evidence of structure document for a regulatory submission. Data generated by an LTQ-Orbitrap are capable of high resolution and high accuracy mass measurements at multiple charge states. Instead of using mass spectrometry to generate average molecular masses through the deconvolution, these mass determination and sequence-confirming fragmentation products were all at less than 5 ppm level.[64]

4.5.2 MALDI–TOF

Matrix-assisted laser desorption/ionization time-of-flight mass spectrometry (MALDI–TOF MS) is another major mass spectrometric technique that has been used in the characterization of OGNs. In MALDI–TOF analysis, a strong UV or IR-absorbing organic matrix is mixed with an analyte. The analyte/matrix then absorbs energy from a pulsed laser to initiating analyte desorption and ionization in the gas phase. Common MALDI matrices are 3-hydroxypicolinic acid (3-HPA) for DNAs,[65] and a mixture of 2,3,4- and 2,4,6-trihydroxy-acetophenone or 3-HPA for RNAs.[66] MALDI as a "soft" ionization technique that typically generates singly charged OGN ions greatly simplifying the data interpretation. The high mass resolution and mass accuracy of the TOF analyzer also makes unambiguous assignment of ladders containing natural mononucleotides and those with various chemical modifications possible solely based on m/z values.

4.5.3 ICP–MS

Inductively coupled plasma mass spectrometry (ICP–MS) coupled with HPLC separation is a great technique for elemental analysis. There are a few applications of HPLC-ICPMS reported for OGN analysis. Limbach group conducted separation and detection of OGNs by HILIC–ICP–MS.[67] With a TSK gel Amide-80 column, separation of dT15, dT20, and dT30 can be achieved with acceptable reproducibility. Due to the high amount of acetonitrile present at the start of HILIC gradient separation, oxygen was added before the ICP torch to prevent carbon buildup at the interface of the ICP–MS. Detection of OGNs was accomplished by reaction of oxygen with phosphodiester linkages, to generate an m/z ratio of 47, which corresponds to $^{31}P^{16}O^+$. LOD and LOQ were established at pmol for the OGNs studied.

More recently, Studzińska et al. researched the separation and detection of phosphorothioate OGNs with IP–HPLC–ICP–MS.[68] Besides the commonly used quadrupole mass analyzer, another mass detector, double focusing sector field mass analyzer (SF-MS) which has higher resolution was also used in the study. Simultaneous monitoring of ^{31}P and ^{32}S were applied to discriminate between target antisense OGNs and unmodified OGN (Figure 4.9). Their results indicated that lower HFIP and TEA concentration gave lower LOD and LOQ for all the OGNs studied. Optimum conditions for good resolution, peak symmetry, and ICP–MS sensitivity were 200 mM HFIP and 2.55 mM TEA with 18% v/v of methanol as mobile phase. In terms of LOD, SP-MS is much more sensitive than the quadrupole mass analyzer. For example, twenty-fold lower concentrations of phosphorothioate OGNs were demonstrated. Finally, application of IP-UHPLC-ICP-SF-MS for analysis of OGNs and their metabolites in serum were attempted. The method showed one order of magnitude of linearity and LOD in the range 0.068–0.125 mg/L.

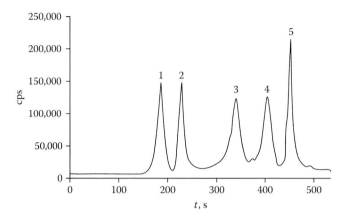

FIGURE 4.9 Separation OL5 and its four metabolites (OL6–OL9) with the use of HR–ICP–MS. Chromatographic conditions: Hypersil GOLD C18 1.9 μm (2.1 × 100 mm) column; mobile phase composed of methanol MeOH and 200 mM HFIP/2.55 mM TEA; gradient elution: 0 min–16.5% v/v MeOH, 5 min–22% v/v MeOH, 7 min–22% v/v MeOH; flow rate 0.1 mL min⁻¹; autosampler temperature 45°C; column temperature 60°C; injection volume 5 μL. Notation: 1 – OL9 (5′-G*T*T*C*T*C*G*C*T*G*G*T*G*A*G*T-3′), 2 – OL8 (5′-G*T*T*C*T*C*G*C*T*G*G*T*G*A*G*T*T-3′), 3 – OL7 (5′-G*T*T*C*T* C*G*C*T*G*G*T*G*A*G*T*T*T-3′), 4 – OL6 (5′-G*T*T*C*T*C*G*C*T*G*G*T*G*A *G*T*T*T*C-3′), 5 – OL5 (5′-G*T*T*C*T*C*G*C*T*G*G*T*G*A*G*T*T*T*C*A-3′). (From Studzińska, S. et al., *Anal. Chim. Acta.*, 855, 13–20, 2015.)

4.6 APPLICATION

4.6.1 SEQUENCING

It is important to confirm the exact sequence of the OGNs to be used in therapeutic applications. Tandem mass spectrometry can be used to reliably confirm the sequence of DNA and RNA, including modified OGNs. Applications include sequence confirmation of known sequences, as well as de novo sequencing and identification of modified residues. Compared with other techniques, the major benefit of LC-MS is the online sequencing of separated OGNs with substitutions or modifications. Several excellent reviews describing OGN sequencing using MS have been published since the 1990s.[49,69] ESI and MALDI are the two most widely used interface techniques employed for MS sequence ladder analysis.

The sequencing of OGNs via ESI–MS is performed by interpreting the OGN fragmentation produced by collision-induced dissociation (CID).[54a and b,70] The nomenclature for OGN fragment ions is proposed by McLuckey based on the peptide fragmentation pattern in CID[54a] at different locations along the phosphodiester backbone to produce a/w, b/x, c/y, and d/z ion series (Figure 4.10). Generally, OGNs are used as a broad concept for both short-chain DNA and RNA. However, ODNs (in DNA) and OGNs (in RNA) have different fragmentation patterns in the gas phase CID. The most abundant ions are (a-B) (a-type ions that have lost their nucleobases) and w-ion series are in DNA for sequencing in forward and reverse directions,[54b] while in RNA, the major ions are the c/y series.[70]

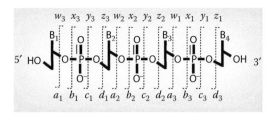

FIGURE 4.10   Fragmentation pattern of oligonucleotides.

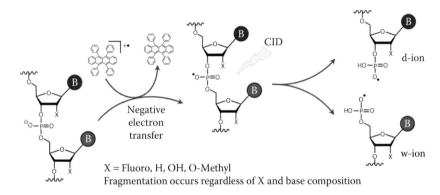

X = Fluoro, H, OH, O-Methyl
Fragmentation occurs regardless of X and base composition

FIGURE 4.11   Negative electron transfer (NET)-CID for oligonucleotides with 2′-position modifications. (From Gao, Y. et al., *Anal. Chem.*, 85, 4713–4720, 2013.)

The resulting fragment ions can be used for construction of mass ladders and OGN sequencing. For OGNs whose backbone has been modified, they will generate the same diagnostic ion series; however, it will differ in the ratios of the ions.[71] Due to the different dissociation patterns of DNA and RNA residues in CID, McLuckey group compared ion trap (IT)-CID and negative electron transfer (NET)-CID for the dissociation patterns of synthetic 21–23-mer OGNs with 2′-position modifications (F, OH, or OMe). (See Figure 4.11 for NET-CID mechanism.) They concluded that NET-CID is a more effective sequencing method for DNA-containing OGNs than conventional IT-CID.

The McLuckey group also studied using CID for sequencing of intact duplexes.[72] Since the direct CID spectra for duplexes were too complicated, the ions corresponding to the dissociated sense and antisense strands were used for sequencing. Two critical steps are applied for the duplex sequencing. First proton-transfer reactions are used to decrease the charge states and simplify the MS/MS spectra to make spectral interpretation easier. Second, the ion trap collisional activation energy is decreased to produce either c/y ions or a-B/w ions to achieve complete sequencing of the sense and antisense strands.

To cope with the massive data processing of the fragmentation, several computer algorithms have been developed since 1996 by different research groups.[73–77] Nyakas et al. introduced a software package containing two algorithms, the OGN mass assembler and the OGN peak analyzer for native and modified nucleic acids in 2013.[78]

Turner and his coworkers described unique methods for the sequence determination of Spiegelmers, which are mirror images of naturally occurring RNA with high resistance to enzymatic degradation.[60] In their methods, either an affinity tag or a dye label was attached to the pre-installed amino-modified Spiegelmer intermediate NOX-E36, NOX-e36 mutant, and NOX-A12. Chemical fragmentation of the labeled molecule followed under mild basic conditions, either isolating or resolving the fragments being produced to create the sequencing ladder for structure identification. They successfully tested the methodology on a mutant NOX-E36 Spiegelmer and were able to generate all desired fragments in one step, which can be resolved in one LC-MS analysis, without MS/MS experiments for subsequent interpretation.

Bahr, Aygun, and Karas demonstrated that the use of strong acid at pH 1–2 lead to rapid hydrolysis of the phosphodiester bonds followed by MALDI to analyze the degradation products for sequencing both single- and double-stranded RNA.[79] Coupling with an Orbitrap high-resolution mass spectrometer, Bahr was able to sequence both strands of the duplex by simultaneously using mass ladders from both sides of the nucleotides without interfering fragments from base losses or internal fragments. However, this method did not work for the RNA with a hydroxyl group on the 2′ position (such as 2′ O-methyl or for 2′ ribose modifications) because the OH group is required for hydrolysis reaction. Farand and his coworkers used the chemical approach combined with accurate mass and tandem mass to successfully sequence multiple 2′ fluoro and 2′ O-methyl modifications siRNAs.[80]

4.6.2 CHARACTERIZATION OF SYNTHETIC IMPURITIES IN OGNs

Although most synthetic methods for OGN can achieve yields 99.0+% per synthetic cycle, a 20-mer OGN can contain as much as 10% impurities prior to purification.[81] When these impurities are present in therapeutic OGNs, they can influence not only the drug efficacy and toxicity but also raise questions for a regulatory submission. Common impurities observed are failure sequences carrying 3′-terminal phosphate and phosphorothioate monoesters or incomplete backbone sulfurization and desulfurization products for phosphorothioate OGNs, acid or base reaction impurities, and degradation impurities.[82] LC-MS has been shown to be applicable for QC and impurity identification during OGN synthesis.[49]

Ion pairing-reversed-phase (IP–RP) HPLC has been employed for the separation of synthetic OGNs and their impurities. One potential drug candidate reported by GlaxoSmithKline (a 20-mer 2′-O-methyl phosphorothioate OGN) was characterized by the use of negative ion electrospray with a combination of high resolution and high mass accuracy (LTQ-Orbitrap).[64] In this paper, accurate mass of this 20-mer OGN and its three impurities were obtained for molecular mass determinations and sequence-confirming fragmentation products with less than 5 ppm mass error. Limbach utilized IP–RP coupling with two mass analyzers (linear ion trap and Fourier transform mass spectrometry [FTMS]) to identify a crude 24-mer phosphorothioate OGN.[82a] Figure 4.12 shows the UV and MS chromatograms for the 24-mer OGN synthesis reaction mixture, which is representative of the complexity of the sample impurities

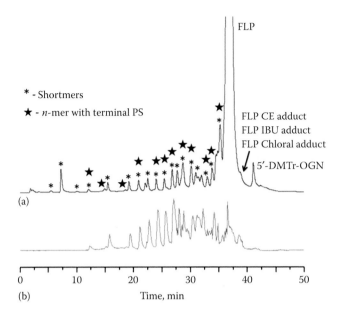

(a)

(b)

0   10   20   30   40   50

(b)

Time, min

FIGURE 4.12 Separation of crude sample containing 24-mer phosphorothioate OGN and its impurities: (a) total ion chromatogram from LC–MS analysis and (b) UV trace at 260 nm. Mobile phase consisted of 400 mM HFIP/16.3 mM TEA in water (buffer A) or in methanol (buffer B). (From Nikcevic, I. et al., *Int. J. Mass Spectrom.*, 304, 98–104, 2011.)

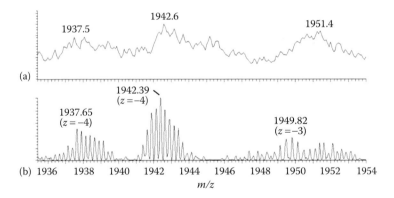

FIGURE 4.13 Comparison of (a) LC–MS spectrum and (b) LC–FTMS spectrum of low-level impurities. The use of high-resolution FTMS enables isotopes to be resolved so that the charge state can be determined for the impurity. (From Nikcevic, I. et al., *Int. J. Mass Spectrom.*, 304, 98–104, 2011.)

and separation challenges. The comparison of low mass resolution and high mass resolution spectrometers for low-level impurities is shown in Figure 4.13. The key advantage of using FTMS is its ability to accurately assign higher ion charge state for those low-level impurities, in comparison this is very difficult to do when using a low resolving power mass analyzer.

4.6.3 QUANTIFICATION OF OGN AND ITS METABOLITES

OGNs can be degraded *in vivo* by exonucleases that cleave the phosphodiester linkage at the ends of the strand and proceed sequentially inward, or cleave internally by endonucleases, or a combination of both. Knowing the exact location where the OGN is susceptible to enzymatic degradation can provide important guidance for incorporating chemical modifications in the OGN sequence to improve stability. Approaches for the identification of OGN metabolites is very similar to those for the sequencing of OGNs with the goal of recognizing the truncated OGN strands. However, the analysis of OGN drug metabolism usually involves analyzing very low concentrations in a biological matrix. Several excellent reviews have extensively covered the use of LCMS for the identification and characterization of the metabolism of siRNA as RNA *in vitro* and *in vivo*.[3,49,83] To evaluate the PK/PD performance of OGNs in biological matrices to support their preclinical and clinical development, it is critical to have reliable bioanalytical methods with high selectivity, sensitivity, accuracy, and precision for measuring OGNs quantitatively *in vivo*. The first comprehensive and validated LC/MS/MS method for a model phosphorothioate backbone OGN drug (7692 amu) from rat plasma was reported by Zhange in 2007.[21] In this paper, an ion-pairing buffer (7 mM TEA and 3 mM ammonium formate)/methanol, 50:50 (v/v), was used as an ESI–MS infusion solvent to enhance the resolution of multiple charge-state distribution. The combination of a phenol/chloroform LLE and SPE steps produced the absolute recovery to >70%. The method was validated in the range of 5–2000 ng/mL and had precision (percent relative standard deviation) <10.1% and accuracy (percent relative error) <11.4% using Applied Biosystem MDS triple-quadrupole mass spectrometer in MRM mode. This methodology was quickly adopted by Deng et al. and used for the determination of antisense OGN and its chained shortened metabolites in rat plasma.[53] The lower limit of quantification (LLQ) was 4.0 ng/mL for AS-ODN and its four metabolites. The intermediate precision was less than 12.0% and the accuracy ranged from 9.6% to 6.0% across validation runs.

However, the LOQ (4 ng/ml) still did not meet the sensitivity required to monitor the elimination phase of OGNs in clinically relevant doses. In recent years, several quantitative LC–MS methods were developed and validated for antisense OGNs in mouse tissue,[24] rat, and human plasma[23,50,51,84] to simplify the extraction procedure or optimize the chromatographic conditions to enhance the sensitivity. Hemsley and his coworkers developed an LC-MS/MS for quantification of a 15-mer siRNA at sub-ng/ml (~50 pg/ml) concentration using online SPE system.[51] Most recently, a 18-mer antisense PT-OGN for transforming growth factor $\beta2$ (TGF-$\beta2$) and its six metabolites (5′n-1, 5′n-2, 5′n-3, 3′n-1, 3′n-2, and 3′n-3) were independently quantified by LLE and SPE with UHPLC–MS/MS from human plasma by Trabedersen.[50] In their method, seven individual analytes in human plasma were quantified within a single assay with a calibration range of 2–1000 ng/ml and the accuracy, precision, and selectivity to meet regulatory expectations. In this particular case, two n-2 metabolites and the two n-3 metabolites were isobaric pairs, which are indistinguishable by MS alone, but were able to be separated under the chromatographic conditions to reduce the risk of two interfering molecules eluting together. Figure 4.14 shows the overlay of all 7 analytes and internal standard chromatograms, in which, only 3′ n-1 and 5′ n-2 are coeluted at 14.5 min. Table 4.2 is the LC and MS/MS parameters for this method, and in MS/

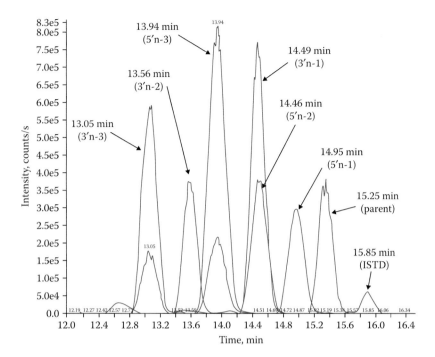

FIGURE 4.14 Raw data (unprocessed) chromatogram showing all analytes at the ULOQ (1000 ng/ml) and the IS. The chromatographic window has been zoomed to 12.0–16.5 min for clarity. Only the 3´n-1 and 5´n-2 analytes could not be separated at all, however these showed no crossed-ion interference with each other. The n-3 metabolites show two traces at their retention times as they are analyzed using two different transitions (different charge states of the same mass), which are shared by both analytes but in different relative abundances. (From Ewles, M. et al., *Bioanalysis*, 6, 447–464, 2014.)

MS parameter table, the two coeluted analytes (3´ n-1 and 5´ n-2) have different MRM patterns so they can be monitored without interference. On the other hand, the two analytes (5´ n-2 and 3´ n-2) have the same MRM patterns, but different elution times (at 14.46 and 13.56 min, respectively) so that they can also be monitored separately.

4.6.4 OGN Aptamers in Biomarker Discovery

Aptamers are synthesized OGNs with high affinity toward specific protein targets. Over the past decade, there are several interesting reports in the application of aptamers as bimolecular devices for binding target molecules, such as protein nanopores,[85] SPE,[86] and liquid chromatography.[87] Due to its high affinity and specificity, an aptamer binds to a specific target. Many applications of aptamers in bioanalytical MS have been reported either as the immobilized surface for MS being used as a read-out probe, or as aptamer-ligand complex for MS to be directly characterized.[88]

Recently, a novel solid-phase microextraction (SPME) coating with an anti-thrombin DNA aptamer for enrichment and determination of thrombin from human plasma is reported by Du et al.[86b] The detection was carried out using LC-MS/MS

TABLE 4.2
LC and MS Parameters

LC Parameter	Details
LC system	Waters® Acquity system
Analytical column	Waters Acquity BEH C18 100 × 2.1 mm 1.7 μm particle size
Column temperature	60°C
Autosampler temperature	5°C
Mobile phase A	Water/HFIP/TEA (100/1/0.1 v/v/v) (freshly prepared)
Mobile phase B	Methanol/HFIP/TEA (100/1/0.1 v/v/v) (freshly prepared)
Gradient	0.00–18.00 min 10%–19.5% B, 0.2 ml/min
	18.00–18.05 min 19.5%–90% B, 0.2 to 0.5 ml/min
	18.05–20.00 min 90% B, 0.5 ml/min
	20.00–20.05 min 90% to 10% B, 0.5 ml/min
	20.05–21.50 min 10% B, 0.5 ml/min
	21.50–21.55 min 10% B, 0.5 to 0.2 ml/min
	21.55–22.00 min 10% B, 0.2 ml/min
Autosampler weak wash	Water/methanol/TEA (60/40/0.2 v/v/v), 4000 μl
Autosampler strong wash	Acetonitrile/water/TEA (60/40/1 v/v/v), 4000 μl
Injection mode	Partial loop with needle overfill (3 μl)
Loop volume	50 μl
Typical injection volume	15 μl

MS/MS Parameter	Details
MS/MS system	AB SCIEX API 5000™ in negative turbo ionspray
Gas and voltages	CAD = 6; CUR = 20; GS1 = 80; GS2 = 80; ISV = −4500; source temperature = 500°C
Resolution	Q1 = High, Q3 = Unit

Ions Monitored

Analyte	Transition	Dwell	DP	EP	CE	CXP
Trabedersen	639.9–319.1	80	−130	−10	−34	−22
5′n-1	779.5–319.1	80	−145	−10	−46	−22
3′n-1	678.9–319.1	80	−130	−10	−37	−22
5′n-2/3′n-2	730.2–319.1	80	−140	−10	−43	−22
5′n-3	595.5–319.1	80	−125	−10	−32	−22
3′n-3	680.9–319.1	80	−130	−10	−39	−22
IS	697.3–319.1	80	−130	−10	−40	−22

Notes: HFIP: 1,1,1,3,3,3-hexafluoro-isopropanol; TEA: Triethylamine.
Pause time = 5 ms; CAD gas = Nitrogen; Analysis time = 18 min.
CAD: Collision gas; CE: Collision energy; CUR: Curtain gas; CXP: Collision cell exit potential; DP: Declustering potential; EP: Entrance potential; GS1: Nebuliser gas; GS2: Heated turbo gas; ISV: Ionspray voltage.

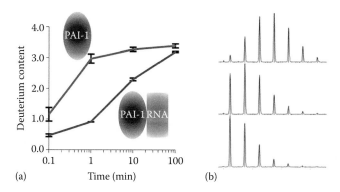

FIGURE 4.15 Example data of a deuterium-labeled PAI-1 peptide (a) and deuterium uptake curves of all analyzed peptides from active PAI-1 (b). (From Trelle, M. B. et al., *ACS Chem. Biol.*, 9, 174–182, 2014.)

and provided a good linear range of 0.5–50 nM for thrombin detection in diluted samples with the detection limit at 0.3 nM. Mass spectrometry can also be used for direct characterization of aptamer–ligand interactions. Besides the well-known three "S" advantages of mass spectrometry (specificity, sensitivity, and speed), using MS for noncovalent interactions can add another "S" (Stoichiometry) as its unique benefit.[88] Guo and his coworkers used ESI-FTMS to calculate the binding affinity of L-argininamide aptamers,[89] and their results are contradictory to what was reported by Brodbelt group where a quadrupole ion trap was used.[90] It is noted that the harsh ESI interface in ion trap is not suitable for preserving bimolecular complexes in their native state which could explain the differences in the results.

In the past decade there have been limited publications utilizing hydrogen–deuterium (H/D) exchange mass spectrometry (MS) to study protein–ligand interactions.[91] The Gross group modified the traditional H/D exchange protocol by using a strong anion exchange column (SAX) to rapidly remove the OGNs from solution before MS analysis. They successfully studied the protein–OGN system (thrombin with the thrombin-binding aptamer).[92] In 2014, Trelle and his coworkers used H/D exchange MS to study the RNA aptamer-binding dynamics of plasminogen activator inhibitor 1 (PAI-1).[93] 0.3 µM PAI-1 in the absence and presence of 0.6 µM RNA aptamer was diluted 20-fold in deuterated PBS-buffer for the hydrogen/deuterium-exchange reaction. At various time points, labeled PAI-1 samples were digested to peptides using pepsin and the deuterium content measured using a cooled LC/MS setup. Deuterium uptake curves of complex and PAI-1 only are shown in Figure 4.15. Their results show that the two aptamers in this study were potent inhibitors of the structural transition of PAI-1 from the active state to the inactive.

4.7 CONCLUSIONS AND FUTURE PERSPECTIVES

The use of OGNs as a drug modality encompasses a vast array of approaches based on their mechanism of action. This field could create drugs that affect protein targets, RNA transcript alteration, upregulation or replacement, biology manipulation, and even genome correction/alteration. Similar to the development of small-molecule drugs, OGN candidates should possess significant potency, stability in biological systems, favorable PKs, distribution and PD, acceptable specificity, and safety. In addition, OGNs also create a new set of considerations since they fall somewhere in the spectrum between small molecules and biologics, such as the target tissue, size of the drug molecule, and its mechanism of action drive requirements for chemical modification and delivery strategies. Therefore, the demand for reliable bioanalytical methods to test OGNs and their metabolites is quickly accelerating.

Unlike other bioanalytical methods (PCR and ELISA), LC-MS allows for the separation, identification, and quantification of an OGN; its impurities; and its metabolites. To meet the regulatory requirements, MS techniques can also be used to provide the accurate mass and the sequence of OGN. With the decreasing LOQ levels to sub-ng/ml range, LC-MS has become more applicable to clinical studies and several LC-MS methods have been validated for the quantification of OGNs and their metabolites in biological fluids. With the development of novel instrumentation such as nano-UPLC, chip technology and modified ionization methods, and mass analyzers, the performance of LC-MS and sensitivity could be greatly improved. Future efforts for LC-MS will focus on the quantitative analysis of intact siRNA, OGN complexes as well as generate complete PK/PD, metabolic stability, and bioavailability data.

REFERENCES

1. Goodchild, J. (ed.), Therapeutic oligonucleotides. In Therapeutic Oligonucleotides: *Methods and Protocols*. Humana Press: New York, **2011**; vol. 764, pp. 1–15.
2. (a) Fichou, Y.; Férec, C., The potential of oligonucleotides for therapeutic applications. *Trends in Biotechnology* **2006**, 24(12), 563–570; (b) Wilson, C.; Keefe, A. D., Building oligonucleotide therapeutics using non-natural chemistries. *Current Opinion in Chemical Biology* **2006**, 10(6), 607–614; (c) Spurgers, K. B.; Sharkey, C. M.; Warfield, K. L.; Bavari, S., Oligonucleotide antiviral therapeutics: Antisense and RNA interference for highly pathogenic RNA viruses. *Antiviral Research* **2008**, 78(1), 26–36.
3. (a) van Dongen, W. D.; Niessen, W. M. A., Bioanalytical LC–MS of therapeutic oligonucleotides. *Bioanalysis* **2011**, 3(5), 541–564; (b) Lin, Z. J.; Li, W.; Dai, G., Application of LC–MS for quantitative analysis and metabolite identification of therapeutic oligonucleotides. *Journal of Pharmaceutical and Biomedical Analysis* **2007**, 44(2), 330–341.
4. Kole, R.; Krainer, A. R.; Altman, S., RNA therapeutics: beyond RNA interference and antisense oligonucleotides. *Nature Reviews Drug Discovery* **2012**, 11(2), 125–140.
5. Mansoor, M.; Melendez, A. J., Advances in antisense oligonucleotide development for farget identification, validation, and as novel therapeutics. *Gene Regulation and Systems Biology* **2008**, 2, 275–295.
6. (a) Tremblay, G. A.; Oldfield, P. R., Bioanalysis of siRNA and oligonucleotide therapeutics in biological fluids and tissues. *Bioanalysis* **2009**, 1(3), 595–609; (b) Devi, G. R., siRNA-based approaches in cancer therapy. *Cancer Gene Therapy* **2006**, 13(9), 819–829.
7. Batkai, S.; Thum, T., Analytical approaches in microRNA therapeutics. *Journal of Chromatography B: Analytical Technologies in the Biomedical and Life Sciences* **2014**, 964, 146–152.
8. Singh, J.; Kaur, H.; Kaushik, A.; Peer, S., A review of antisense therapeutic interventions for molecular biological targets in various diseases. *International Journal of Pharmacology* **2011**, 7(3), 294–315.

9. Dallas, A.; Vlassov, A. V., RNAi: a novel antisense technology and its therapeutic potential. *Medical Science Monitor* **2006**, 12(4), RA67–RA74.

10. (a) Ambros, V., The functions of animal microRNAs. *Nature* **2004**, 431(7006), 350–355; (b) Bartel, D. P., MicroRNAs: genomics, biogenesis, mechanism, and function. *Cell* **2004**, 116(2), 281–297.

11. (a) Bauersachs, J.; Thum, T., Biogenesis and regulation of cardiovascular microRNAs. *Circulation Research* **2011**, 109(3), 334–347; (b) Gupta, S. K.; Bang, C.; Thum, T., Circulating microRNAs as biomarkers and potential paracrine mediators of cardiovascular disease. *Circulation: Cardiovascular Genetics* **2010**, 3(5), 484–488.

12. Mastroyiannopoulos, N. P.; Uney, J. B.; Phylactou, L. A., The application of ribozymes and DNAzymes in muscle and brain. *Molecules* **2010**, 15, 5460–5472.

13. Jain, A.; Wang, G.; Vasquez, K. M., DNA triple helices: Biological consequences and therapeutic potential. *Biochimie* **2008**, 90(8), 1117–1130.

14. Keefe, A. D.; Cload, S. T., SELEX with modified nucleotides. *Current Opinion in Chemical Biology* **2008**, 12, 448–456.

15. Keefe, A. D.; Pai, S.; Ellington, A., Aptamers as therapeutics. *Nature Reviews Drug Discovery* **2010**, 9(7), 537–550.

16. Eulberg, D.; Klussmann, S., Spiegelmers: biostable aptamers. *European Journal of Chemical Biology* **2003**, 4(10), 979–983.

17. Vater, A.; Klussmann, S., Turning mirror-image oligonucleotides into drugs: the evolution of Spiegelmer® therapeutics. *Drug Discovery Today* **2015**, 20(1), 147–155.

18. Wang, H.; Rayburn, E. R.; Wang, W.; Kandimalla, E. R.; Agrawal, S.; Zhang, R., Immunomodulatory oligonucleotides as novel therapy for breast cancer: pharmacokinetics, in vitro and in vivo anticancer activity, and potentiation of antibody therapy. *Molecular Cancer Therapeutics* **2006**, 5(8), 2106–2114.

19. Krieg, A. M., Mechanisms and applications of immune stimulatory CpG oligodeoxynucleotides. *Biochimica et Biophysica Acta* **1999**, 1489, 107–116.

20. (a) Warncke, S.; Gegout, A.; Carell, T., Phosphorothioation of oligonucleotides strongly influences the inhibition of bacterial (M.HhaI) and human (Dnmt1) DNA methyltransferases. *European Journal of Chemical Biology* **2009**, 10(4), 728–734; (b) Canello, T.; Ovadia, H.; Refael, M.; Zrihan, D.; Siegal, T.; Lavon, I., Antineoplastic effect of decoy oligonucleotide derived from MGMT enhancer. *PLoS One* **2014**, 9(12), e113854/1–e113854/18, 18 pp; (c) Kortylewski, M. T.; Swiderski, P. M., STAT3 inhibitors comprising of STAT3 decoy oligodeoxyribonucleotides linked to TLR9 ligand and uses thereof in cancer treatment. WO2015077657A2, 2015; (d) Nakazawa, T., Points-to-consider on the research and development of oligonucleotide therapeutics. *Iyakuhin Iryo Kiki Regyuratori Saiensu* **2014**, 45(5), 387–391.

21. Zhang, G.; Lin, J.; Srinivasan, K.; Kavetskaia, O.; Duncan, J. N., Strategies for bioanalysis of an oligonucleotide class macromolecule from rat plasma using liquid chromatography–tandem mass spectrometry. *Analytical Chemistry* **2007**, 79(9), 3416–3424.

22. Beverly, M.; Hartsough, K.; Machemer, L.; Pavco, P.; Lockridge, J., Liquid chromatography electrospray ionization mass spectrometry analysis of the ocular metabolites from a short interfering RNA duplex. *Journal of Chromatography B: Analytical Technologies in the Biomedical and Life Sciences* **2006**, 835(1–2), 62–70.

23. Chen, B.; Bartlett, M., A one-step solid phase extraction method for bioanalysis of a phosphorothioate oligonucleotide and its 3′ n-1 metabolite from rat plasma by uHPLC–MS/MS. *American Association of Pharmaceutical Scientists Journal* **2012**, 14(4), 772–780.

24. Turnpenny, P.; Rawal, J.; Schardt, T.; Lamoratta, S.; Mueller, H.; Weber, M.; Brady, K., Quantitation of locked nucleic acid antisense oligonucleotides in mouse tissue using a liquid-liquid extraction LC-MS/MS analytical approach. *Bioanalysis* **2011**, 3(17), 1911–1921.

25. Wu, L.-X.; Lu, D.-D.; Zhou, Z.; Zhang, H.-Y.; Zhang, Y.-L.; Wang, S.-Q., A combined solid phase extraction/capillary gel electrophoresis method for the determination of phosphorothioate oligodeoxynucleotides in biological fluids, tissues and feces. *Journal of Chromatography B: Analytical Technologies in the Biomedical and Life Sciences* **2009**, 877(4), 361–368.

26. Wheller, R.; Summerfield, S.; Barfield, M., Comparison of accurate mass LC-MS and MRM LC-MS/MS for the quantification of a therapeutic small interfering RNA. *International Journal of Mass Spectrometry* **2013**, 345–347, 45–53.

27. (a) Willems, A. V.; Deforce, D. L.; Van Peteghem, C. H.; Van Bocxlaer, J. F., Analysis of nucleic acid constituents by on-line capillary electrophoresis-mass spectrometry. *Electrophoresis* **2005**, 26(7–8), 1221–1253; (b) Friedecky, D.; Tomkova, J.; Maier, V.; Janost'akova, A.; Prochazka, M.; Adam, T., Capillary electrophoretic method for nucleotide analysis in cells: application on inherited metabolic disorders. *Electrophoresis* **2007**, 28(3), 373–380; (c) Kleparnik, K.; Bocek, P., DNA Diagnostics by capillary electrophoresis. *Chemical Reviews* **2007**, 107(11), 5279–5317; (d) Willems, A.; Deforce, D.; Van Bocxlaer, J., Analysis of oligonucleotides using capillary zone electrophoresis and electrospray mass Spectrometry. In Schmitt-Kopplin, P. (ed.) *Capillary Electrophoresis: Methods and Protocols.* Humana Press: Totowa, NJ, **2008**; Vol. 384, pp. 401–414; (e) Wan, F.; Zhang, J.; Chu, B., Advances in electrophoretic techniques for DNA sequencing and oligonucleotide analysis. *Advances in Chromatography* **2009**, 47, 59–125; (f) In Carmody, J.; Noll, B. (ed.) *Purity and Content Analysis of Oligonucleotides by Capillary Gel Electrophoresis*, CRC Press: Boca Raton, FL, **2011**; pp. 243–264.

28. Chen, B.; Bartlett, M. G., Determination of therapeutic oligonucleotides using capillary gel electrophoresis. *Biomedical Chromatography* **2012**, 26(4), 409–418.

29. Gaspar, A.; Englmann, M.; Fekete, A.; Harir, M.; Schmitt-Kopplin, P., Trends in CE-MS 2005 – 2006. *Electrophoresis* **2008**, 29(1), 66–79.

30. Klampfl, C. W.; Buchberger, W., Recent advances in the use of capillary electrophoresis coupled to high-resolution mass spectrometry for the analysis of small molecules. *Current Analytical Chemistry* **2010**, 6(2), 118–125.

31. Thayer, J. R.; Barreto, V.; Rao, S.; Pohl, C., Control of oligonucleotide retention on a pH-stabilized strong anion exchange column. *Analytical Biochemistry* **2005**, 338(1), 39–47.

32. Cook, K.; Thayer, J., Advantages of ion-exchange chromatography for oligonucleotide analysis. *Bioanalysis* **2011**, 3(10), 1109–1120.

33. (a) Thayer, J. R.; Puri, N.; Burnett, C.; Hail, M.; Rao, S., Identification of RNA linkage isomers by anion exchange purification with electrospray ionization mass spectrometry of automatically desalted phosphodiesterase-II digests. *Analytical Biochemistry* **2010**, 399(1), 110–117; (b) Thayer, J. R.; Wu, Y.; Hansen, E.; Angelino, M. D.; Rao, S., Separation of oligonucleotide phosphorothioate diastereoisomers by pellicular anion-exchange chromatography. *Journal of Chromatography A* **2011**, 1218(6), 802–808.

34. Huber, C. G.; Oefner, P. J.; Bonn, G. K., High-resolution liquid chromatography of oligonucleotides on nonporous alkylated styrene-divinylbenzene copolymers. *Analytical Biochemistry* **1993**, 212(2), 351–358.

35. Apffel, A.; Chakel, J. A.; Fischer, S.; Lichtenwalter, K.; Hancock, W. S., Analysis of oligonucleotides by HPLC-electrospray ionization mass spectrometry. *Analytical Chemistry* **1997**, 69(7), 1320–1325.

36. Gong, L.; McCullagh, J. S. O., Comparing ion-pairing reagents and sample dissolution solvents for ion-pairing reversed-phase liquid chromatography/electrospray ionization mass spectrometry analysis of oligonucleotides. *Rapid Communications in Mass Spectrometry* **2014**, 28(4), 339–350.

37. Studzińska, S.; Buszewski, B., Evaluation of ultrahigh-performance liquid chromatography columns for the analysis of unmodified and antisense oligonucleotides. *Analytical and Bioanalytical Chemistry* **2014**, 406(28), 7127–7136.

38. Biba, M.; Welch, C. J.; Foley, J. P.; Mao, B.; Vazquez, E.; Arvary, R. A., Evaluation of core–shell particle columns for ion-pair reversed-phase liquid chromatography analysis of oligonucleotides. *Journal of Pharmaceutical and Biomedical Analysis* **2013**, 72, 25–32.

39. Biba, M.; Welch, C. J.; Foley, J. P., Investigation of a new core-shell particle column for ion-pair reversed-phase liquid chromatography analysis of oligonucleotides. *Journal of Pharmaceutical and Biomedical Analysis* **2014**, 96, 54–57.

40. Garcia-Gomez, D.; Rodriguez-Gonzalo, E.; Carbias-Martinez, R., *Trends in Analytical Chemistry* **2013**, 47, 111–128.

41. Biba, M.; Jiang, E.; Mao, B.; Zewge, D.; Foley, J. P.; Welch, C. J., Factors influencing the separation of oligonucleotides using reversed-phase/ion-exchange mixed-mode high performance liquid chromatography columns. *Journal of Chromatography A* **2013**, 1304, 69–77.

42. (a) Premstaller, A.; Oberacher, H.; Huber, C. G., High-performance liquid chromatography–electrospray ionization mass spectrometry of single- and double-stranded nucleic acids using monolithic capillary columns. *Analytical Chemistry* **2000**, 72(18), 4386–4393; (b) Huber, C. G.; Premstaller, A.; Xiao, W.; Oberacher, H.; Bonn, G. K.; Oefner, P. J., Mutation detection by capillary denaturing high-performance liquid chromatography using monolithic columns. *Journal of Biochemical and Biophysical Methods* **2001**, 47(1–2), 5–19; (c) Bakry, R.; Huck, C. W.; Bonn, G. K., Recent applications of organic monoliths in capillary liquid chromatographic separation of biomolecules. *Journal of Chromatographic Science* **2009**, 47(6), 418–431; (d) Sharma, V. K.; Glick, J.; Vouros, P., Reversed-phase ion-pair liquid chromatography electrospray ionization tandem mass spectrometry for separation, sequencing and mapping of sites of base modification of isomeric oligonucleotide adducts using monolithic column. *Journal of Chromatography A* **2012**, 1245, 65–74.

43. Engelhardt, H., *Introduction to Microscale High-Performance Liquid Chromatography.* Herausgegeben von D. Ishii (ed.). VCH Verlags-gesellschaft, Weinheim/VCH Publishers, New York, **1988**, XIV, 208 S., geb. DM 118.00.–ISBN 3-527-26636-4; 0-89573-309-9.

44. Bisjak, C. P.; Bakry, R.; Huck, C. W.; Bonn, G. K., Amino-functionalized monolithic poly(glycidyl methacrylate-co-divinylbenzene) ion-exchange stationary phases for the separation of oligonucleotides. *Chromatographia* **2005**, 62(13), s31–s36.

45. Wieder, W.; Bisjak, C. P.; Huck, C. W.; Bakry, R.; Bonn, G. K., Monolithic poly(glycidyl methacrylate-co-divinylbenzene) capillary columns functionalized to strong anion exchangers for nucleotide and oligonucleotide separation. *Journal of Separation Science* **2006**, 29(16), 2478–2484.

46. Kawamura, K.; Ikoma, K.; Maruoka, Y.; Hisamoto, H., Separation behavior of short oligonucleotides by ion-pair reversed-phase capillary liquid chromatography using a silica-based monolithic column applied to simple detection of SNPs. *Chromatographia* **2015**, 78(7–8), 487–494.

47. Fenn, J. B.; Mann, M.; Meng, C. K.; Wong, S. F.; Whitehouse, C. M., Electrospray ionization for mass spectrometry of large biomolecules. *Science* **1989**, 246(4926), 64–71.

48. Huber, C. G.; Oberacher, H., Analysis of nucleic acids by on-line liquid chromatography-mass spectrometry. *Mass Spectrometry Reviews* **2001**, 20(5), 310–343.

49. Basiri, B.; Bartlett, Michael G., LC–MS of oligonucleotides: applications in biomedical research. *Bioanalysis* **2014**, 6(11), 1525–1542.

50. Ewles, M.; Goodwin, L.; Schneider, A.; Rothhammer-Hampl, T., Quantification of oligonucleotides by LC–MS/MS: the challenges of quantifying a phosphorothioate oligonucleotide and multiple metabolites. *Bioanalysis* **2014**, 6(4), 447–464.

51. Hemsley, M.; Ewles, M.; Goodwin, L., Development of a bioanalytical method for quantification of a 15-mer oligonucleotide at sub-ng/ml concentrations using LC–MS/MS. *Bioanalysis* **2012**, 4(12), 1457–1469.

52. Liu, Z.; Wu, J.; Xie, Z.; Liu, S.; Fan-Havard, P.; Huang, T. H.; Plass, C.; Marcucci, G.; Chan, K. K., Quantification of regional DNA methylation by liquid chromatography/tandem mass spectrometry. *Analytical Biochemistry* **2009**, 391, 106.

53. Deng, P.; Chen, X.; Zhang, G.; Zhong, D., Bioanalysis of an oligonucleotide and its metabolites by liquid chromatography–tandem mass spectrometry. *Journal of Pharmaceutical and Biomedical Analysis* **2010**, 52(4), 571–579.

54. (a) McLuckey, S. A.; Van Berkel, G. J.; Glish, G. L., Tandem mass spectrometry of small, multiply charged oligonucleotides. *Journal of the American Society for Mass Spectrometry* **1992**, 3(1), 60–70; (b) Wu, J.; McLuckey, S. A., Gas-phase fragmentation of oligonucleotide ions. *International Journal of Mass Spectrometry* **2004**, 237(2–3), 197–241.

55. Goodenough, A. K.; Schut, H. A. J.; Turesky, R. J., Novel LC-ESI/MS/MS(n) method for the characterization and quantification of 2′-deoxyguanosine adducts of the dietary carcinogen 2-amino-1-methyl-6-phenylimidazo[4,5-b]pyridine by 2-D linear quadrupole ion trap mass spectrometry. *Chemical Research in Toxicology* **2007**, 20(2), 263–276.

56. Easter, R.; Barry, C.; Caruso, J.; Limbach, P., Separation and identification of phosphorothioate oligonucleotides by HILIC-ESIMS. *Analytical Methods* **2013**, 5(11), 2657–2659.

57. Tretyakova, N.; Villalta, P. W.; Kotapati, S., Mass spectrometry of structurally modified DNA. *Chemical Reviews* **2013**, 113(4), 2395–2436.

58. (a) Gilar, M.; Fountain, K. J.; Budman, Y.; Holyoke, J. L.; Davoudi, H.; Gebler, J. C., Characterization of therapeutic oligonucleotides using liquid chromatography with on-line mass spectrometry detection. *Oligonucleotides* **2003**, 13(4), 229–243; (b) Gong, L. Z.; McCullagh, J. S. O., Analysis of oligonucleotides by hydrophilic interaction liquid chromatography coupled to negative ion electrospray ionization mass spectrometry. *Journal of Chromatography A* **2011**, 1218, 5480.

59. (a) Ivleva, V. B.; Yu, Y.-Q.; Gilar, M., Ultra-performance liquid chromatography/tandem mass spectrometry (UPLC/MS/MS) and UPLC/MSE analysis of RNA oligonucleotides. *Rapid Communications in Mass Spectrometry* **2010**, 24(17), 2631–2640; (b) Chen, B.; Bartlett, M. G., Evaluation of mobile phase composition for enhancing sensitivity of targeted quantification of oligonucleotides using ultra-high performance liquid chromatography and mass spectrometry: application to phosphorothioate deoxyribonucleic acid. *Journal of Chromatography A* **2013**, 1288, 73–81.

60. Turner, J. J.; Hoos, J. S.; Vonhoff, S.; Klussmann, S., Methods for L-ribooligonucleotide sequence determination using LCMS. *Nucleic Acids Research* **2011**, 39(21), e147–e147.

61. Makarov, A.; Denisov, E.; Kholomeev, A.; Balschun, W.; Lange, O.; Strupat, K.; Horning, S., Performance evaluation of a hybrid linear ion trap/orbitrap mass spectrometer. *Analytical Chemistry* **2006**, 78(7), 2113–2120.

62. Makarov, A.; Denisov, E.; Lange, O., Performance evaluation of a high-field orbitrap mass analyzer. *Journal of the American Society for Mass Spectrometry* **2009**, 20(8), 1391–1396.

63. Michalski, A.; Damoc, E.; Hauschild, J.-P.; Lange, O.; Wieghaus, A.; Makarov, A.; Nagaraj, N.; Cox, J.; Mann, M.; Horning, S., Mass spectrometry-based proteomics using Q exactive, a high-performance benchtop quadrupole orbitrap mass spectrometer. *Molecular and Cellular Proteomics* **2011**, 10(9), M111.011015.

64. Smith, M., Characterization of a modified oligonucleotide together with its synthetic impurities using accurate mass measurements. *Rapid Communications in Mass Spectrometry* **2011**, 25(4), 511–525.

65. Nordhoff, E.; Cramer, R.; Karas, M.; Hillenkamp, F.; Kirpekar, F.; Kristiansen, K.; Roepstorff, P., Ion stability of nucleic acids in infrared matrix-assisted laser desorption/ionization mass spectrometry. *Nucleic Acids Research* **1993**, 21(15), 3347–3357.

66. Zhu, Y. F.; Chung, C. N.; Taranenko, N. I.; Allman, S. L.; Martin, S. A.; Haff, L.; Chen, C. H., The study of 2,3,4-trihydroxyacetophenone and 2,4,6-trihydroxyacetophenone as matrices for DNA detection in matrix-assisted laser desorption/ionization time-of-flight mass spectrometry. *Rapid Communications in Mass Spectrometry* **1996**, 10, 383.

67. Easter, R. N.; Kröning, K. K.; Caruso, J. A.; Limbach, P. A., Separation and identification of oligonucleotides by hydrophilic interaction liquid chromatography (HILIC)–inductively coupled plasma mass spectrometry (ICPMS). *The Analyst* **2010**, 135(10), 2560–2565.

68. Studzińska, S.; Mounicou, S.; Szpunar, J.; Łobiński, R.; Buszewski, B., New approach to the determination phosphorothioate oligonucleotides by ultra high performance liquid chromatography coupled with inductively coupled plasma mass spectrometry. *Analytica Chimica Acta* **2015**, 855, 13–20.

69. (a) Rulka, A.; Markiewicz, W. T., Sequencing of combinatorial libraries with mass spectrometry. *Collection Symposium Series* **2011**, 12 (Chemistry of Nucleic Acid Components), 388–390; (b) Fisher, H. C.; Smith, M.; Ashcroft, A. E., De novo sequencing of short interfering ribonucleic acids facilitated by use of tandem mass spectrometry with ion mobility spectrometry. *Rapid Communications in Mass Spectrometry* **2013**, 27(20), 2247–2254; (c) Schuette, J. M.; Pieles, U.; Maleknia, S. D.; Srivatasa, G. S.; Cole, D. L.; Moser, H. E.; Afeyan, N. B., Sequence analysis of phosphorothioate oligonucleotides via matrix-assisted laser desorption ionization time-of-flight mass spectrometry. *Journal of Pharmaceutical and Biomedical Analysis* **1995**, 13(10), 1195–1203; (d) Haff, L.; Juhasz, P.; Martin, S.; Roskey, M.; Smirnov, I.; Stanick, W.; Vestal, M.; Waddell, K., Oligonucleotide analysis by MALDI-MS. *Analusis* **1998**, 26(10), M26–M30; (e) Polo, L. M.; Limbach, P. A., Analysis of oligonucleotides by electrospray ionization mass spectrometry. *Current Protocols in Nucleic Acid Chemistry* **2001**, Chapter 10, Unit 10.2; (f) Hayashizaki, Y.; Ono, T. MALDI-TOF-MS analysis and/or sequencing of oligonucleotides using modified ribonucleotides. WO2002046468A2, 2002; (g) Hail, M. E.; Elliott, B.; Anderson, K., High-throughput analysis of oligonucleotides using automated electrospray ionization mass spectrometry. *American Biotechnology Laboratory* **2004**, 22(1), 12–14; (h) Castleberry, C. M.; Rodicio, L. P.; Limbach, P. A., Electrospray ionization mass spectrometry of oligonucleotides. *Current Protocols in Nucleic Acid Chemistry* **2008**, Chapter 10, Unit 10.2.

70. Huang, T.-y.; Kharlamova, A.; Liu, J.; McLuckey, S. A., Ion trap collision-induced dissociation of multiply deprotonated RNA: c/y-ions versus (a-B)/w-ions. *Journal of the American Society for Mass Spectrometry* **2008**, 19(12), 1832–1840.

71. (a) Gao, Y.; McLuckey, S. A., Collision-induced dissociation of oligonucleotide anions fully modified at the 2′-position of the ribose: 2′-F/-H and 2′-F/-H/-OMe mix-mers. *Journal of Mass Spectrometry* **2012**, 47(3), 364–369; (b) Gao, Y.; Yang, J.; Cancilla, M. T.; Meng, F.; McLuckey, S. A., Top-down interrogation of chemically modified oligonucleotides by negative electron transfer and collision induced dissociation. *Analytical Chemistry* **2013**, 85(9), 4713–4720.

72. Huang, T.-y.; Liu, J.; Liang, X.; Hodges, B. D. M.; McLuckey, S. A., Collision-induced dissociation of intact duplex and single-stranded siRNA anions. *Analytical Chemistry* **2008**, 80(22), 8501–8508.

73. Ni, J.; Pomerantz, C.; Rozenski, J.; Zhang, Y.; McCloskey, J. A., Interpretation of oligonucleotide mass spectra for determination of sequence using electrospray ionization and tandem mass spectrometry. *Analytical Chemistry* **1996**, 68, 1989.

74. Rozenski, J.; McCloskey, J. A., SOS: a simple interactive program for ab initio oligo-nucleotide sequencing by mass spectrometry. *Journal of the American Society for Mass Spectrometry* **2002**, 13, 200.

75. Oberacher, H.; Wellenzohn, B.; Huber, C. G., Comparative sequencing of nucleic acids by liquid chromatography–tandem mass spectrometry. *Analytical Chemistry* **2002**, 74(1), 211–218.

76. Oberacher, H.; Mayr, B. M.; Huber, C. G., Automated de novo sequencing of nucleic acids by liquid chromatography-tandem mass spectrometry. *Journal of the American Society for Mass Spectrometry* **2004**, 15(1), 32–42.

77. Liao, Q.; Shen, C.; Vouros, P., GenoMass–a computer software for automated identification of oligonucleotide DNA adducts from LC-MS analysis of DNA digests. *Journal of Mass Spectrometry* **2009**, 44(4), 549–560.

78. Nyakas, A.; Blum, L.; Stucki, S.; Reymond, J.-L.; Schürch, S., OMA and OPA— software-supported mass spectra analysis of native and modified nucleic acids. *Journal of the American Society for Mass Spectrometry* **2013**, 24(2), 249–256.

79. Bahr, U.; Aygün, H.; Karas, M., Sequencing of single and double stranded RNA oligo-nucleotides by acid hydrolysis and MALDI mass spectrometry. *Analytical Chemistry* **2009**, 81(8), 3173–3179.

80. (a) Farand, J.; Beverly, M., Sequence confirmation of modified oligonucleotides using chemical degradation, electrospray ionization, time-of-flight, and tandem mass spectrometry. *Analytical Chemistry* **2008**, 80(19), 7414–7421; (b) Farand, J.; Gosselin, F., De novo sequence determination of modified oligonucleotides. *Analytical Chemistry* **2009**, 81(10), 3723–3730.

81. McCarthy, S. M.; Gilar, M.; Gebler, J., Reversed-phase ion-pair liquid chromatography analysis and purification of small interfering RNA. *Analytical Biochemistry* **2009**, 390(2), 181–188.

82. (a) Nikcevic, I.; Wyrzykiewicz, T. K.; Limbach, P. A., Detecting low-level synthesis impurities in modified phosphorothioate oligonucleotides using liquid chromatography–high resolution mass spectrometry. *International Journal of Mass Spectrometry* **2011**, 304(2–3), 98–104; (b) Capaldi, D. C.; Gaus, H. J.; Carty, R. L.; Moore, M. N.; Turney, B. J.; Decottignies, S. D.; McArdle, J. V.; Scozzari, A. N.; Ravikumar, V. T.; Krotz, A. H., Formation of 4,4′-dimethoxytrityl-C-phosphonate oli-gonucleotides. *Bioorganic and Medicinal Chemistry Letters* **2004**, 14(18), 4683–4690; (c) Gaus, H.; Olsen, P.; Sooy, K. V.; Rentel, C.; Turney, B.; Walker, K. L.; McArdle, J. V.; Capaldi, D. C., Trichloroacetaldehyde modified oligonucleotides. *Bioorganic and Medicinal Chemistry Letters* **2005**, 15(18), 4118–4124; (d) Oberacher, H.; Niederstätter, H.; Parson, W., Characterization of synthetic nucleic acids by electrospray ionization quadrupole time-of-flight mass spectrometry. *Journal of Mass Spectrometry* **2005**, 40(7), 932–945; (e) Kurata, C.; Bradley, K.; Gaus, H.; Luu, N.; Cedillo, I.; Ravikumar, V. T.; Sooy, K. V.; McArdle, J. V.; Capaldi, D. C., Characterization of high molecular weight impurities in synthetic phosphorothioate oligonucleotides. *Bioorganic and Medicinal Chemistry Letters* **2006**, 16(3), 607–614.

83. Beverly, M. B., Applications of mass spectrometry to the study of siRNA. *Mass Spectrometry Reviews* **2011**, 30(6), 979–998.

84. Erb, R.; Leithner, K.; Bernkop-Schnürch, A.; Oberacher, H., Phosphorothioate oligo-nucleotide quantification by μ-liquid chromatography-mass spectrometry. *American Association of Pharmaceutical Scientists Journal* **2012**, 14(4), 728–737.

85. Ding, S.; Gao, C.; Gu, L.-Q., Capturing single molecules of immunoglobulin and ricin with an aptamer-encoded glass nanopore. *Analytical Chemistry* **2009**, 81(16), 6649–6655.

86. (a) Zhao, Q.; Li, X.-F.; Shao, Y.; Le, X. C., Aptamer-based affinity chromatographic assays for thrombin. *Analytical Chemistry* **2008**, 80(19), 7586–7593; (b) Du, F.; Alam, M. N.; Pawliszyn, J., Aptamer-functionalized solid phase microextraction–liquid chromatography/tandem mass spectrometry for selective enrichment and determination of thrombin. *Analytica Chimica Acta* **2014**, 845, 45–52.

87. Cho, S.; Lee, S.-H.; Chung, W.-J.; Kim, Y.-K.; Lee, Y.-S.; Kim, B.-G., Microbead-based affinity chromatography chip using RNA aptamer modified with photocleavable linker. *Electrophoresis* **2004**, 25(21–22), 3730–3739.

88. Gulbakan, B., Oligonucleotide aptamers: emerging affinity probes for bioanalytical mass spectrometry and biomarker discovery. *Analytical Methods* **2015**, 7(18), 7416–7430.

89. Chiang, C.-K.; Chen, W.-T.; Chang, H.-T., Nanoparticle-based mass spectrometry for the analysis of biomolecules. *Chemical Society Reviews* **2011**, 40(3), 1269–1281.

90. Keller, K. M.; Breeden, M. M.; Zhang, J.; Ellington, A. D.; Brodbelt, J. S., Electrospray ionization of nucleic acid aptamer/small molecule complexes for screening aptamer selectivity. *Journal of Mass Spectrometry* **2005**, 40(10), 1327–1337.

91. Wales, T. E.; Engen, J. R., Hydrogen exchange mass spectrometry for the analysis of protein dynamics. *Mass Spectrometry Reviews* **2006**, 25(1), 158–170.

92. Sperry, J. B.; Wilcox, J. M.; Gross, M. L., Strong anion exchange for studying protein-DNA interactions by H/D exchange mass spectrometry. *Journal of the American Society for Mass Spectrometry* **2008**, 19(6), 887–890.

93. Trelle, M. B.; Dupont, D. M.; Madsen, J. B.; Andreasen, P. A.; Joergensen, T. J. D., Dissecting the effect of RNA aptamer binding on the dynamics of plasminogen activator inhibitor 1 ssing hydrogen/deuterium exchange mass spectrometry. *ACS Chemical Biology* **2014**, 9(1), 174–182.

5 Uncertainty Evaluation in Chromatography

Veronika R. Meyer

CONTENTS

5.1   INTRODUCTION

Analysts are charged with the weighty responsibility for generating high-quality and reliable results. In the case of quantitative analysis this data is usually a concentration, expressed in g/g, g/L, %, ppm, ppb, or other non-SI units (SI units such as kg/kg or kg/m^3 are neither memorable nor popular in this field). Common sense leads to the insight that the data should come together with additional data which give information about the *quality* of the analytical result, sometimes termed as its *fitness for purpose*. Measurement uncertainty (MU) is a widely accepted concept which can describe this *quality* [1]; the closer a result is to an end user, such as the buyer of merchandise or the consumer of foods, the more important the disclosure of this type of information. It is also needed for the comparison of analytical results found by different laboratories, for example, headquarters versus subsidiary, producer versus customer or versus legal control laboratory, and the like. Figure 5.1 shows the presence or absence of result conformity between laboratories.

Many analytical (and other scientific) data come with a ± sign followed by a number. This number can be interpreted as standard deviation but it could be a standard uncertainty as well; therefore it is desirable—although this wish seems to be a pie in the sky—that the ± sign is no longer used. Instead, the abbreviations SD for standard deviation, RSD for relative standard deviation (or CV for coefficient of variation), u_c for combined standard uncertainty, or U for expanded standard uncertainty should be used and the relevant data declared. Of all these possibilities, the standard uncertainty u_c or U is the most scientific because it is found by a comprehensive investigation of the analytical problem. We can expect that it is determined with quite a bit of evaluation instead of calculating the standard deviation of n consecutive experiments (n being as low as 3 in the worst case).

The aim of this chapter is to explain how standard uncertainties can be evaluated. The concept arose first in physics as *error analysis* which led to the *Guide to the Expression of Uncertainty in Measurement* (GUM) [2]. Its first version was published in 1995. Although some parts may be hard to understand without a physicist's background in mathematics, the GUM is the reference document for all

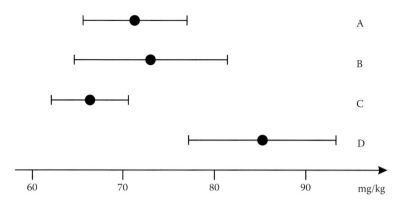

FIGURE 5.1 Mean analytical results (dots) and measurement uncertainties (bars) for an identical sample as obtained by four different laboratories. Laboratories A, B, and C found coincident results within the span of their respective uncertainties; as a consequence, company A can sell its products to companies B and C. The result of laboratory D does not match with A and C, but B and D may accept each other's data although the conformity range is narrow.

questions of MU. For its better understanding, an introductory text was also published [3]. Another document, based on the GUM, describes the evaluation of MU for analytical chemistry results: the guide *Quantifying Uncertainty in Analytical Measurement* (QUAM), first published in 1995 and by now available in the third edition [4]. Any laboratory, which produces quantitative analytical results, should have knowledge of the QUAM, should have at least one person who is trained in the evaluation of the MU, and should follow the guidelines and recommendations of the QUAM.

There is an *official* definition of MU to be found in the *International Vocabulary of Metrology*: Measurement uncertainty is a "non-negative parameter characterizing the dispersion of the quantity values being attributed to a measurand, based on the information used" [5]. This definition is hard to understand, and the one given by both the GUM and the QUAM may be better: "A parameter, associated with the result of a measurement, that characterizes the dispersion of the values that could reasonably be attributed to the measurand" [2,4]. (The quantity of interest, e.g., the concentration of an analyte, is the *measurand* in terms of MU.) The indication to *reasonably* is important; it is a matter of fact that results obtained by repeated scientific measurements show some variability if the measuring system has sufficient resolution, even if the measuring instrument is well maintained and properly used and if the operator is skilled. For a vendor–customer relationship it may be detrimental that the term *measurement uncertainty* is connoted with a negative meaning. But it is an established and scientific term and the vendor is coerced to explain it to the customer.

An uncertainty is not an error. If the true value of the measurand would be known, the difference between the experimental data and the true value would be the error. However, the true value is unknown by definition. Even with some kind of *molecule counting machine*, the true value cannot be determined. Besides the questions of *true* sampling (what is a representative sample and what size is needed?), the counting process would be falsified by adsorption processes, decomposition, contamination, or electronic disturbances in the counting system. In terms of MU, errors are the unwanted mistakes which may occur during the laboratory work: using the wrong measuring flask, resulting in a wrong dilution; using the wrong reference standard; mixing up manually recorded data such as 759 instead of 579; and so on. These errors are sometimes termed as blunder.

In principle, the standard uncertainty of an analysis can be evaluated by two different approaches. The first one, which is explained in this chapter, is the bottom–up method which looks at every influence parameter of the method, determines its uncertainty, and adds them all to the combined standard uncertainty of the method. For a simple example see the paper by Rösslein [6]. The second one is the top–down method which starts from the repeatability or reproducibility of the procedure, that is, from a standard deviation, and then considers some additional parameters. More explanations can be found in Section 5.14.

For the sound evaluation of the MU of an analytical method, a good knowledge of the topic is a prerequisite. Quick and easy recipes cannot be found in this chapter. They are not forbidden, but they must only be used by experienced analysts who know what they do and who can explain it.

5.2 STANDARD UNCERTAINTIES

5.2.1 GENERAL

If a measuring process (such as an analytical method) is composed from a number of operations, the uncertainties of all these operations will sum up to a combined uncertainty, correctly: to the combined standard uncertainty u_c (you may be familiar with this treatment from your practical course of physics where it was referred to as the *propagation of errors*):

$$u_c = \sqrt{u_{\text{weighing}}^2 + u_{\text{diluting}}^2 + u_{\text{chromatography}}^2} \tag{5.1}$$

as an example for a simple method which needs weighing and diluting of the sample before the chromatography can be performed.

By squaring the individual uncertainties, their sign becomes positive in any case and Equation 5.1 yields a realistic value. (If, e.g., all data of the three operations are too high by 1%, their simple addition results in $u_c = 3\%$. If only one data is too high and the other two too low, $u_c = 1\%$. Squaring and extracting the root of the sum gives the correct result of $u_c = 1.7\%$.)

The individual u's can be standard deviations but they can also belong to another class of probability distribution. In analytical chemistry, the common distribution functions are the normal, rectangular, and triangular distribution. Nonsymmetric distributions could occur but they are usually not needed in analytical chemistry. All standard uncertainties represent the second moment of their distribution function if it is symmetric:

$$u(x) = x^2 f(x) dx \tag{5.2}$$

5.2.2 THE NORMAL DISTRIBUTION (GAUSS)

If an experiment is repeated under identical conditions, the obtained results will show some scatter. In the absence of trends, such as a continuous increase of temperature or pH, the distribution of data is described by the equation:

$$f(x, \mu, \sigma) = \frac{1}{\sigma\sqrt{2\pi}} e^{-(x-\mu)^2/2\sigma^2} \tag{5.3}$$

with x = axis of possible results, μ = mean, σ = standard deviation, (and σ^2 = variance). The normal distribution is shown in Figure 5.2. Note that its both ends are located at $-\infty$ and $+\infty$, that is, results far from the mean may occur, although their probability decreases strongly by their distance from the mean. The range embraced by $\pm1\sigma$ covers 68.2% of all values (remember: two-thirds), $\pm2\sigma$ covers 95.4% (remember: 19 out of 20), and $\pm3\sigma$ covers 99.7% (remember: 3 out of 1000). The standard uncertainty of normally distributed data is the standard deviation calculated as follows:

$$\sigma(x) = \sqrt{\frac{1}{n-1} \sum_{i=1}^{n} (x_i - \mu)^2} \tag{5.4}$$

with n = number of results. However, the symbols σ and μ should only be used in the case of an infinitely large number of experiments, which is never possible; therefore, the symbols s and $\bar{x}$ are to be preferred, and the abbreviation SD is

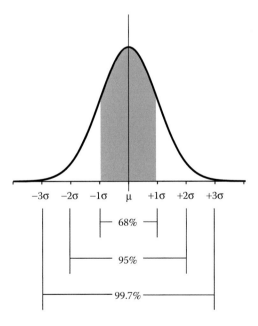

FIGURE 5.2   The normal distribution function. The gray area covers the range of ±1 standard deviation on both sides of the mean or 68% of the whole area under the function.

frequently used. SDs obtained by scientific calculators or by the function STDEV in Excel use Equation 5.4. The RSD is

$$RSD = \frac{s}{\bar{x}}$$

(5.5)

or, expressed in percentage for easier perception:

$$RSD = 100\frac{s}{\bar{x}}[\%]$$

(5.6)

A related term is the standard deviation of the mean $s(\bar{x})$:

$$s(\bar{x}) = \frac{s(x)}{\sqrt{n}}$$

(5.7)

which yields a lower number than SD is, thus there is a temptation to use it as a MU data. However, it must only be used if a certain, well-defined sample was investigated n times; for example, the concentration of the pesticide endosulfan in a single can of apple puree, determined by 10 analyses. But $s(\bar{x})$ is not allowed for the characterization of the method *Endosulfan in apple puree* as performed in a laboratory; its number of merit is s, RSD, or u_c as will be discussed below.

The term *identical conditions* as a prerequisite for the determination of SDs is not unambiguous, therefore it is necessary to explain how an SD was found. It may refer to immediately repeated measurements by the same person using the same instrument on the one hand, or to measurements performed by different persons with different instruments and on different days but using the same standard operating procedure (SOP or method description) as the other extreme. In the first case, the SD will be (much) lower than in the latter one [7]. Especially, n consecutive injections of the same sample solution into a GC or an LC instrument will only yield the repeatability of the combined injection, chromatography, and detection processes and not of the analytical method. (Exceptions may be the direct injection of an untreated water sample in ion chromatography or of an untreated gasoline sample in gas chromatography.)

5.2.3   The Rectangular Distribution

In many cases, an uncertainty cannot be determined by repeated experiments but some data is known from experience or can be found in tables, data sheets, and so on. If no further information is available, its uncertainty function is a rectangle of width $2a$, presented in Figure 5.3a. The range embraced by $\pm a$ covers 58% of all values. A typical example is the temperature in a room without air conditioning. It is, for example, known that the room is never colder than 18°C (in winter time) and never warmer than 28°C (in summer) but it is unknown if a certain temperature (23°C?) is dominant over the year. Therefore, one is forced to assign the same probability to all possible temperatures between the extremes. The standard uncertainty of the rectangular distribution is

$$u(x)_{\text{rectangular}} = \frac{a}{\sqrt{3}} = 0.577a \approx 0.6a$$

(5.8)

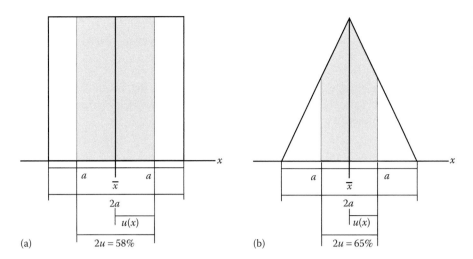

(a)   2u = 58%   (b)   2u = 65%

FIGURE 5.3   The rectangular (a) and triangular (b) distribution functions and their standard uncertainties.

which results in $u(T) = 0.6°C \cdot 5°C = 3°C$ for the example given above because $2a = 10°C$. It is not important whether the correct relationship $u(x) = a/\sqrt{3}$ or the approximation $u(x) = 0.6a$ is used. The correct equation gives $2.89°C$ which is largely the same as $3.0°C$ considering the inherent uncertainty of all uncertainty determinations.

5.2.4 THE TRIANGULAR DISTRIBUTION

If it is known that the mean value of a certain data range is the most probable one, its uncertainty function is a triangle of width $2a$, see Figure 5.3b. The range embraced by $\pm a$ covers 65% of all values. In a room with air conditioning, the most probable temperature will be the target value, for example 25°C. Depending on the performance of the air conditioning system and the frequency of external disturbances, the temperature may vary between, for example, 23°C and 27°C. The standard uncertainty of the triangular distribution is

$$u(x)_{\text{triangular}} = \frac{a}{\sqrt{6}} = 0.408\, a \approx 0.4\, a \tag{5.9}$$

which results in $u(T) = 0.4°C \cdot 2°C = 0.8°C$ for this example.

5.2.5 THE CASE OF UNSUFFICIENT INSTRUMENTAL RESOLUTION

If an instrument has a poor resolution d or if its proper resolution is not utilized, the MU becomes:

$$u(x) = \frac{d/2}{\sqrt{3}} = \frac{d}{\sqrt{12}} = 0.289\, d \approx 0.3\, d \tag{5.10}$$

because this case represents a special type of the rectangular distribution. A common application of Equation 5.10 occurs if a weighing accuracy is requested which is considerably lower than the technical performance of the balance. If, for example, a balance displays 1/100 of the mg but the SOP requires *weighing with an accuracy of 1 mg*, then its technical performance (see Section 5.7) is not important. Instead, the required accuracy represents a rectangular distribution of width $d = 1$ mg. The uncertainty of this weighing operation is thus 0.3 mg. Similarly, the length determination of a table with a coarse measuring tape with only 0.1 m marks yields an uncertainty of 3 cm (if the centimeters are not estimated optically).

5.3 CALCULATION RULES

5.3.1 The Equation of the Measurand

Equation 5.1 which describes a simple analytical method is too simplified. For the evaluation of the MU, the detailed equation of the measurand is needed. Usually the measurand is a concentration such as the content of endosulfan $c_{\text{Endosulfan}}$ in the apple puree. A simple procedure could consist of sample weighing, extraction, dilution, and gas chromatography with the direct comparison of the peak areas of sample and reference. The variables of the equation are mass m_S of the puree sample, recovery (*Rec*) of the extraction, volume V_S of the sample dilution, concentration of the reference solution c_R (assuming that it could be purchased as a ready-made solution which does not need further dilution), sample solution peak area A_S, and reference solution peak area A_R. Their mathematical relationship is

$$c_{\text{Endosulfan}} = \frac{c_R \, A_S V_S}{A_R \, m_S \, Rec} \tag{5.11}$$

if the injection volumes of sample and reference solutions are identical. Because of the inherent injection problems of GC, an internal standard (IS) is frequently used for quantitative analyses, therefore its peak area and amount must also be included into the equation of the measurand. Similarly, if the calibration is done with several points of different concentration (linear regression), the equation includes additional or other parameters. Even Equation 5.11 is not really complete because the purity of the reference is not included (see Section 5.8).

In addition, it may be necessary to consider other influence parameters with value = 1 and uncertainty $\neq 0$, such as Hom for homogeneity of the sample (with $u[\text{Hom}]$ perhaps being rather high). Another, maybe very important contribution to the overall uncertainty is the representativeness of the sample and the sampling process, including the storage. This topic is not presented here. In 2007, an extra guide discussing the uncertainty of sampling was published [8].

Analogously, in cases where the analytical result is not corrected by the recovery, the variable *Rec* must be considered in the MU calculation; that is, *Rec* is included in the equation of the measurand with $Rec = 1$ and $u(Rec) \neq 0$.

For the proper evaluation of a MU by the bottom–up approach the equation of the measurand must be set up properly. It must be one single equation which includes all variables (such as the mass of sample weighted in and its dilution but not the concentration of the injected sample solution). The equation can only be set up if the analyst understands the method fully. It must be noted in the SOP.

For a beginner in the field of MU, the most demanding task may be the correct setup of the equation of the measurand. The subsequent calculation or estimation of the uncertainties belonging to the discrete variables can be tedious but not really difficult.

5.3.2 General Calculation Rule for the Combined Uncertainty

Once the equation of the measurand is set up, the next step is to find out the standard uncertainties of all its variables (Equation 5.11 has six of them). This can be done by experiments, estimations, or by looking for data in user manuals, technical descriptions, or handbooks. The standard uncertainties of common laboratory operations or data are discussed in Sections 5.6 through 5.12. The combined standard uncertainty of the measurand $u_c(M)$ is the square root of the sum of all squared partial differentials $u^2(x_i)$ of M with respect to all the individual uncertainty sources x_i which are present in the equation of the measurand. The mathematical form of this definition is easier to understand than the sentence:

$$u_c(M) = \sqrt{\sum_{i=1}^{n} \left(\frac{\partial M}{\partial x_i} \right)^2 u^2(x_i)} \qquad (5.12)$$

Equation 5.12 is the general calculation rule for $u_c(M)$. Fortunately, in almost all cases which occur in analytical chemistry, simplified equations can be used as discussed in Sections 5.3.3 through 5.3.5. In cases where Equation 5.12 must be used, the resulting equation for $u_c(M)$ can be tedious. An example is the two-point calibration by the bracketing method (no fit and without inclusion of the zero point). The equation of the measurand c_{Sample} is

$$c_{\text{Sample}} = \frac{S_S - S_{R1}}{S_{R2} - S_{R1}} (c_{R2} - c_{R1}) + c_{R1} \qquad (5.13)$$

with $S =$ signal ($S_S =$ signal of the sample, $S_{Rn} =$ signal of reference n), and analogously $c =$ concentration. This equation cannot be simplified and both terms S_{R1} and c_{R1} occur twice. Therefore the procedure of Equation 5.12 must be performed which, after factoring out, leads to

$$u_c(c_{\text{Sample}}) = \frac{1}{S_{R2} - S_{R1}} \qquad (5.14)$$

$$\sqrt{\begin{array}{l} (c_{R2} - c_{R1})^2 \cdot u^2(S_S) + \left(\dfrac{(S_S - S_{R2})(c_{R2} - c_{R1})}{S_{R2} - S_{R1}} \right)^2 \cdot u^2(S_{R1}) + \\[2em] \left(\dfrac{(S_{R1} - S_S)(c_{R2} - c_{R1})}{(S_{R2} - S_{R1})} \right)^2 \cdot u^2(S_{R2}) + (S_{R2} - S_S)^2 \cdot u^2(c_{R1}) + (S_S - S_{R1})^2 \cdot u^2(c_{R2}) \end{array}}$$

5.3.3 EQUATIONS WITH SUMMANDS ONLY

Equations which contain only sums or differences lead to a simple expression if the procedure according to Equation 5.12 is performed. The uncertainty of M in the equation

$$M = p + q - r \tag{5.15}$$

is calculated as follows:

$$u_c(M) = \sqrt{u^2(p) + u^2(q) + u^2(r)} \tag{5.16}$$

It is a matter of course that all u's must have the same unit. An example is the loss on drying:

$$\text{Loss} = m_{\text{wet}} - m_{\text{dry}} \tag{5.17}$$

with m = mass of the sample. The combined standard uncertainty of the loss is

$$u_c(\text{Loss}) = \sqrt{u^2(m_{\text{wet}}) + u^2(m_{\text{dry}})} \tag{5.18}$$

Numerical coefficients are also squared. The uncertainty of, for example, $M = 3p + 7q$ is

$$u_c(M) = \sqrt{9u^2(p) + 49u^2(q)} \tag{5.19}$$

5.3.4 EQUATIONS WITH PRODUCTS ONLY

The most frequent equation type in many analytical methods, including chromatography, is the one which is composed only from products and quotients:

$$M = \frac{pq}{r} \tag{5.20}$$

In this case $u_c(M)$ becomes:

$$u_c(M) = M\sqrt{\left(\frac{u(p)}{p}\right)^2 + \left(\frac{u(q)}{q}\right)^2 + \left(\frac{u(r)}{r}\right)^2} \tag{5.21}$$

for the absolute uncertainty (with a unit). The relative uncertainty, which is often of higher interest, is of course:

$$\frac{u_c(M)}{M} = \sqrt{\left(\frac{u(p)}{p}\right)^2 + \left(\frac{u(q)}{q}\right)^2 + \left(\frac{u(r)}{r}\right)^2} \tag{5.22}$$

and if the result is multiplied by 100 one gets the uncertainty in percentage. The endosulfan concentration in apple puree was described in Equation 5.11. Its combined standard uncertainty is

$$\frac{u_c(c_{\text{Endosulfan}})}{c_{\text{Endosulfan}}} = \sqrt{\left(\frac{u(c_R)}{c_R}\right)^2 + \left(\frac{u(A_S)}{A_S}\right)^2 + \left(\frac{u(V)}{V}\right)^2 + \left(\frac{u(A_R)}{A_R}\right)^2 + \left(\frac{u(m)}{m}\right)^2 + \left(\frac{u(Rec)}{Rec}\right)^2}$$

$$\tag{5.23}$$

Numerical coefficients do not appear in the equation of $u_c(M)$. Squared variables v^2 which are in the numerator appear as $(2 \cdot u(v)/v)^2$. For other cases (v^2 in the denominator, higher powers) the general rule with the partial derivatives is to be applied. If all uncertainties are known as relative data, such as percentage, Equation 5.22 becomes very simple; $u_c(M)$ of a *product type* equation then looks like Equation 5.16. The result is also in percentage in this case:

$$u_c(M)[\%] = \sqrt{u^2(p)[\%] + u^2(q)[\%] + u^2(r)[\%]} \tag{5.24}$$

5.3.5 MIXED EQUATIONS, ALL VARIABLES OCCUR ONLY ONCE

This rule must only be applied to equations (or parts of equations) where every variable occurs only once. The equation is subdivided into parts with summands only and products/quotients only. Every part is then treated in accordance to Equations 5.16 or 5.21, respectively:

$$M = \frac{p+q}{r-s} = \frac{N}{D} \tag{5.25}$$

$$u_c(N) = \sqrt{u^2(p) + u^2(q)} \qquad u_c(D) = \sqrt{u^2(r) + u^2(s)} \tag{5.26}$$

$$u_c(M) = M\sqrt{\left(\frac{u_c(N)}{N}\right)^2 + \left(\frac{u_c(D)}{D}\right)^2} = M\sqrt{\frac{u^2(p) + u^2(q)}{(p+q)^2} + \frac{u^2(r) + u^2(s)}{(r-s)^2}} \tag{5.27}$$

An equation of the type $M = (a-c)/(b-c)$ must not be treated in this way because the term c occurs twice (the c's are correlated) and the equation cannot be simplified by factoring out. It is necessary to use the general rule of Equation 5.12.

5.3.6 MONTE CARLO METHOD

The uncertainty propagation rules as presented above are based on a Taylor series which includes only the first order. The calculated results are not correct if a function is nonlinear. The Monte Carlo process is an elegant alternative to avoid complicated equations [9]. It varies all variables (or *throws the dice*) in accordance to their uncertainty and calculates the measurand. A software program repeats this procedure a great number of times; 10^5 or more calculations may be necessary for measurands with many variables. Finally, the mean of the measurand (or the median if this number is needed), its uncertainty, and the true distribution of the data are calculated. Monte Carlo simulations can be performed not only with specialized software but also with common spreadsheet calculations [10].

Advantages of the Monte Carlo method are as follows:

- The equation of the measurand can be of any complexity, differentiation is not necessary
- The correct uncertainty is obtained in any case
- Correlations are considered without extra effort
- In cases with linear regression all uncertainties are considered correctly

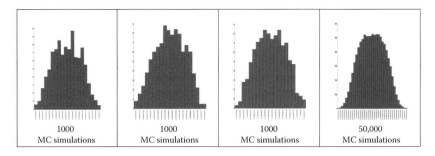

FIGURE 5.4 Monte Carlo simulations of a combined standard uncertainty function which is composed from a normal and a rectangular distribution function. With 1000 simulations, no stable distribution is obtained but 50,000 simulations yield a distinct function.

The correct uncertainty distribution is obtained also in cases where it is of nontrivial shape or is asymmetrical. An example is shown in Figure 5.4 with a function which is composed from a normal and a rectangular distribution; the Gaussian part comes from the handling repeatability of an instrument, the rectangular part is due to the temperature influence on the result if the instrument is used in a laboratory without air conditioning (as discussed in Section 5.2.3). It can be seen that 1000 simulations do not yield a stable distribution but that 50,000 runs result in a function which clearly shows its composition from normal and rectangular parts. With enough computer power available, 10^5 simulations or more would give an even smoother function.

5.4 EXPANDED UNCERTAINTY

5.4.1 The Coverage Factor

The combined standard uncertainty u_c, as calculated according to the rules presented in Section 5.3, covers approximately only two-thirds of the expected results. The approach is based on the concept of the normal distribution with the standard deviation σ or s defined according to Equation 5.4. The range of ± 1 s covers not more than 68.3% of the experimental results, that is, one-third of the data will lie outside of these limits. However, one is usually interested in a 95% coverage, and in some cases an even higher statistical certainty is needed (e.g., in industrial production where a failure rate of 1 item out of 1000 is way too large). Therefore the calculated standard uncertainty can be expanded by the coverage factor k:

- $k = 2$ for normally distributed data, covering 95.45% of the data
- $k = 1.65$ for data with rectangular distribution, covering 95% of the data
- $k = 1.93$ for data with triangular distribution, covering 95% of the data
- For higher demands: $k = 3$ with the normal distribution, $k = 1.73$ for the rectangular distribution, and $k = 2.4$ for the triangular distribution covering 99.7% of the data

If, for example, the combined relative standard uncertainty of normally distributed experimental data was found to be $u_c = 4.7\%$, it is multiplied by $k = 2$, resulting in an expanded uncertainty $U = 9.4\%$ with a confidence interval of 95%. In fact, since every uncertainty data has its own uncertainty, this value should be rounded to 10%. Note that the symbol for the expanded uncertainty is an upper case U without subscript. The coverage range, usually 95%, should be noted, too.

5.4.2 COVERAGE IN THE CASE OF A LIMITED NUMBER OF EXPERIMENTS

Experimental data are usually assumed to follow a normal distribution. The coverage factor for 95% confidence of this function is 2.0, as mentioned in Section 5.4.1. However, this factor is valid only for a large number of experiments (theoretically for $n = \infty$) [11]. For a limited number, larger coverage factors are needed based on Student's seminal paper of 1908 about the t-distribution [12]. These are shown in Table 5.1. The example below shows how to use them.

Determination of caffeine in beverages by high-performance liquid chromatography (HPLC). After filtration the beverage can be injected directly without further sample preparation. Five repeated injections of the same sample result in a standard deviation of ±1.8% of the peak areas. The concentration uncertainty of the caffeine reference solution is 0.5% (as a combination from weighing, dilution, and purity of the caffeine). In accordance to Equation 5.24, we get the combined uncertainty:

$$u_c = \sqrt{u^2(c) + u^2(HPLC)} = \sqrt{0.5^2 + 1.8^2}\,\% = \sqrt{0.25 + 3.24}\,\% = 1.9\% \quad (5.28)$$

This combined standard uncertainty is clearly dominated by the HPLC term. Its uncertainty was determined by not more than five measurements. Therefore, the coverage factor from Table 5.1 is $k = 2.8$ and the expanded measurement uncertainty $U = u_c \cdot 2.8 = 1.9\% \cdot 2.8 = 5.3\%$ (95%).

Uncertainty parameters with a rectangular or triangular distribution do not need such a treatment because they are not based on experiments but on knowledge (or assumptions), therefore they come with their constant coverage factors.

TABLE 5.1
95% Coverage Factors for a Limited Number of Experiments

Number of Experiments n	Statistical Degree of Freedom ν	Coverage Factor k
2	1	12.7
3	2	4.3
4	3	3.2
5	4	2.8
6	5	2.6
7	6	2.5
8	7	2.4
9	8	2.3
10	9	2.3
20	19	2.1
50	49	2.01
∞	∞	2.00

5.4.3 COMBINATION OF SEVERAL INFLUENCE PARAMETERS

The selection of the correct coverage factor can be less simple in the case of several influence parameters involved.

- A single element is dominating the equation of the combined standard uncertainty, that is, its size is three or more times larger than the next one. Depending on the type of distribution to which it belongs, the appropriate coverage factor k is used
- The elements are of similar magnitude but they belong to different types of distribution. For such a problem, the GUM [2] must be consulted (paragraph G.4.2 on p. 62). Luckily the combination (convolution) of three or more uncertainty elements quickly leads to a normal distribution; this approximation is closer to the reality the more elements are present. In such a case the coverage factor for 95% confidence is $k = 2$
- All elements are more or less equal in size and follow the normal distribution (because they all were determined by repeated measurements of the respective phenomenon). The effective degree of freedom v_{eff} is calculated as follows:

$$v_{eff} = u_c(M)^4 / \sum_{i=1}^{n} \frac{u(x_i)^4}{v_i} \qquad (5.29)$$

with $u_c(M)$ = combined standard uncertainty of the measurand M, $u(x_i)$ = standard uncertainty of influence element x_i, and v_i = degree of freedom of influence element x_i (i.e., number of experiments −1). The reader can test this equation with the following data: $M = a \cdot b$; $u(a) = 1.8\%$ (RSD of five experiments); $u(b) = 2.1\%$ (RSD of three experiments). Results: $u_c(M) = 2.8\%$ (Equation 5.24), $v_{eff} = 4.0$ (Equation 5.29).

5.5 GRAPHICAL TOOLS

5.5.1 FLOWCHART

The equation of the measurand can only be set up correctly if the analytical method is fully comprehended. A clear and detailed description (in many cases an SOP) is needed. It may be helpful, although not mandatory, to draw a flowchart; its most simple design is shown in Figure 5.5 which represents a method with three working steps only. Every step gets a box, the boxes are connected with arrows, and the uncertainty sources may also be noted. If the SOP is not written clearly, the attempt to draw a flowchart will bring its weaknesses forward.

A flowchart usually has more than one branch because it represents the preparation of the reference solution on the one hand and the sample preparation on the other. The branches join wherever this is logical, but at the latest at the step of calculating the analytical result.

The degree of detail of a flowchart is up to the analyst.

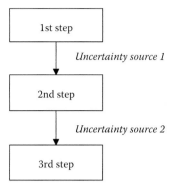

FIGURE 5.5   Principle of a flowchart.

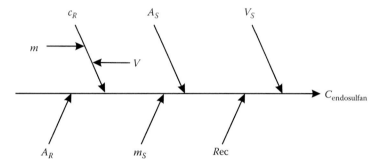

FIGURE 5.6   An Ishikawa diagram representing Equation 5.11 with the influence parameters concentration of the reference solution c_R (obtained from a mass m and a volume V), peak area of the sample A_S, volume of sample solution V_S, peak area of the reference A_R, weighted mass of the sample m_S, and recovery of the analyte during sample preparation. Together they yield the concentration of endosulfan in the apple puree sample. The diagram can or should be drawn with more side arrows of second and third order for a discussion of all possible influence parameters.

5.5.2   Ishikawa Diagram

The Ishikawa diagram (cause and effect diagram, fishbone diagram) is another and mandatory aid to find all uncertainty sources which have an influence on the result of a measurement [13,14]. All variables of the equation are represented in a fishbone-like structure (usually horizontal, but can be vertical as well). The main arrow represents the measurand; side arrows of first order are formed coercively by the variables present in the equation of the measurand. The order or position of the side arrows is not important. In many cases additional side arrows of second order are needed.

Figure 5.6 represents Equation 5.11 with the concentration of endosulfan as *effect*, that is, as the main arrow. There are six side arrows for the six variables in Equation 5.11. One of them carries two additional arrows: The concentration c_R of the reference solution depends on the mass m of compound weighted in and on the volume V of its dilution. Both m and V have their specific influence parameters, as discussed in Sections 5.6 and 5.7, and they can be added to the diagram, too.

Similarly, the other five side arrows can or should be equipped with their influence parameters.

Ishikawa diagrams of the influence parameters which finally yield an HPLC peak with its retention time, height, width, and shape were published by Meyer [15].

5.6   UNCERTAINTY OF VOLUMETRIC OPERATIONS

5.6.1   GENERAL

The uncertainty sources of a volumetric operation are the following:

- The calibration of the volumetric instrument at a certain temperature
- The repeatability of its handling
- The change in liquid volume when the temperature changes

These influence parameters are additive. In the norms for glass instruments the terms calibration and repeatability are combined to the maximum permissible error (MPE) [16–18], reducing the uncertainty sources to two instead of three. Unfortunately, the Eurachem guide [4] does not present the MPE concept but calculates all three uncertainty sources which results in too large combined uncertainties.

The evaluation and concept of $u(V)$ is presented here in detail because it is a plain model of how the bottom–up approach of MU works. However, in chromatographic analyses, the uncertainty of volumetric operations for the preparation of sample and reference solutions can usually be neglected. The important contributions to the overall MU come from the repeatability of all sample preparation steps (or of the recovery, which represents the opposite, but equally valid view on sample preparation operations). Depending on the chromatographic technique used and the signal-to-noise ratio of the peaks, the overall repeatability of the gas or liquid chromatographic separation process will be important as well; see the remarks given in Section 5.12. The *volumetric operation* of injection is not evaluated separately because it is a part of the chromatographic process; for curiosity, the isolated $u(V)$ of an HPLC autosampler was studied by Kehl and Meyer [19], a number which depends highly on the instrument used and its maintenance.

It must be kept in mind that the published technical data of volumetric instruments are usually based on pure water. The performance of an instrument, including the volume itself, can be different or even very poor with more or less viscous fluids, including aqueous solutions with detergents, biological fluids, organic solvents, oils, and so on. Calibration and repeatability can be determined by a series of inter- or intrapersonal handling operations of an instrument with weighing control.

5.6.2   GLASS INSTRUMENTS

Volumetric instruments made from glass, that is, pipets and measuring flasks, come with two imprinted numbers such as 10 mL ± 0.02 mL (20°C). The first one is the volume and the second one the MPE. The MPE is defined as the "extreme value of measurement error, with respect to a known reference quantity value, permitted by specifications or regulations for a given measurement, measuring instrument, or measuring system" in paragraph 4.26 of Ref. [5]. This statement means that that the combined effect of manufacturing tolerances and instrument handling by a trained person will not be higher than the indicated MPE (if the liquid is water). It can be assumed that the MPE is characterized by a triangular distribution because both the manufacturer and the user try to hit the nominal volume. Therefore, $u(\text{MPE})$ is $\text{MPE}/\sqrt{6} = 0.4 \text{ MPE}$.

The detailed equation for the temperature effect is as follows:

$$u(V_T) = V(\gamma_{\text{liquid}} - \gamma_{\text{instrument}})u(T) \tag{5.30}$$

with V = nominal volume and γ = cubic expansion coefficient of the liquid and the material of the volumetric instrument, respectively. The expansion coefficients are:

- $2 \cdot 10^{-4}/°C$ for water in the vicinity of $20°C$
- Approximately $1 \cdot 10^{-3}/°C$ for organic solvents, that is, much larger than for water
- $10 \cdot 10^{-6}/°C$ for borosilicate glass (e.g., Pyrex) used for measuring flasks and burets
- $25 \cdot 10^{-6}/°C$ for soda lime glass used for pipets
- $1.6 \cdot 10^{-6}/°C$ for quartz glass
- $4.5 \cdot 10^{-4}/°C$ for polypropylene used for transparent measuring flasks and pipets
- $3.5 \cdot 10^{-4}/°C$ for polymethylpentene used for clear measuring flasks

The different combinations of liquid and instrument can be considered for very detailed MU evaluations for methods like titration. In most other situations, the contribution of a glass instrument γ_{glass} can be neglected because it is too small and Equation 5.30 uses only γ_{liquid}. With plastic instruments and water, $u(V_T)$ becomes nearly zero and the temperature term of $u(V)$ does not need to be considered at all. With plastic instruments and organic solvents it may be necessary to calculate $\gamma_{\text{liquid}} - \gamma_{\text{instrument}}$.

The temperature uncertainty $u(T)$ is handled as a rectangular or triangular distribution as discussed in Sections 5.2.3 and 5.2.4. What is important in the context of MU is the possible temperature drift in the laboratory over the time needed for a working step or an analysis. If the relevant working steps are completed within 1 h, it is this time interval and the temperature drift ΔT within it which needs to be considered. With other scenarios it is necessary to know (or to estimate) ΔT over 24 h or over a year. It is good working practice to equilibrate the instruments and liquids at the same temperature, then ΔT is negligible and the temperature uncertainty according to Equation 5.30 is non-existing. On the other hand, it can be high if a solution is stored at low temperature and used before it has reached the room temperature.

$u(\text{MPE})$ and $u(V_T)$ are then combined to $u(V)$ according to Equation 5.15:

$$u(V) = \sqrt{u^2(V_{\text{MPE}}) + u^2(V_T)} \tag{5.31}$$

The reader can practice its calculation skills in MU with the following data: 10 mL glass pipet of MPE = ±0.02 mL used with water, rectangular temperature uncertainty of ±3°C. The results are: $u(V_{MPE}) = 0.008$ mL, $u(V_T) = 0.0036$ mL, rounded $u(V) = 0.01$ mL. In this case the main uncertainty comes from the MPE.

5.6.3 Piston-Operated Instruments

For such instruments the three effects of calibration, repeatability, and temperature are considered separately:

$$u(V) = \sqrt{u^2(V_{cal}) + u^2(V_{rep}) + u^2(V_T)} \qquad (5.32)$$

$u(V_{cal})$ is a triangular distribution, $u(V_{rep})$ is a standard deviation, and with regard to $u(V_T)$ the function depends on whether the temperature is controlled or not, as mentioned above.

5.7 UNCERTAINTY OF WEIGHING OPERATIONS

5.7.1 General

Mass determinations are the most precise operations in the analytical laboratory if the weighing good is not volatile, hygroscopic, electrically charged, or reactive (in such cases it is necessary to use special techniques such as an evaporation trap or the back weighing of a syringe). In many cases the contribution of a weighing operation is the smallest one of the complete uncertainty budget of an analysis. Usually this contribution cannot be influenced by the analyst as long as the balance is in a good condition and gets a regular service.

Similar to the case of unproblematic volumetric operations, the uncertainty of mass determinations does not influence the MU budget of a chromatographic analysis. Weighing can be performed in an uncertainty range of 10–100 ppm of the sample mass. Therefore, usually all weighing operations can be omitted in the uncertainty evaluation of GC or HPLC methods. Nevertheless, this chapter first presents the background of weighing uncertainty (the strict approach) and then a *forbidden* simple method (the pragmatic approach) if one does not want to omit mass in a combined MU budget.

5.7.2 The Strict Approach

The uncertainty sources of a weighing operation with an analytical balance are the repeatability (Rep) (which also includes the resolution of the readability or of the digital display), the nonlinearity (NL) of the characteristic curve, the sensitivity tolerance (ST) of the characteristic curve, the temperature coefficient (TC) of the characteristic curve, and the uncertainty of the buoyancy (BU) [20]. The first four terms depend on the technical data of the balance in use, whereas BU depends on the density of the weighing good and of the air. Rep refers to the gross weight (tare and net weight) but NL, ST, and TC refer to the net weight. NL must be considered twice, because the determination of the tare (which is often done by setting the balance display to zero) is also a weighing operation. For details of BU see Section 5.7.3. The combination of the five factors gives the equation for the uncertainty of the mass $u_c(m)$:

$$u_c(m) = \sqrt{u^2(Rep) + 0.67\,NL^2 + u^2(ST) + u^2(TC) + u_c^2(BU)} \qquad (5.33)$$

The technical data are specified for a certain balance type and in some cases also for the weighed mass (e.g., if the mass is <10 g or <200 g when a 200 g balance is used). Unfortunately, the mutual influences of the atmospheric conditions, the density uncertainty of the weighing good, the technical data of the balance, and the masses of sample and tare are rather complex. Therefore it is not possible to define simple rules of thumb, and the use of Equation 5.33 becomes tedious, although it can be handled with a spreadsheet. BU, the buoyancy of the weighing good in the surrounding air, is usually markedly larger than the combined standard uncertainty of the other four parameters, especially with materials of low density. Therefore it is inscrutable why this term is not considered and discussed in the Eurachem guide [4].

5.7.3 THE PRAGMATIC APPROACH BASED ON AIR BUOYANCY

Because buoyancy is an important parameter, a pragmatic (and conservative) approach for the implementation of $u_c(m)$ in an uncertainty budget can be based on this effect. The extent of BU depends on the air density ρ_A (i.e., the height above sea level and the current atmospheric pressure, temperature, and air humidity [21]) and on the density ρ_W of the weighing good with regard to the density of the reference (calibration) weight ρ_R (8.006 kg/dm³ in the widespread laboratory balances with internal reference weight) of the balance. Therefore, the uncertainty of BU is composed from the density uncertainties of air $u(\rho_A)$, weighing good $u(\rho_W)$, and reference weight $u(\rho_R)$, resulting in a complex equation not shown here.

Although metrologically wrong, a possible approach is to consider BU not as a bias but as an uncertainty source. In common analytical balances, the weighing value (the displayed mass) of goods with a density <8 kg/dm³ is too low, whereas it is (slightly) too high for goods with a density >8 kg/dm³. The relative buoyancy BU_{rel}, which refers to the reference weight, is shown in Figure 5.7. It can be looked at as the range of a one-sided rectangular distribution (therefore the division is only by $\sqrt{3}$, not $2 \cdot \sqrt{3}$), and the pragmatic uncertainty of a mass becomes:

$$\frac{u_m}{m} = \frac{BU_{rel}}{\sqrt{3}} = 0.6\,BU_{rel} \qquad (5.34)$$

with the absolute value of BU_{rel} taken from Figure 5.7.

It must be stressed again that this procedure is wrong from a metrological point of view, but it can be used by analysts who are not afraid of pragmatic concepts. For most weighing goods (namely, the ones with a density of approximately <5 kg/dm³, also depending on the uncertainty of this density), Equation 5.34 yields a higher uncertainty of a mass determination than Equation 5.33, therefore the pragmatic approach is a conservative one.

FIGURE 5.7 The relationship between relative buoyancy and weighing good density if a balance with an internal reference weight of $\rho_R = 8$ kg/dm³ is used. This figure is needed for the pragmatic determination of the uncertainty of a weighing operation according to Equation 5.34.

5.8 UNCERTAINTY OF THE PURITY OF REFERENCE COMPOUNDS

In quantitative chromatographic analysis, it is necessary to use a reference standard for calibration. The purity of this compound must be included in the equation of the measurand. Therefore, Equation 5.11 which describes the concentration of endosulfan in apple puree is not complete but must be completed with the purity term P_R of the endosulfan reference:

$$c_{\text{Endosulfan}} = \frac{c_R \, P_R A_S V_S}{A_R \, m_S \, Rec} \tag{5.35}$$

As a consequence, the uncertainty of purity of this compound must be considered in the calculation of the measurand's combined uncertainty. Many reference compounds are commercially available with a defined purity such as $a \pm$ data (whereby the question remains open if this symbol marks a standard deviation, a rectangular or a triangular distribution), $a \geq$ data, or, in the best and most expensive case, with a true uncertainty data u_c or $U_{95\%}$ of certified reference materials.

The problem of chemicals with a defined minimum purity such as $\geq 98\%$ (which means ≥ 0.98 in a mathematical sense) was discussed by van Look and Meyer [22]; it has lost part of its nuisance because nowadays many companies offer more detailed information on the Internet if the lot number of a certain product is known. If not, a rectangular distribution with a range $1 - 0.xy = \pm a = 2a$ must be assumed and the uncertainty of the purity becomes:

$$u(P_R) = (1 - 0.xy)/\left(2\sqrt{3}\right) = 0.3(1 - 0.xy) \tag{5.36}$$

Some pharmaceutical reference compounds which are sold with a content of 100.0% by definition [23] represent a special case. This approach is a possible one, but nevertheless, even this content data must come with an uncertainty which is often not the case [24].

The purity of the reference is an influence parameter which must never be forgotten. In multiple-point calibration (linear regression, see Section 5.11), it represents a bias in x direction of the calibration points which should be considered in the respective equation of the measurand. If the top–down approach is used for the MU evaluation of a method, it is not necessary to know all influence parameters mentioned above, such as $u(m)$ and $u(V)$. However, in the standard deviation of a method obtained from validation, the uncertainty of the purity or content of the reference is not included. As a consequence, the equation of the combined standard uncertainty of the measurand M becomes:

$$\frac{u_c(M)}{M} = \sqrt{\left(\frac{u(M_{\text{Val}})}{M_{\text{Val}}}\right)^2 + \left(\frac{u(P_R)}{P_R}\right)^2} \tag{5.37}$$

with M_{Val} = the mean value of the measurand (i.e., the analytical result) as determined during validation.

5.9   UNCERTAINTY OF ATOMIC WEIGHTS

The uncertainty of atomic weights is not relevant in most chromatographic analyses because these weights do not appear in the equation of the measurand. These are (theoretically) important in titration or in ion chromatography where, for example, bromide is the analyte which calls for the usage of a potassium bromide reference. The ratio of Br in KBr is 0.6715, and for the calculation of the uncertainty of this data, the atomic weight uncertainties of K and Br must be known. So far, the IUPAC published atomic weight data with their uncertainties circa every 2 years. The last paper in this series brought a paradigm shift with pragmatic values for 12 elements, including bromine [25]. Their atomic weights are "not so constant after all" [26]. For some elements, especially including the 21 monoisotopic ones, it is no longer difficult to determine their atomic weights with extremely low uncertainty by mass spectrometry (the best-investigated weight is known for sodium with a relative uncertainty of $1 \cdot 10^{-9}$). However, many elements show a high isotopic abundance variation in nature or their isotopic composition was and is corrupted by human activities (enrichment of certain isotopes for energy production, transformation of atomic numbers by processes in nuclear reactors, etc.). Twelve elements are especially affected by such influences, therefore the uncertainty of their atomic weights is rather high (the worst case is boron with a relative uncertainty of $1 \cdot 10^{-3}$), therefore the IUPAC commission provides now *conventional* atomic weight values without specific uncertainties for them [25]. Their uncertainty is ± 1 of the last published digit by definition.

In cases where atomic weights appear in the equation of the measurand, the most recent IUPAC values should be used. The equation for the uncertainty calculation of the atomic mass fraction in a molecule can be found elsewhere [27]. The relative uncertainty of the amount of Br in KBr is thus 5.4 ppm; both atomic weights have an absolute uncertainty of ± 0.001 (by chance). In reality, the uncertainty of purity of a chemical (see Section 5.8) may be markedly higher than any uncertainty coming from atomic weights, thus the latter can be neglected. Nevertheless, it is a sin of omission that the topic of atomic weight uncertainty is not mentioned at all in the Eurachem guide [4].

5.10 UNCERTAINTY OF THE RECOVERY DURING SAMPLE PREPARATION

Incomplete (or, in some cases of trace analysis, a higher than 100%) recovery is a bias which must be considered in uncertainty evaluation [28,29], therefore it should show up in the equation of the measurand. Many analyses require a multistep sample preparation which is prone to variability and errors. The analyst is primarily interested in the overall recovery of the analyte. The uncertainty of recovery can only be evaluated experimentally, therefore this is a classical example of a normally distributed result, and its standard deviation is $u(Rec)$. It is a characteristic of a certain method. Therefore it should be determined by a large number of independent experiments in order to keep the necessary coverage factor low (see Section 5.4 and the example of Section 5.13).

It can be assumed that the recovery is overestimated in cases where the analyte is added purposefully to the reference material in order to perform the calibration. In this case it is perhaps not chemically bound to the matrix but only adsorbed, therefore it can be extracted easily. The best way out of this problem is to use a (expensive) certified matrix reference material if such a thing is available. For the problem mentioned in Section 5.3.1 this would be a material *endosulfan in apple puree.* If no reference material of this type can be found in the market, one might try an endosulfan standard in another matrix, such as apple chips or a vegetable. In this case, the experimental uncertainty of recovery (the standard deviation) should be enlarged mathematically, but it is impossible to recommend a certain factor. In any case, the analyst must fully disclose their considerations and method of uncertainty evaluation.

The notation of $u(Rec)$ must be clear. If, for example, the recovery was determined as 0.88 ± 0.057, such a result is usually presented in a percentage notation. It is recommended to write $(88 \pm 5.7)\%$ instead of $88\% \pm 5.7\%$ because the latter version might indicate 5.7% of 88% which gives 5.0%.

Note that $u(Rec)$ is the standard deviation according to Equations 5.4 through 5.6 because it is the number of merit of a procedure. It must never be calculated as the standard deviation of the mean according to Equation 5.7.

Since the uncertainty of recovery can be the largest influence parameter of the MU of an analytical method, one should try to improve (minimize) it, although this can be very demanding or impossible. Such an investigation would involve a detailed examination of all sample preparation steps. May be a lower-than-desirable recovery is easier to accept than a poor repeatability of the whole sample preparation procedure. In addition, it will be virtually impossible to identify the contribution of the chromatographic analysis on the overall uncertainty of the sample preparation; in most cases of chromatographic methods, the recovery itself will be determined by chromatography.

It may be a good idea to perform the sample preparation in such a way that the resulting MU is minimized [30] although *green* aspects such as the low consumption of organic solvents must be considered, too.

A valuable investigation of random and systematic errors due to sample preparation (besides the ones originating from sampling and the chemical analysis) was published by Lyn et al. [31]. It deals with the GC-MS quantification of pesticide residues in strawberries.

5.11 UNCERTAINTY OF A RESULT OBTAINED BY A LINEAR REGRESSION FUNCTION

In almost all quantitative chromatographic analyses, the relation between signal size (peak area or height) and sample concentration (e.g., in HPLC with UV detection) or sample amount (e.g., in GC with flame ionization detection) is linear. An exception is often found in quantitative thin layer chromatography (TLC) with relations of the quadratic type. The linear function is established by several calibration points defined by their x and y values:

$$y_R = a + bx_R \qquad (5.38)$$

with y_R = signal of the reference, x_R = concentration of the reference solution (or amount as well), a = intercept of the calibration function, and b = slope of the calibration function. The unknown data is x_S, the concentration of the injected sample solution, calculated from its signal y_S:

$$x_S = \frac{y_S - a}{b} \qquad (5.39)$$

Strictly speaking, the uncertainty $u(x_S)$ can only be determined with differential equations which, in most cases, do not have closed solutions but numerical calculation is needed. The uncertainty can be calculated with good approximation with the assumption that the uncertainty of the x_R values is low compared to the uncertainty of the y_R values. This assumption is usually valid, because the calibration solutions x_R were prepared by weighing and dilution, operations with low uncertainty as discussed in Sections 5.6 and 5.7. The uncertainty of the signal, that is, of a chromatographic peak area or height is larger by orders of magnitude (Section 5.12). With this assumption the following equation is valid [4]:

$$u(x_S) = \frac{s_R}{b} \sqrt{\frac{1}{m} + \frac{1}{n} + \frac{(x_S - \bar{x}_R)^2}{S_{xx}}} \qquad (5.40)$$

with m = number of measurements (injections) of the sample solution, n = number of data points used for the determination of the calibration function, $\bar{x}_R$ = mean of all x_R values, S_{xx} = sum of all values $(x_R - \bar{x}_R)^2$, and s_R = standard deviation of the determined y_R values around the calibration function according to the equation:

$$s_R = \sqrt{\frac{\sum_{R=1}^{n} (y_R - y_{CF})^2}{n-2}} \tag{5.41}$$

with y_{CF} = the y value of a calibration point x_R, as calculated with the calibration function.

When using Equation 5.40 it is assumed that the variabilities of both the y_R and the y_S values are identical. This assumption is true if the calibration solutions and the sample solution have identical properties, especially if the sample chromatogram does not show higher noise, higher baseline drift, or overlapping peaks. In addition, the calibration should be established in a close time frame before or after the analysis of the sample; the analytical instrument must not undergo any change in the meantime.

If the validation of the method has already been done, the uncertainty of the sample signal $u(y_S)$ is known as its standard deviation $s(y_S)$. Then the equation for $u(x_S)$ has a different form:

$$u(x_S) = \frac{1}{b} \sqrt{\frac{s^2(y_s)}{m} + s_R^2 \left(\frac{1}{n} + \frac{(x_S - \bar{x}_R)^2}{S_{xx}} \right)} \tag{5.42}$$

In this case, $s(y_S)$ is the *absolute* standard deviation of the *sample signal* from validation (1σ value, not RSD, and not standard deviation of the mean). m is the number of measurements of the sample solution in use.

There are some approximate equations which can replace Equation 5.40 but which yield too low values for $u(x_S)$:

$$u(x_S) = \frac{s_R}{b} \tag{5.43}$$

if n and m are large and if x_S is close to $\bar{x}_R$.

$$u(x_S) = \frac{s_R}{b} \sqrt{\frac{1}{m} + \frac{1}{n}} \tag{5.44}$$

if x_S is close to $\bar{x}_R$.

$$u(x_S) = \frac{s_R}{b} \sqrt{\frac{1}{m} + \frac{4}{n}} \tag{5.45}$$

if x_S is close to the lower or upper limit of the calibration interval.

From a practical point of view, it may be allowed and even recommended to use the repeatability of the *instrument response*, here of the chromatographic peak size, as the uncertainty of $u(x_S)$ [32,33].

The uncertainty of purity of the reference compound, as discussed in Section 5.8, is not a subject of all these equations because it is a bias which influences all x_R values in the same way. It must be considered later in the overall equation of the measurand as explained in Section 5.8.

In addition, Monte Carlo simulation can be an excellent tool to investigate the variability of a calibration function [34].

5.12   UNCERTAINTY OF THE CHROMATOGRAPHIC PROCESS

The chromatographic process within the column can be described mathematically, see for example, several chapters in Guiochon and Guillemin [35] for GC and Chapter 2 in Moldoveanu and David [36] for HPLC. The knowledge of the relevant equations is important for the understanding of chromatography but it is not helpful with regard to the uncertainty of the analytical results obtained. The complexity of HPLC was visualized by Ishikawa diagrams [15], and these figures show that it is hopeless to calculate the combined uncertainty of a peak area from the uncertainties of all its influence parameters. For the sake of curiosity, the variability of the HPLC injection process proper was studied [19] but in reality, it is the interplay of injection, chromatographic separation, detection, and integration which yields the variability (standard deviation) of a certain quantitative chromatographic process. It is worse if the signal-to-noise ratio is low [37].

In a routine laboratory, it is almost impossible to keep the repeatability of an HPLC method lower than 1% for a longer period of time. Chromatography is not a highly precise technique, and usually its performance cannot be separated from the (in many cases even worse) performance of the sample preparation. Nevertheless, its precision should be better than the variability of the sample composition to be studied [38].

The uncertainty sources of GC and HPLC methods were discussed in a review by Barwick [39]. A special case is TLC. A paper by Prošek et al. from 2002 presents a comprehensive list of influence parameters on the quantitative result [40]. It is interesting to note that the calculated and validated uncertainty data, expressed as RSDs, matched well. Since then, the instrumentation for quantitative TLC was improved markedly and thus also the reliability of the analyses [41]. A discussion about the validation of TLC and HPTLC methods, including the MU, was published by Renger et al. [42].

It is a matter of fact that chromatographic peaks must be resolved well for quantitative analysis. If not, the peak heights and/or areas will be wrong [43] which does not directly contribute to uncertainty but to bias, that is, to a wrong analytical result. The resolution can be physical (real) or it can be obtained by selective detection.

5.13   AN EXAMPLE: DRUG CONCENTRATION IN SERUM BY HPLC

5.13.1   GENERAL

The serum concentration of a drug D which needs individual dosage and monitoring is determined by HPLC. For the quantitative analysis an IS is used.

Remark concerning analyses with an IS: All volumes are unimportant with the exception of the ones shown in the flow diagram. In principle, a quantitative HPLC procedure leads to the determination of concentrations; however, as soon

TABLE 5.2
Calibration Data of the Drug Analysis

Calibration Solution	1	2	3	4	5
m_D (ng)	10.4	26.0	52.0	78.0	104.0
m_{IS} (ng)	48.6	48.6	48.6	48.6	48.6
m_D/m_{IS}	0.214	0.535	1.070	1.605	2.140
A_D/A_{IS}	0.308	0.881	1.952	2.910	3.984

Notes: Weighted mass of reference = 104.0 mg, weighted mass of IS = 48.6 mg. Results of the two sample injections are A_D/A_{IS} = 1.741 and 1.658, mean A_D/A_{IS} = 1.700, giving $(m_D/m_{IS})_{sample}$ = 0.95.

as the IS has been added it becomes a procedure for the determination of analyte amounts. Chromatographic analyses with IS use the *ratios* of peak heights or areas and not directly the peak sizes. The calibration relies on these ratios, too. The effectively weighted mass and the purity of the IS are not needed for the calculation of the analytical result because equal amounts are added to the sample solution and to the reference solutions. Analogously, identical amounts of the reference and sample solutions must be injected into the chromatographic system; thus the accuracy of the autosampler is not important and its repeatability is part of the repeatability of the chromatographic process.

Summary of the procedure: 1 mL of serum is mixed with 2 mL of buffer solution and 50 ng of IS. The mixture is centrifuged. 2 mL of the supernatant is loaded onto a solid-phase cartridge, washed, and eluted. The eluate is dried under a stream of nitrogen. The residue is reconstituted in 100 µL of mobile phase and centrifuged. The HPLC injection volume is 10 µL.

Some experimental data (see also Table 5.2):

Recovery of D in serum: (87 ± 4.0)% at 50 µg/L (three determinations)
Calibration function (ratio of peak areas): $A_D/A_{IS} = -0.115 + 1.907\ m_D/m_{IS}$
Correlation coefficient: $r = 0.9998$
Repeatability of the HPLC analysis: 1.2% (10 injections from the same vial with serum sample, performed during the validation studies)
Purity of the drug reference: ≥ 99.5%

5.13.2 PROCEDURE

The simplified SOP is given below (in reality it should also include the detailed composition of the solvents, buffers, etc. used).

Internal standard solution: 50 mg IS is dissolved to 100 mL. 1 mL of this solution is diluted to 100 mL, and 1 mL thereof is diluted to 100 mL (=50 ng/mL).
Drug solutions for calibration: 100 mg D is dissolved to 100 mL (gives the reference stock solution with 100 ng/mL). From this solution 10, 25, 50, and 75 mL are diluted to 100 mL (gives calibration dilutions with 10, 25, 50, and 75 ng/mL).
Determination of the calibration function: 1 mL each of the calibration dilutions and 1 mL of the reference stock solution are individually mixed with 2 mL of

ammonium carbonate buffer and with 1 mL of IS, then worked up as described
below, and injected. Thus five calibration points with D/IS mass ratios of 0.2,
0.5, 1.0, 1.5, and 2.0 are obtained.

Sample preparation: 1 mL of serum is mixed with 2 mL of ammonium carbonate
buffer and 1 mL of IS. The IS is added with the same pipette as used above.

Work-up: The mixed solutions (total volume = 4 mL each) are centrifuged.
2 mL of the supernatant is loaded on a C_{18} cartridge, washed with ammonium carbonate buffer, and eluted with methanol/acetic acid 0.5 M (9:1).
The eluates are evaporated under nitrogen, reconstituted in 100 µL of
mobile phase, and centrifuged. 10 µL is injected.

Chromatography: Stationary phase: RP-18, 4 µm. Mobile phase: 0.05 M
ammonium formate buffer pH 3.0/acetonitrile 3:1. Detection: MS (single
ion monitoring). Double determination (two injections from each vial).

The result is corrected by the recovery.

5.13.3 FLOWCHART

The flowchart of the procedure outlined above is shown in Figure 5.8. It consists of
several branches. The ones for the sample (serum) and the reference meet only at the

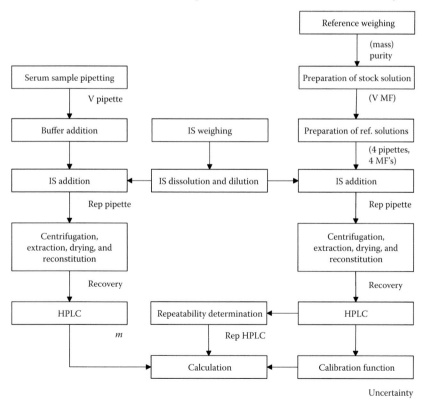

FIGURE 5.8   Flowchart of the quantitative analysis of a drug in serum by HPLC.

box for the calculation. The IS preparation is used for the sample and all reference solutions, therefore it is arranged in the middle of the chart.

The working steps for the IS preparation have no uncertainty at all because, in fact, the IS solution can be prepared in a very sloppy way. The purity of the IS is not important. It is not even necessary to know the precise amount of weighted IS (although m_{IS} is a variable in the equation of the measurand, see Equation 5.46 and discussion below), and the dilution can be made without much care because identical amounts of the IS solution are added to all solutions which are processed further. Note that the same pipet (probably a piston-operated one) is used for all six IS addition steps. With good laboratory practice, the additions of the IS are made immediately one after the other, thus eliminating any temperature effect on the delivered volume.

5.13.4   Equation of the Measurand

Although the SOP of the procedure is a rather long text, the equation of the measurand is surprisingly simple in this case:

$$c_D = \frac{1}{Rec} \cdot \frac{m_D}{m_{IS}} \cdot m_{IS} \cdot \frac{1}{V_{Ser}} \cdot P_R \tag{5.46}$$

with c_D = concentration of the drug in serum, Rec = recovery, m_D/m_{IS} = mass ratio of drug reference and IS as found by the multipoint calibration function, m_{IS} = mass of IS added in step *IS addition* of the flowchart, V_{Ser} = volume of the serum sample, and P_R = purity of the drug reference.

The equation is composed from no more than five variables. The recovery is dealt with as described in Section 5.10. m_D/m_{IS} is found from the multipoint calibration function as discussed in Section 5.11. The volume of serum and the purity of the reference are dealt with as shown in Sections 5.6 and 5.8, respectively.

m_{IS} is a special case because its true mass need not be known. The IS is weighted and diluted, then it is added with a 1 mL piston pipet to the sample and reference solutions. It is not important how much IS is added although the weighted amount, the dilution, and the resulting concentration are noted in the laboratory notebook. Its value is used in Equation 5.46 whether it is accurate or not but mathematically it disappears by cancellation. However, what must be known for the uncertainty evaluation is its uncertainty which is nothing else than the repeatability of the 1 mL pipet.

5.13.5   Ishikawa Diagram

Once the equation of the measurand is defined, it is a simple task to draw the main backbone of the Ishikawa diagram. It consists of the horizontal arrow which designates the sought-after concentration and the five side arrows which represent the variables of Equation 5.46. The latter carry side arrows of second and third order which give detailed insight into the vast number of influence parameters. This diagram is shown in Figure 5.9.

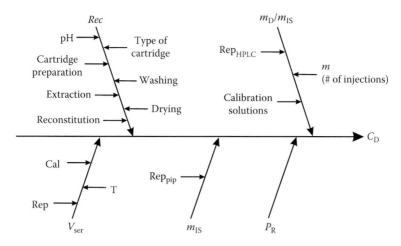

FIGURE 5.9 Ishikawa diagram for the quantitative analysis of a drug in serum by HPLC.

5.13.6 UNCERTAINTIES OF THE INDIVIDUAL PARAMETERS

The variables of Equation 5.46 are linked together in a multiplicative manner, therefore all uncertainties can be evaluated as relative values. Then the combined standard uncertainty of the drug concentration in serum is calculated as defined by a member of the Equations 5.21, 5.22, and 5.24 family.

Recovery (Section 5.10). Its uncertainty is the repeatability as found during the validation studies. 4% of 87% gives a relative uncertainty $u_{rel}(Rec)$ of 0.046.

Calibration function (Section 5.11). Because the calibration points were established with pure reference solutions, whereas the original sample was serum, it is not recommended to use Equation 5.40 but Equation 5.42 instead. The values of its parameters were found during validation or with the sample analyses (Table 5.2) and are as follows: $b = 1.907$; repeatability (HPLC) = 1.2%, yielding (together with the mean ratio of $A_D/A_{IS} = 1.700$ of the sample chromatograms) an absolute standard deviation of the sample signal $s(y_s) = 0.0204$; $m = 2$; $s_R = 0.0322$ (validation data used for Equation 5.41); $n = 5$; $x_S - \bar{x}_R = -0.16$ (experimental data); $S_{xx} = 2.44$ (experimental data). b, m, and n are directly noted in Section 5.13.1, the values of the other parameters can be calculated from the data given in Table 5.2. The uncertainty coming from the calibration function is therefore:

$$u(m_D/m_{IS}) = \frac{1}{1.907}\sqrt{\frac{0.0204^2}{2} + 0.0322^2\left(\frac{1}{5} + \frac{-0.16^2}{2.44}\right)} = 0.0108 \qquad (5.47)$$

Together with the experimental result $(m_D/m_{IS})_{sample} = 0.95$, the relative uncertainty of the data obtained by the calibration function is $u_{rel}(m_D/m_{IS}) = 0.0114$. Note that the correlation coefficient of the calibration function is not important with regard to its uncertainty although it was mentioned in Section 5.13.1.

IS. Here, only the repeatability of the 1 mL pipet used is relevant which is claimed in the data sheet to be 0.004 mL, giving a relative uncertainty $u_{rel}(m_{IS}) = 0.004$.

Volume of the serum sample (Section 5.6). A 1 mL piston-operated pipet is used, its uncertainty parameters are: 0.008 mL calibration uncertainty with a triangular distribution, therefore $u(V_{cal}) = 0.003$ mL; 0.004 mL repeatability as standard deviation, therefore $u(V_{rep}) = 0.004$ mL; estimated temperature range $= \pm 3°C$ (the serum was stored at low temperature and it may be not fully equilibrated with the laboratory temperature), therefore $u(T) = 1.8°C$ with a rectangular distribution. The cubic expansion coefficient of water is $2 \cdot 10^{-4}/°C$, and in lack of the data for serum this value is used. According to Equation 5.32 we get:

$$u(V_{Ser}) = \sqrt{0.003^2 + 0.004^2 + \left(1 \cdot 2 \cdot 10^{-4} \cdot 1.8\right)^2} \text{ mL} = 0.0050 \text{ mL} \qquad (5.48)$$

Together with the volume of 1 mL, the relative uncertainty $u_{rel}(V_{Ser})$ is 0.005.

Purity of the drug reference (Section 5.8). Its purity is claimed to be $\geq 99.5\%$. Thus $P_R = 0.9975$ is used in the equation of the measurand, that is, the mean between the lower limit and 100%. The range of the rectangular distribution is $1 - 0.995 = 0.005$ (which means $\pm a = 2a$). According to Equation 5.36 we get

$$u(P_R) = 0.3(1 - 0.995) = 0.0015 \qquad (5.49)$$

which is directly the relative data $u_{rel}(P_R)$.

5.13.7 COMBINED UNCERTAINTY OF THE ANALYTICAL RESULT

The analytical result, according to Equation 5.46 is

$$c_D = \frac{1}{0.87} \cdot 0.95 \cdot 48.6 \text{ ng} \cdot \frac{1}{1 \text{ mL}} \cdot 0.9975 = 53 \text{ ng/mL} \qquad (5.50)$$

Its uncertainty equation is of the multiplicative type, and the squared relative uncertainties can be added:

$$u_{rel}(c_D) = \sqrt{u_{rel}^2(Rec) + u_{rel}^2(m_D/m_{IS}) + u_{rel}^2(m_{IS}) + u_{rel}^2(V_{Ser}) + u_{rel}^2(P_R)} \qquad (5.51)$$

In numbers:

$$u_{rel}(c_D) = \sqrt{0.046^2 + 0.0114^2 + 0.004^2 + 0.005^2 + 0.0015^2} = 0.048 = 4.8\% \qquad (5.52)$$

As an absolute value, these 4.8% correspond to 2.5 ng/mL. The percentages of the different contributions are: recovery 93.3%, calibration function 5.3%, amount of IS 0.4%, sample volume 0.9%, and purity 0.1%. It is obvious that it is not necessary to consider also the uncertainty of the calibration solutions, that is, the weighting of the drug reference and the dilution steps.

5.13.8 EXPANSION TO 95% CONFIDENCE

The dominating contribution to the uncertainty is the recovery. Since it was determined three times, the coverage factor for 95% of all analyses is 4.3 (Section 5.4.2 and Table 5.1). Therefore the expanded relative uncertainty U is 4.8% $\cdot$ 4.3 = 21%

which can be rounded to 20%. In concentration units it means 11 ng/mL. This value is the within-laboratory uncertainty (determined in one laboratory). It could be lower if the recovery was determined more than three times only.

Note that MU data itself has an uncertainty of ±20% or more.

5.14   THE TOP–DOWN APPROACH

The bottom–up approach is not the only possibility for the evaluation of the MU [44]. One can use validation data [45,46], and if this is done properly, the within-laboratory repeatability s_R (a standard deviation) can be declared as the MU of a given method performed in a given laboratory. The investigation can be enlarged, for example, to all laboratories of a company [47], which usually results in a higher uncertainty than the above-mentioned s_R. A reliable MU is obtained by interlaboratory comparisons because different persons, different instruments, and different circumstances are involved [48–50]. It can be assumed that all possible settings contributed to the between-laboratory reproducibility S_R. However, even in this scenario, the purity of the reference materials used (Section 5.8) must not be forgotten. Its uncertainty, which may not be a constant if different references (from different manufacturers) were used, needs to be added to S_R, just in analogy to Equation 5.37:

$$\frac{u_c(M)}{M} = \sqrt{S_R^2 + \left(\frac{u(P_R)}{P_R}\right)^2} \tag{5.53}$$

The temptation to use the top–down approach comes from the fact that it looks so simple. If the analyst has a good knowledge of the general concepts of MU, including the bottom–up approach, and of the analytical method, there is nothing to be said against this way of uncertainty evaluation. It cannot be recommended for beginners. In fact, the Eurolab Technical Report of 1/2007 notes in its conclusions that "the following are required: devotion, competence in measuring techniques, a sound knowledge about the test item, basic know-how about measurement uncertainty" [49].

The drawback of the top–down approach is the fact that it does not disclose the fractions of the different influence parameters to the overall uncertainty $u_c(M)$, thus it cannot be used directly for the improvement of a method.

5.15   DARK UNCERTAINTY

In 2011, Thompson and Ellison pointed to the fact that standard uncertainties, as evaluated by the bottom–up approach, may be too small [51]. This suspicion is based on the fact that the top–down approach frequently yields higher combined standard uncertainties than the detailed evaluation according to the GUM [2] or the QUAM [4], especially if the between-laboratory S_R is compared to $u_c(M)$ according to Equations 5.12 or 5.23 (the latter being just an example). The authors think that there exists some *dark uncertainty* which is not covered by the Ishikawa diagram or similar considerations. In a later paper they state: "Dark uncertainty seems to be not only ubiquitous but almost inevitable in chemical measurement" [52]. It is a fact that the GUM [2] was designed for measurements in physics, and that many chemical processes are not understood in full detail; the best example is perhaps the broad field of sample preparation procedures.

Nevertheless, it is hard to understand why the careful investigation of an analytical method according to the bottom–up approach does not lead to a sound MU data, at least if compared with the within-laboratory uncertainty s_R. At present, there is a lack of investigations which might elucidate this problem. One of its aspects may be the overestimation of sample homogeneity mentioned in Section 5.3.1 (although a sample must be scrupulously homogenized before it is sent to the participants of an inter-laboratory comparison). In addition, Thompson assumes that sporadic errors and out-of-scope test materials may play a role [52].

Considering the fact that something like *dark uncertainty* seems to exist, even the use of the between-laboratory standard deviation S_R as the *true measurement uncertainty* creates some misgivings. Thompson in his 2014 paper proposes how this uncertainty about MU can be handled [53]. There is no question that the problem of *dark uncertainty* deserves special efforts for its better understanding.

5.16   UNCERTAINTY IN THE CHROMATOGRAPHIC LITERATURE

Uncertainty in budgets and evaluations are not often reported in the chromatographic literature, compared to the enormous number of papers with new methods which are published every year. A review by Konieczka and Namieśnik includes a list of 27 papers, published before 2009, which present not only GC and HPLC methods but also their uncertainty budgets [54]. The list may be incomplete because the authors refer to it as of *examples*.

As a complement, Table 5.3 lists a number of more recent papers [55–99], including one single publication dealing with TLC [90]. It is not comprehensive since it is restricted to the *Journals of Chromatography A and B*. The papers were found in www.sciencedirect.com with the keyword *uncertainty*, starting in 2009 (search performed on March 12, 2015). The paper by Quintela et al. presents the detailed uncertainty evaluation, including flowcharts, Ishikawa diagrams, and chromatograms of an HPLC method (chlorides by ion exchange) and of a GC method (palmitic and stearic acid, flame ionization detection [FID]) [80].

TABLE 5.3
Some Papers Presenting Chromatographic Analyses with Uncertainty Data

Analyte(s)	Matrix	Method	Detection	Ref.
Pesticides, PCBs	Wine	GC	TOF-MS	[55]
Pesticides	Vegetables	GC	NPD, ECD	[56]
PAHs	Air particulate matter	GC	MS	[57]
Organometallic compounds	Water	GC	MS/MS	[58]
Pesticides	Wine	HPLC	MS/MS	[59]
Polychlorinated compounds	Incinerator emissions	GC	MS	[60]
Pesticides	Honeybees	GC	MS/MS	[61]
Anti-inflammatory drugs	Bovine milk	HPLC	MS/MS	[62]
Quinolones	Animal feed	HPLC	FLU, UV	[63]
Firocoxib	Bovine milk	HPLC	MS/MS	[64]

(Continued)

TABLE 5.3 (*Continued*)
Some Papers Presenting Chromatographic Analyses with Uncertainty Data

Analyte(s)	Matrix	Method	Detection	Ref.
Retinol, α-tocopherol	Human serum	HPLC	FLU, UV	[65]
Oxytetracycline	Nasal secretions	HPLC	MS/MS	[66]
Dioxins, dioxin-like PCBs	Food	GC	ID-MS	[67]
Pesticides	Tea, infusions, leaves	HPLC	MS/MS	[68]
Pesticides, POPs	Grape, wine	2D-GC	TOF-MS	[69]
Cardiac glycosides	Digitalis plant extract	HPLC	MS^n	[70]
Pesticides, carbamates	Food	HPLC	MS/MS	[71]
Pesticides	Water	GC	MS	[72]
Tiloronoxim, tilorone	Human blood	HPLC	MS/MS	[73]
Aldehydes	Human urine	HPLC	MS/MS	[74]
Urea	Milk, milk powder	GC	ID-MS	[75]
Ototoxic solvents	Saliva	GC	MS	[76]
Geosmin, 2-methylisoborneol	Water, wine	GC	MS	[77]
Pesticides, PCBs, PAHs	Water-based commodities	GC	TOF-MS	[78]
Alkylphenols, bisphenol A	Seawater	HPLC	MS/MS	[79]
Chloride	Lixiviate	HPLC	COND	[80]
Fatty acids	Magnesium stearate	GC	FID	[80]
Volatile compounds	Roasted hazelnuts	2D-GC	MS	[81]
Accelerants	Fire debris	GC	MS	[82]
Alkylphenols, bisphenol A	Bivalve molluscs	HPLC	MS/MS	[83]
Veterinary drugs	Fish	HPLC	MS/MS	[84]
Chlorpyrifos + metabolite	Wine	GC	ECD	[85]
Fluoroquinolones	Meat	HPLC	ID-MS/MS	[86]
Impurities	Aldrin	2D-GC	various	[87]
Alkylphenols	Water	HPLC	MS/MS	[88]
Pharmaceuticals	Water	HPLC	MS/MS	[89]
Drugs	Tablet formulations	TLC	DENS	[90]
Antibiotics	Seawater	HPLC	MS/MS	[91]
PCBs, PAHs, pesticides	Atmosphere	GC	MS, ECD	[92]
Marine biotoxins	Mussels	HPLC	MS	[93]
Glycerol	Biodiesel	GC	FID	[94]
Retinoic acid	Plasma	HPLC	UV	[95]
Caffeine + metabolites	Human plasma	HPLC	MS/MS	[96]
Bisphenol A	Foodstuffs	GC	MS/MS	[97]
Alternaria toxins	Cereal-based foodstuffs	HPLC	MS/MS	[98]
Dioxins, polychlorobiphenyls	Feed	GC	MS/MS	[99]

Abbreviations: COND = conductivity; DENS = densitometry; ECD = electron capture detector; FID = flame ionization detector; FLU = fluorescence; ID = isotope dilution; MS = mass spectroscopy; MS/MS = tandem mass spectroscopy; NPD = nitrogen–phosphorous detector; PAH = polycyclic aromatic hydrocarbon; PCB = polychlorinated biphenyls; POP = persistent organic pollutant; TOF = time of flight; UV = ultraviolet.

5.17   DO NOT BE AFRAID OF MEASUREMENT UNCERTAINTY

In a literature survey, Ruiz-Angel et al. found that MU is not a characteristic number that is published as a matter of course [100]. They checked the reports, published between 2004 and 2013, dealing with HPLC method validation in the life and health sciences. Of the 12,668 articles found they investigated a random selection of 200 papers. *Uncertainty*, as one of the possible validation parameters, was reported explicitly in only 3.5% of them (note that this is the mean of 10 years). The authors think that uncertainty "is usually expressed as the standard deviation, a multiple of the standard deviation, or the width of a confidence interval. This parameter is usually referred in the calculus of accuracy or precision, thus making sense not to study it separately." It is, however, possible that the percentage of papers which disclose a true MU data is now higher than it was in 2004.

This study [100] leads to the assumption that the knowledge and awareness of MU is still underdeveloped. For the readers of this whole chapter it should be clear that MU is more than just a standard deviation. As explained in Section 5.14, the standard deviation obtained by an interlaboratory comparison, expanded by the purity contribution of the reference, may represent something like a sound MU. However, an assessment is only possible by an experienced and trained analyst who has a good knowledge of the bottom–up approach.

The evaluation and knowledge of a method's MU opens the mind for its possible pitfalls. In some cases a method can be improved and simplified when it is reinvestigated for the purpose of MU evaluation; an example from titrimetry was published by Brix et al. [101]. It is always necessary to set up the proper equation of the measurand in an SOP, as explained in Section 5.3.1. Starting from there, it is not too difficult to look for the possible influence parameters of the MU. What may be more demanding is the evaluation of hidden variables such as sample homogeneity. The analyst must be aware of the fact that these hidden contributions to the combined MU may be present and needs to consider them in the first draft of the MU budget.

Despite of the negative connotation of a data termed *measurement uncertainty*, its knowledge leads to more certainty about the quality of an analytical result.

REFERENCES

1. S.K. Kimothi, *The Uncertainty of Measurements*, ASQ Quality Press, Milwaukee, WI. 2002.
2. Evaluation of measurement data – Guide to the expression of uncertainty in measurement. 2008. Joint Committee for Guides in Metrology, JGCM 100:2008. Free download: http://www.bipm.org/en/publications/guides (accessed March 18, 2015).
3. Evaluation of measurement data – An introduction to the "Guide to the expression of uncertainty in measurement" and related documents, JCGM 104:2009. Free download: http://www.bipm.org/en/publications/guides (accessed March 18, 2015).
4. Quantifying uncertainty in analytical measurement. 3rd ed. 2012. Eurachem, ISBN 0-948926-15-5. Free download: http://eurachem.org/index.php/publications/guides/quam (accessed March 18, 2015).
5. International vocabulary of metrology – Basic and general concepts and associated terms (VIM). 3rd ed. 2012. Joint Committee for Guides in Metrology, JCGM 200:2012. Free download: http://www.bipm.org/en/publications/guides (accessed March 18, 2015).

6. M. Rösslein, *Accred. Qual. Assur.*, 5:88 (2000).
7. V.R. Meyer, *J. Chromatogr. A*, 1158:15 (2007).
8. Measurement uncertainty arising from sampling – A guide to methods and approaches. 2007. Eurachem, ISBN 978-0-948926-26-6. Free download: http://www.eurachem.org/images/stories/Guides/pdf/UfS_2007 (accessed March 18, 2015).
9. Evaluation of measurement data – Supplement 1 to the "Guide to the expression of uncertainty in measurement" – Propagation of distributions using a Monte Carlo method. Joint Committee for Guides in Metrology, JCGM 101:2008. Free download: http://www.bipm.org/utils/common/documents/jcgm/JCGM_101_2008_E.pdf (accessed March 18, 2015).
10. G. Chew and T. Walczyk, *Anal. Bioanal. Chem.*, 402:2463 (2012).
11. V.R. Meyer, *LC GC Eur.*, 25:417 (2012).
12. Student (W.S. Gosset), *Biometrika*, 6:1 (1908).
13. K. Ishikawa, *Introduction to Quality Control*, Kluwer, Dordrecht, 1991, Chapter 4.7.4.
14. S.L.R. Ellison and V.J. Barwick, *J. Accred. Qual. Assur.*, 3:101 (1998).
15. V.R. Meyer, *J. Chromatogr. Sci.*, 41:439 (2003).
16. EN ISO 648:2009-01, *Laboratory Glassware – Single-volume Pipettes*. Beuth, Berlin, 2009.
17. EN ISO 1042:1999-08, *Laboratory Glassware – One-mark Volumetric Flasks*. Beuth, Berlin, 1999.
18. V.R. Meyer, J. Pfohl and B. Winter, *Accred. Qual. Assur.*, 15:705 (2010).
19. K.G. Kehl and V.R. Meyer, *Anal. Chem.*, 73:131 (2001).
20. A. Reichmuth, S. Wunderli, M. Weber and V.R. Meyer, *Microchim. Acta*, 148:133 (2004).
21. M. Pozivil, W. Winiger, S. Wunderli and V.R. Meyer, *Microchim. Acta*, 154:55 (2006).
22. G. van Look and V.R. Meyer, *Analyst*, 127:825 (2002).
23. USP certificate of palmitic acid as just one example: http://www.usp.org/pdf/EN/referenceStandards/certificates/1492007-L0K054.pdf (accessed March 18, 2015).
24. R. Borges and V.R. Meyer, *J. Pharm. Biomed. Anal.*, 77:40 (2013).
25. M.E. Wieser et al., *Pure Appl. Chem.*, 85:1047 (2013).
26. W.A. Brand, *Anal. Bioanal. Chem.*, 405:2755 (2013).
27. V.R. Meyer, *Anal. Bioanal. Chem.*, 377:775 (2003).
28. V.J. Barwick and S.L.R. Ellison, *Analyst*, 124:981 (1999).
29. T.P.J. Linsinger, *Trends Anal. Chem.*, 27:916 (2008).
30. V.R. Meyer, *LC GC North Am.*, 20:106 (2002); also *LC GC Eur.*, 15:398 (2002).
31. J.A. Lyn, M.H. Ramsey, R.J. Fussell and R. Wood, *Analyst*, 128:1391 (2003).
32. D.B. Hibbert, *Analyst*, 131:1273 (2006).
33. D.B. Hibbert, *Analyst*, 132:587 (2007) (correction of equation 16 in ref. 32).
34. V.R. Meyer, *LC GC Eur.*, 28:204 (2015).
35. G. Guiochon and C.L. Guillemin, *Quantitative Gas Chromatography*, Elsevier, Amsterdam, 1988.
36. S.C. Moldoveanu and V. David, *Essentials in Modern HPLC Separations*, Elsevier, Amsterdam, 2013.
37. C. Meyer, P. Seiler, C. Bies, C. Cianculli, H. Wätzig and V.R. Meyer, *Electrophoresis*, 33:1509 (2012).
38. B. Renger, *J. Chromatogr. B*, 745:167 (2000).
39. V.J. Barwick, *J. Chromatogr. A*, 849:13 (1999).
40. M. Prošek, A. Golc-Wondra and I. Vovk, *J. Chrom. Sci.*, 40:598 (2002).
41. P. Jazbec Krizman, K. Cernelic, A. Golc Wondra, Z. Rodic, M. Prosek and M. Prosek, *J. Planar Chromatogr.*, 26:299 (2013).
42. B. Renger, Z. Végh and K. Ferenczi-Fodor, *J. Chromatogr. A*, 1218:2712 (2011).
43. V.R. Meyer, *J. Chromatogr. Sci.*, 33:26 (1995).
44. W. Horwitz, *J. AOAC Int.*, 86:109 (2003).
45. V.J. Barwick and S.L.R. Ellison, *Accred. Qual. Assur.*, 5:47 (2000).

46. V.J. Barwick, S.L.R. Ellison, M.J.Q. Rafferty and R.S. Gill, *Accred. Qual. Assur.*, 5:104 (2000).
47. S. Populaire and E.C. Giménez, *Accred. Qual. Assur.*, 10:485 (2006).
48. I.R. Bertoni Olivares and F.A. Lopes, *Trends Anal. Chem.*, 35:109 (2012).
49. Measurement uncertainty revisited: Alternative approaches to uncertainty evaluation. Eurolab Technical Report 1/2007. Free download: http://www.eurolab.org/documents/1-2007.pdf (accessed March 18, 2015).
50. Handbook for calculation of measurement uncertainty in environmental laboratories. Nordtest Technical Report 537, Oslo 2012. Free download: http://www.nordtest.info/images/documents/nt-technical-reports/nt_tr_537_ed3_1_English_Handbook%20for%20Calculation%20of%20Measurement%20uncertainty%20in%20environmental%20laboratories.pdf (accessed March 18, 2015).
51. M. Thompson and S.L.R. Ellison, *Accred. Qual. Assur.*, 16:483 (2011).
52. M. Thompson and S.L.R. Ellison, *Anal. Methods*, 4:2609 (2012).
53. M. Thompson, *Anal. Methods,* 6:8454 (2014).
54. P. Konieczka and J. Namieśnik, *J. Chromatogr. A*, 1217:882 (2010).
55. S.H. Patil, K. Banerjee, S. Dasgupta, D.P. Oulkar, S.B. Patil, M.R. Jadhav, R.H. Savant, P.G. Adsule and M.B. Deshmukh, *J. Chromatogr. A*, 1216:2307 (2009).
56. E.G. Amvrazi and N.G. Tsiropoulos, *J. Chromatogr. A*, 1216:2789 (2009).
57. B.L. van Drooge, I. Nikolova and P.P. Ballesta, *J. Chromatogr. A*, 1216:4030 (2009).
58. E. Beceiro-González, A. Guimaraes and M.F. Alpendurada, *J. Chromatogr. A*, 1216:5563 (2009).
59. A. Economou, H. Botitsi, S. Antoniou and D. Tsipi, *J. Chromatogr. A*, 1216:5856 (2009).
60. K. Martínez, J. Rivera-Austrui, M.A. Adrados, M. Abalos, J.J. Llerena, B. van Bavel, J. Rivera and E. Abad, *J. Chromatogr. A*, 1216:5888 (2009).
61. S. Walorczyk and B. Gnusowski *J. Chromatogr. A*, 1216:6522 (2009).
62. G. Dowling, P. Gallo, E. Malone and L. Regan, *J. Chromatogr. A*, 1216:8117 (2009).
63. R. Galarini, L. Fioroni, F. Angelucci, G.R. Tovo and E. Cristofani, *J. Chromatogr. A*, 1216:8158 (2009).
64. G. Dowling, P. Gallo and L. Regan, *J. Chromatogr. B*, 877:541 (2009).
65. A. Semeraro, I. Altieri, M. Patriarca and A. Menditto, *J. Chromatogr. B*, 877:1209 (2009).
66. M.A. Bimazubute, E. Rozet, I. Dizier, J.C. Van Heugen, E. Arancio, P. Gustin, J. Crommen and P. Chiap, *J. Chromatogr. B*, 877:2349 (2009).
67. G. Eppe and E. De Pauw, *J. Chromatogr. B*, 877:2380 (2009).
68. B. Kanrar, S. Mandal and A. Bhattacharyya, *J. Chromatogr. A*, 1217:1926 (2010).
69. S. Dasgupta, K. Banerjee, S.H. Patil, M. Ghaste, K.N. Dhumal and P.G. Adsule, *J. Chromatogr. A*, 1217:3881 (2010).
70. R.D. Josephs, A. Daireaux, S. Westwood and R.I. Wielgosz, *J. Chromatogr. A*, 1217:4535 (2010).
71. S.W.C. Chung and B.T.P. Chan, *J. Chromatogr. A*, 1217:4815 (2010).
72. E. Passeport, A. Guenne, T. Culhaoglu, S. Moreau, J.M. Bouyé and J. Tournebize, *J. Chromatogr. A*, 1217:5317 (2010).
73. X. Zhang, J. Duan, S. Zhai, Y. Yang and L. Yang, *J. Chromatogr. B*, 878:492 (2010).
74. C.E. Baños and M. Silva, *J. Chromatogr. B*, 878:653 (2010).
75. X. Dai, X. Fang, F. Su, M. Yang, H. Li, J. Zhou and R. Xu, *J. Chromatogr. B*, 878:1634 (2010).
76. M. Gherardi, A. Gordiani and M. Gatto, *J. Chromatogr. B*, 878:2391 (2010).
77. C. Cortada, L. Vidal and A. Canals, *J. Chromatogr. A*, 1218:17 (2011).
78. S. Dasgupta, K. Banerjee, S. Utture, P. Kusari, S. Wagh, K. Dhumal, S. Kolekar and P.G. Adsule, *J. Chromatogr. A*, 1218:6780 (2011).
79. N. Salgueiro-González, E. Concha-Graña, I. Turnes-Carou, S. Muniategui-Lorenzo, P. López-Mahía and D. Prada-Rodríguez, *J. Chromatogr. A*, 1223:1 (2012).

80. M. Quintela, J. Báguena, G. Gotor, M.J. Blanco and F. Broto, *J. Chromatogr. A*, 1223:107 (2012).
81. J. Kiefl, C. Cordero, L. Nicolotti, P. Schieberle, S.E. Reichenbach and C. Bicchi, *J. Chromatogr. A*, 1243:81 (2012).
82. P.A.S. Salgueiro, C.M.F. Borges and R.J.N. Bettencourt da Silva, *J. Chromatogr. A*, 1257:189 (2012).
83. N. Salgueiro-González, I. Turnes-Carou, S. Muniategui-Lorenzo, P. López-Mahía and D. Prada-Rodríguez, *J. Chromatogr. A*, 1270:80 (1012).
84. R.P. Lopes, R. Cazorla Reyes, R. Romero-González, J.L. Martínez Vidal and A. Garrido Frenich, *J. Chromatogr. B*, 895–896:39 (2012).
85. F.O. Pelit, L. Pelit, H. Ertaş and F.N. Ertaş, *J. Chromatogr. B*, 904:35 (2012).
86. S. Lee, B. Kim and J. Kim, *J. Chromatogr. A*, 1277:35 (2013).
87. X. Li, X. Dai, X. Yin, M. Li, Y. Zhao, J. Zhou, T. Huang and H. Li, *J. Chromatogr. A*, 1277:69 (2013).
88. N. Salgueiro-González, I. Turnes-Carou, S. Muniategui-Lorenzo, P. López-Mahía and D. Prada-Rodríguez, *J. Chromatogr. A*, 1281:46 (2013).
89. M.R. Boleda, M.T. Galceran and F. Ventura, *J. Chromatogr. A*, 1286:146 (2013).
90. D.H. Shewiyo, E. Kaale, P.G. Risha, B. Dejaegher, J. De Beer, J. Smeyers-Verbeke and Y. Vander Heyden, *J. Chromatogr. A*, 1293:159 (2013).
91. M. Borecka, A. Białk-Bielińska, G. Siedlewicz, K. Kornowska, J. Kumirska, P. Stepnowski and K. Pazdro, *J. Chromatogr. A*, 1304:138 (2013).
92. G. Aslan-Sungur, E.O. Gaga and S. Yenisoy-Karakaş, *J. Chromatogr. A*, 1325:40 (2014).
93. A. Domènech, N. Cortés-Francisco, O. Palacios, J.M. Franco, P. Riobó, J.J. Llerena, S. Vichi and J. Caixach, *J. Chromatogr. A*, 1328:16 (2014).
94. L. Ruano Miguel, M. Ulberth-Buchgraber and A. Held, *J. Chromatogr. A*, 1338:127 (2014).
95. C.M. Teglia, M.D. Gil García, M. Martínez Galera and H.C. Goicoechea, *J. Chromatogr. A*, 1353:40 (2014).
96. A.L. Gassner, J. Schappler, M. Feinberg and S. Rudaz, *J. Chromatogr. A*, 1353:121 (2014).
97. Y. Deceuninck, E. Bichon, S. Durand, N. Bemrah, Z. Zendong, M.L. Morvan, P. Marchand, G. Dervilly-Pinel, J.P. Antignac, J.C. Leblanc and B. Le Bizec, *J. Chromatogr. A*, 1362:241 (2014).
98. J. Walravens, H. Mikula, M. Rychlik, S. Asam, E.N. Ediage, J.D. Di Mavungu, A. Van Landschoot, L. Vanhaecke and S. De Saeger, *J. Chromatogr. A*, 1372:91 (2014).
99. B. L'Homme, G. Scholl, G. Eppe and J.F. Focant, *J. Chromatogr. A*, 1376:149 (2015).
100. M.J. Ruiz-Angel, M.C. García-Alvarez-Coque, A. Berthod and S. Carda-Broch, *J. Chromatogr. A*, 1353:2 (2014).
101. R. Brix, S.H. Hansen, V. Barwick and J. Tjørnelund, *Analyst*, 127:140 (2002).

6 Comprehensive Two-Dimensional Hydrophilic Interaction Chromatography × Reversed-Phase Liquid Chromatography (HILIC × RP–LC)

Theory, Practice, and Applications

André de Villiers and Kathithileni Martha Kalili

CONTENTS

6.1   INTRODUCTION

Ever-increasing demands being placed on accurate analytical data for complex samples by diverse fields such as biotechnology, metabolomics, proteomics, lipidomics, petroleomics, and natural product research continue to provide the impetus for the development of more powerful analytical techniques and instrumentation. In this context, mass spectrometry (MS) undoubtedly represents one of the most important instrumental techniques, as evidenced by the rapid growth in MS instrumentation and applications during especially the past decade and a half. For the complex samples dealt with in each of the fields listed above, however, separation plays an equally important role, due in part to limitations of MS in terms of the identification and accurate quantification of similar analytes in complex mixtures.

High-performance liquid chromatography (HPLC) in particular is an indispensable tool due to the wide application range of the technique, which includes important nonvolatile and/or thermally labile compounds such as proteins and peptides, carbohydrates, polymers, phenolics, and so on. In contrast to gas chromatography (GC), where the use of long open tubular capillary columns provides exceptional separation performance, HPLC suffers from limited chromatographic efficiency, a consequence of its reliance on relatively short packed columns dictated by slow diffusion of analytes in the liquid phase and the associated pressure constraints. Recognition of the complexity of the samples being investigated has prompted extensive development in the field of HPLC, and especially in the past 15 years several notable achievements have provided a quantum leap in the performance of the technique. Important developments include the use of ultra high pressures in combination with packed beds comprising ever-decreasing particle sizes (ultra high-pressure liquid chromatography [UHPLC]) (MacNair et al. 1997; Patel et al. 2004), high temperature LC columns and instrumentation (Heinisch and Rocca 2009; Teutenberg 2009), alternative stationary phases such as monoliths (Tanaka 2002) and superficially porous particles (Wang et al. 2006a; Hayes et al. 2014), new detection strategies (Gamache et al. 2005), and of course developments in LC–MS instrumentation (Makarov 2000; Makarov and Scigelova 2010).

Despite these important advances, increasing application to highly complex samples in combination with the latest generation of very sensitive detectors has served to emphasize the limitations of one-dimensional (1D) HPLC. For example, proteomic samples may contain up to 20,000 proteins, corresponding to in excess of 400,000 peptides following enzymatic digestion (Regnier et al. 2001; Anderson and Anderson 2002); this number far exceeds the number of compounds that can be separated using even highly optimized HPLC methods. It is in this context that multidimensional liquid chromatographic (MD–LC) separations have grown in significance. In MD–LC, two or more complementary separation modes are used to provide the requisite increase in selectivity and resolving power.

MD–LC can be performed in either heart-cutting or comprehensive modes. In the former case, selected fractions of the first dimension (^{1}D) effluent are transferred to a second column for further analysis. The benefit of this, provided that the second separation provides an alternative separation mechanism, is that additional information may be obtained about a portion of the sample. Heart-cutting two-dimensional (2D) separation is particularly useful when selected target analytes cannot be separated using a single column/mobile phase combination. The performance gains are however relatively small compared to ^{1}D chromatography.

In contrast, comprehensive 2D liquid chromatography (abbreviated LC × LC) involves transfer of fractions from ^{1}D to the second dimension (^{2}D) column, such that complete sample is submitted to separation in both dimensions. This approach therefore provides more information about the entire sample, and the performance gain is significant compared to 1D HPLC methods (refer to Section 6.2.2). This has been the main impetus behind research and development in the field. Especially during the past decade, important advances in the fundamental and instrumental aspects of LC × LC have coincided with much more widespread application of the technique; for recent overviews of the field of LC × LC, the reader is referred to (Shalliker and Gray 2006; Dugo et al. 2008; François et al. 2009; Stoll 2010; Carr et al. 2012).

At this point, it is worth revisiting the requirements that a MD separation should fulfill in order to be considered comprehensive, as outlined by Schoenmakers et al. (2003) (we use the original wording in this definition): "(i) every part of the sample is subjected to two different separations, (ii) equal percentages (either 100% or lower) of all sample components pass through both columns and eventually reach the detector, and (iii) the separation (resolution) in the first dimension is essentially maintained" (Marriot et al. 2012). Each of these criteria is important, as MD separations that do not meet any one of them cannot be considered comprehensive (although this does not mean that they are not fit for purpose). As will be pointed out further, many 2D separations reported in literature fail to meet one or more of these requirements, most often due to failure to satisfy criterion (iii) above, and therefore are not comprehensive by nature.

Several different modes of HPLC can, and have been combined in LC × LC: ion-exchange chromatography (IEC) × reversed-phase liquid chromatography (RP–LC) or size exclusion chromatography (SEC) × RP–LC have been used for many years in the analysis of proteins and peptides (Opiteck and Jorgenson 1997; Wagner et al. 2000; Washburn et al. 2001; Wolters et al. 2001), while the comprehensive hyphenation of SEC and RP–LC, liquid chromatography under critical conditions (LCCC) or normal phase liquid chromatography (NP–LC) have found extensive application in polymer analysis (Balke and Patel 1983; Malik and Pasch 2014). For low to moderate molecular weight (MW) neutral and weakly acidic or basic organic molecules, however, NP–LC × RP–LC (Dugo et al. 2004; François et al. 2006) and especially RP–LC × RP–LC (François et al. 2008) are becoming more popular because of the molecular properties of the target analytes and the improved separation performance provided by these modes compared to IEC and SEC. In recent years, hydrophilic interaction chromatography has also found increasing application in LC × LC, especially in combination with RP–LC. Although hydrophilic interaction liquid chromatography (HILIC) has been in use since the 1970s (Linden and Lawhead 1975; Neue 1997), the term HILIC was only coined in 1990 (Alpert 1990). HILIC is an aqueous variant of NP–LC, where compounds are separated based on their partitioning into a water-rich layer immobilized on the surface of polar column packings (Neue 1997; McCalley and Neue 2008); retention therefore increases with analyte hydrophilicity. In fact, the suitability of HILIC for the analysis of highly polar molecules, which are often insufficiently retained in RP–LC, is considered one of its most attractive features. HILIC has also been shown to provide improved sensitivity for some analytes in electrospray ionization (ESI)–MS (Periat et al. 2013). The fact that HILIC offers an alternative separation mechanism, and therefore selectivity, compared to RP–LC, makes HILIC × RP–LC such a promising approach for organic analysis. It is therefore not surprising that the number of reports utilizing HILIC × RP–LC has grown significantly during the past decade (Figure 6.1). The technique has found especially widespread application in the fields of proteomics and natural product analysis.

In this chapter, we aim to present a short overview of the current state-of-the-art in HILIC × RP–LC analysis. Particular emphasis will be placed on the fundamental aspects relevant to maximizing performance gains in HILIC × RP–LC separation. These will be used as basis for a proposed systematic approach for HILIC × RP–LC method development and the evaluation of "practical" performance. Second, practical aspects particular to the hyphenation of HILIC and RP–LC will be addressed, and recent instrumental trends pertaining to the hyphenation of these modes will be discussed. Finally, a brief overview of recent applications of HILIC × RP–LC in diverse research fields where the technique has found most use will be presented.

For the purposes of this work, the focus is solely on the "comprehensive" hyphenation of HILIC and RP–LC. Techniques such as heart-cutting, serial coupling, and parallel analysis using these modes will therefore not be addressed. Furthermore, the abbreviation HILIC × RP–LC is used in the current work in a transposable sense, to refer to their comprehensive combination with either mode used in ^{1}D or ^{2}D.

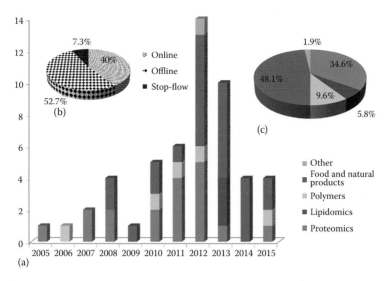

(a)

FIGURE 6.1   (a) Number of papers utilizing HILIC × RP–LC that have appeared in the literature over the past 10 years, (b) the proportion of these papers using on-line, off-line and stop-flow hyphenation modes and (c) the same papers grouped according to application areas. Data based on a Scopus search (October 4, 2015) over this period where HILIC and RP–LC were hyphenated in a comprehensive manner (offline, online, or stop-flow; both HILIC × RP–LC and RP–LC × HILIC are included). Reviews are excluded.

6.2   THEORY

6.2.1   PERFORMANCE OF 1D HPLC METHODS

Various metrics can be used to evaluate the performance of chromatographic methods. Most of these are limited to a specific pair of analyte peaks analyzed using a particular set of conditions, such as for example the resolution (R_s) or selectivity factor (α), or are limited to isocratic analysis, such as the number of theoretical plates (N). As a more generic metric of separation performance, the concept of peak capacity has gained significant traction. The peak capacity of a separation is defined as the number of compounds that can "theoretically" be separated using a given set of experimental conditions (column, gradient, flow rate, temperature, etc.) (Giddings 1967). The concept is equally valid under isocratic or gradient conditions, although it is mostly used in the latter context. The peak capacity of a 1D gradient separation ($n_{c,1D}$) can be calculated using the following relationship (Neue 2005):

$$n_{c,1D} = 1 + \frac{t_g}{\dfrac{1}{n}\displaystyle\sum_1^n w_b} \tag{6.1}$$

where:

   t_g is the gradient time (typically the time up to the flushing step of the gradient is used)

   w_b is the width of analyte peaks at the baseline (the latter averaged for n number of peaks)

For gradient HPLC separations, n_c values typically vary between 50 and 300, although peak capacities in excess of a thousand have been measured for long analysis times under optimized conditions (Shen et al. 2005).

It is important to note that peak capacity, despite its utility in comparing separation systems, is a theoretical concept—no real-life separation will provide separation of a number of components equal to the peak capacity of that separation, since analyte peaks are not equally distributed across the separation space. In what can be considered a worst-case scenario, Davis and Giddings (1983) showed that for complex samples, where analyte retention is approximated by a random distribution, the statistical theory of component overlap indicates that the peak capacity should exceed the number of components to be separated by a significant factor. These authors showed that for such "random" chromatograms, the number of pure (i.e., separated) chromatographic peaks will equal only 18% of the peak capacity of that separation (Davis and Giddings 1983). Put in another way, in order to resolve with 98% certainty n randomly distributed peaks, n_c should be $n \times 100$ (Giddings 1995). While it should be kept in mind that an experienced chromatographer should be capable of tuning analyte retention through judicious choice of experimental parameters, this is no longer possible for samples containing a very large number of components. For such samples, the findings of Davis and Giddings clearly provide incentive for improving the performance of chromatographic separations—in particular also for the interest in multidimensional separations.

In the context of 1D separations, it is relevant to briefly address how peak capacity may be optimized. In the case of gradient separations, peak capacity increases with the gradient time, although this increase does not follow a linear trend. For relatively short gradient times, higher peak capacities are obtained at higher flow rates (typically at or above the optimal mobile phase linear velocity). Furthermore, the peak production rate (i.e., the peak capacity divided by the analysis time) generally increases with a reduction in particle size (up to a point, the maximum peak capacity may not do so due to pressure constraints limiting the column length) and higher temperatures. Figure 6.2 illustrates examples of the dependence of 1D peak capacity on gradient time for several of the latest generation C18 columns. Values for $n_{c,1D}$ will vary as a function of column packing and dimensions, flow rate, temperature, and of course the analytes under investigation (De Villiers et al. 2009b), but the general shape of these curves is valid for 1D HPLC separations.

It is also worth noting that, in general, HILIC provides lower chromatographic efficiency, and therefore peak capacity, compared to RP–LC (Ikegami et al. 2008). The reasons for this are complex and likely related to the different retention mechanisms involved and their effect on mass transfer properties (Gritti et al. 2009, 2010; Gritti and Guiochon 2013a, 2013b). The lower efficiency of HILIC columns is partially offset by the fact that less viscous phases are commonly used; as a result the kinetic performance of HILIC and RP–LC phases may be similar (Song et al. 2014). Also, analogous to RP–LC, significant performance gains may be realized when working under UHPLC conditions in HILIC (Song et al. 2014).

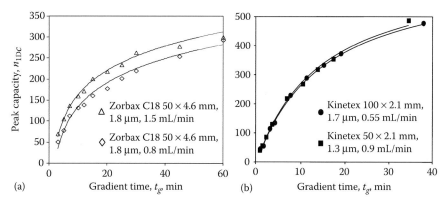

FIGURE 6.2 Variation in one-dimensional peak capacity ($n_{c,1D}$) as a function of gradient time (t_g) for the RP–LC separation of flavonoids. (a) Data obtained for cocoa procyanidins on a C18 column (50 × 4.6 mm, 1.8 μm) at 50°C. (From Kalili, K. M. and de Villiers, A., *J. Chromatogr. A*, 1289, 69–79, 2013a. With permission.) (b) Data obtained for anthocyanins on two superficially porous columns. Mobile phases: 7.5% formic acid in water (A) and 7.5% formic acid in acetonitrile (B); column temperature 60°C (unpublished results). Peak capacities were determined according to Equation 6.1 and fit to Equation 6.9 for both (a) and (b).

6.2.2 Performance of LC × LC Separations

Interest in LC × LC is driven by the gain in peak capacity that may "theoretically" be obtained. This can be conceptualized by considering the 2D space afforded by comprehensive 2D separations, from which the following oft-quoted relationship between comprehensive 2D ($n_{c,2D}$) and 1D chromatographic peak capacities in each dimension (denoted ^{1}D and ^{2}D, respectively) is derived (Karger et al. 1973; Guiochon et al. 1982, 1983; Giddings 1984, 1991)[*]:

$$n_{c,2D} = {}^1n_c \times {}^2n_c \qquad (6.2)$$

From this equation, the significant advantages of LC × LC are evident; what is less appreciated, however, is how hard this ideal is to obtain in practice. Two important constraints have to be considered in this regard: "orthogonality" and the sampling rate.

6.2.2.1 Orthogonality

The term orthogonality as used in MD chromatography refers to the degree of mutually exclusive information that may be obtained from each separation dimension. Simply put, if separation in two dimensions is based on different physicochemical properties, analytes not separated in ^{1}D may ideally be separated in the other, and vice versa. On the other hand, if the separation mechanisms used in both dimensions

---

[*] Throughout this work the nomenclature suggested by Schoenmakers et al. (2003); Marriot et al. (2012) is used to define terms in LC × LC, with superscripts indicating the relevant dimension.

are correlated to a significant degree, strongly retained compounds in ^{1}D will also be highly retained in the other, resulting in distribution of analyte peaks close to the diagonal of the 2D space, and thereby poor utilization of the potential benefit of LC × LC. A high degree of orthogonality is therefore required to effectively exploit the 2D separation space provided by LC × LC operation, and thereby provide maximum gains in resolution.

The concept of orthogonality is graphically illustrated in Figure 6.3. In Figure 6.3a, a hypothetical weakly orthogonal (or highly correlated) 2D system is shown, with peaks distributed along the diagonal. In contrast, in Figure 6.3b representing an orthogonal system, most of the separation space is utilized, and compounds co-eluting in ^{1}D are separated in the other. Figure 6.3c shows an example of an inversely correlated system, where weak retention in ^{1}D corresponds with strong retention in the other. Similar to the highly correlated system, this represents poor utilization of the 2D separation space. This example is shown here because, for some analytes and column/mobile phase combinations, HILIC × RP–LC can provide such an elution pattern, which is not surprising considering the respective retention mechanisms, but is clearly unwanted in the context of improved resolution.

While the concept of orthogonality intuitively makes sense, finding the means of accurately quantifying orthogonality in MD chromatography has proved challenging. This is evident from the large number and diversity of methods used to assess orthogonality in literature: geometric approaches to factor analysis (Liu and Patterson 1995), information theory metrics (Slonecker et al. 1996), various "box-counting" (Gilar et al. 2005; Davis et al. 2008a) and convex hull procedures (Burgman and Fox 2003; Semard et al. 2010), fractal dimensionality (Schure 2011), correlation coefficients (Al Bakain et al. 2011), selectivity approaches based on gradient parameters (D'Attoma et al. 2012), vector approaches (Dück et al. 2012), and others. Each of these metrics has certain inherent limitations, for example when faced with "nonrandom" analyte distribution in two dimensions, and there is as yet little consensus in literature regarding the most accurate metric to estimate orthogonality (this, together with the fact that orthogonality is not universally considered in the estimation of practical performance, severely hampers the comparison of LC × LC performance between studies).

Nevertheless, recent studies comparing several of these metrics for real-life and simulated analyte distributions provide some indication of the most reliable

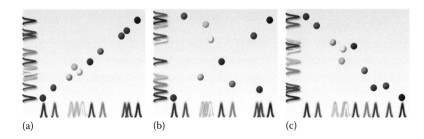

(a)   (b)   (c)

FIGURE 6.3 Graphical representation of the distribution of analyte peaks in the two-dimensional space for each of the following cases: (a) high correlation, (b) high orthogonality, and (c) inverse correlation.

techniques (Gilar et al. 2012; Rutan et al. 2012). In particular, approaches based on determining the effective area of the MD separation space that may be occupied by peaks, the so-called surface coverage (SC) or fractional coverage metrics, have been shown to be the most widely applicable and accurate metrics. One of the attractive features of these approaches is that, in addition to providing a numerical value to quantify orthogonality, they may also be used to quantify the effect of correlation between the two dimensions to determine practical 2D peak capacity, $n'_{c,2D}$

$$n'_{c,2D} = {}^1n_c \times {}^2n_c \times f_c \qquad (6.3)$$

where f_c is the fractional SC of the 2D separation space. f_c can be determined in a number of ways; in this work we will briefly outline the minimum convex hull (CH) approach (Semard et al. 2010), since this has been suggested to be suitable for the purposes of determining practical peak capacities (Rutan et al. 2012). In essence, a CH is the minimum polygon surrounding all data points in the 2D separation space, and is obtained by the Delaunay triangulation (Semard et al. 2010). The fractional SC is then obtained by dividing the area of the CH by the total area of the separation space. This procedure can easily be programmed in a suitable software program such as MATLAB®, and requires as input only the range-scaled retention times of compounds in both dimensions. As an example, Figure 6.4 graphically compares SC values obtained using the CH and "box counting" approaches for the HILIC × RP–LC analysis of anthocyanins (Willemse et al. 2014).

HILIC × RP–LC has elicited significant attention in large part due to the apparent high degree of orthogonality of this combination. While this is generally valid (Gilar et al. 2005), it is relevant to note that the separation mechanisms for these modes have some aspects in common (i.e., both dominated by partitioning of analytes into a phase of similar polarity). This may result in the opposite elution order in HILIC compared to RP–LC for particular compound classes. In such cases, compound retention show

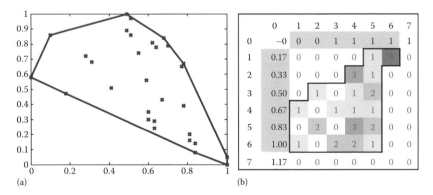

FIGURE 6.4 Examples of surface coverage (SC) plots obtained using (a) the convex hull method (SC_{CH}) and (b) the "box-counting" (SC_{BC}) method for the HILIC × RP–LC analysis of grape skin anthocyanins. The respective surface coverage values are SC_{CH}: 0.482 and SC_{BC}: 0.639. (From Willemse, C. M. et al., *J. Chromatogr. A*, 1359, 189–201, 2014. With permission.)

an inverse correlation between the ^{2}Ds (illustrated exaggeratedly in Figure 6.3c), and the orthogonality is clearly not very high. Yet, certain orthogonality metrics, such as correlation coefficients or the geometric approach to factor analysis (Liu and Patterson 1995), will still indicate a high degree of orthogonality. Caution should therefore be exercised in applying these metrics to evaluate orthogonality in HILIC × RP–LC.

6.2.2.2 ^{1}D Sampling Rate

The ^{1}D sampling rate refers to the time used to collect each ^{1}D fraction relative to the peak width in this dimension. This aspect is essential to satisfy the third criterion for comprehensive separations, that is, that the separation in the ^{1}D be "essentially" maintained (Schoenmakers et al. 2003; Marriot et al. 2012). Note that this definition is purposefully vague, although the authors did qualify the definition somewhat by suggesting that the reduction in ^{1}D peak capacity should not exceed 10% in practice. As will be shown below, extensive recent research has provided the means to quantify the effect of undersampling in comprehensive chromatography, which allows more objective evaluation of the degree to which MD methods meet this criterion to be considered comprehensive.

In simple terms, ^{1}D undersampling refers to the situation where sampling times are too long relative to the ^{1}D peak widths—this results in the recombination of peaks that were partially separated in ^{1}D during the fraction collection time. The performance (peak capacity) of the ^{1}D separation is therefore sacrificed under these conditions. Clearly, this phenomenon should be taken into account in the determination of the practical performance of a 2D separation (cf. Equation 6.2).

Murphy and coworkers (1998) were the first to study the effect of the sampling rate on resolution in comprehensive chromatography. Their conclusion that maximum resolution is achieved when each ^{1}D peak is sampled at least three times has become the standard in the field. Subsequent work by Seeley (2002) confirmed this conclusion. In later work, Horie et al. (2007) concluded that the best compromise between sampling rate and ^{2}D analysis time (and by extension peak capacity) in online LC × LC is achieved when each ^{1}D peak is sampled at least twice. Lower optimal sampling rates have since been found in several studies (Potts et al. 2010; Gu et al. 2011).

Davis and co-workers took this work further, by deriving an empirical undersampling correction factor, β, based on the application of statistical overlap theory to comprehensive 2D data (Davis et al. 2008b; Li et al. 2009). The β value can be used to account for ^{1}D undersampling in the calculation of practical 2D peak capacity according to

$$n'_{c,2D} = \frac{{}^1n_c \times {}^2n_c}{\beta} \tag{6.4}$$

where β is defined as

$$\beta = \sqrt{1 + 3.35 \frac{{}^1t_s \times {}^1n_c}{{}^1t_g}} \tag{6.5}$$

^1t_s is the sampling time and ^1t_g the ^{1}D gradient time. Note that in the original work (Li et al. 2009), which dealt with online LC × LC, the ^{2}D cycle time, ^2t_c, was used instead of ^1t_s in the above equation. Equation 6.5 is more generally applicable, since ^1t_s determines the degree of undersampling in all modes of LC × LC (in online operation, $^1t_s = {^2t_c}$).

By combining Equations 6.2 through 6.4, the practical peak capacity for a 2D separation ($n'_{c,2D}$) taking into account both the effects of finite orthogonality and ^{1}D undersampling may be determined (Stoll et al. 2008):

$$n'_{c,2D} = \frac{^1n_c \times {^2n_c} \times f_c}{\beta} \tag{6.6}$$

This equation therefore provides an estimate of the peak capacity achievable in practice. While it should always be kept in mind that the numerical value for peak capacity does not equal the number of compounds that will be separated, it is imperative that estimates of the 2D peak capacity account for both finite orthogonality and undersampling. Practical peak capacity values obtained in this manner are commonly much lower than theoretical values obtained using Equation 6.2. It is therefore unfortunate that 2D peak capacities reported in literature are determined by correction for one, both or none of these factors—a consequence of which is that it is very hard to compare practical performance between methods.

6.3 PRACTICAL ASPECTS AND HYPHENATION MODES OF LC × LC

The performance gains of LC × LC come at the cost of increased operational complexity; the degree to which both performance and complexity increase is dependent on the hyphenation mode. LC × LC separations can be performed in one of three different modes: offline, online, and stop-flow (also referred to as stop-and-go). The practical aspects relevant to the operation of LC × LC in general for each of these modes will be briefly outlined below.

6.3.1 OFFLINE LC × LC

The simplest way to perform an LC × LC experiment is by using the offline approach. In offline LC × LC, fractions from the ^{1}D are collected and subsequently injected onto the ^{2}D column. Because fraction collection and ^{2}D analyses are performed independently, the time allotted to the ^{2}D separation is independent of the sample collection time. As a result, longer separations may be used in ^{2}D to provide higher peak capacities in this dimension (*cf.* Figure 6.2). At the same time, short sampling times may be used to avoid ^{1}D undersampling. For both these reasons, offline LC × LC provides the highest practical peak capacity of all modes. However, this performance comes at the cost of exorbitant analysis times (up to in the order of several days). Another potential drawback of offline LC × LC is the risk of sample alteration and/or degradation during fraction collection or the time that elapses before the analysis of fractions in ^{2}D.

An offline LC × LC experiment can be performed with a single HPLC instrument by executing both ^{1}D and ^{2}D analyses on the same instrument at different times. Fraction collection can be performed manually, using a fraction collector (various degrees of automation are possible in this case), or even using instrumentation that allows collection of fractions in autosampler trays (Sneekes et al. 2007).

6.3.2 ONLINE LC × LC

In online LC × LC, fractions are continuously collected and analyzed in ^{2}D. The analysis of each ^{1}D fraction is completed in ^{2}D while the subsequent fraction is being collected. Since fraction collection times should be minimized to avoid excessive ^{1}D undersampling, this implies that short ^{2}D analyses are essential in online operation. As a consequence, the peak capacity in this dimension, and of online LC × LC separations in general, is normally significantly lower than in offline operation: Guiochon and coworkers estimated the practical limit at a peak capacity of ~10,000 (Horváth et al. 2009b). However, this performance is achieved in a much shorter time, typically commensurate with 1D HPLC analysis times.

Automated collection and transfer of fractions is most commonly performed using a 2-position valve equipped with two storage loops (Figure 6.5a). The valve is switched according to the sampling time, allowing alternate collection of a fraction and analysis of the previously collected fraction in the same time period. The sample loops may be chosen such that their volumes match or exceed the fraction volume. In the latter case, filling of the loop with the initial mobile phase composition to be used in ^{2}D allows some dilution of fractions with weak mobile phase, which might prove beneficial to minimize injection effects (see below). The minimum experimental requirement to perform online LC × LC is therefore one complete HPLC instrument (including as a minimum a pump, injection device, and detector), a valve, and a second pump to perform the ^{2}D analysis.

Alternative configurations are also possible, for example by adding additional ^{2}D column(s) and pump(s) and/or detector(s) (Opiteck and Jorgenson 1997; Opiteck et al. 1998; Unger et al. 2000; Wagner et al. 2000; Wagner et al. 2002; Haefliger 2003; Tanaka et al. 2004; François et al. 2008) (Figure 6.5b). Guiochon's group reported a theoretical study outlining the potential advantages of increasing the number of ^{2}D columns in online LC × LC (Fairchild et al. 2009a). In essence, this has the effect of extending the time available for each ^{2}D analysis and therefore provides higher peak capacities. There is however a severe price to pay in terms of instrumental complexity, which increases with the number of ^{2}D columns. Another approach is to use trapping columns and make-up flow to dilute each fraction before collection on the trapping column to avoid/minimize solvent mismatch effects (see further) (François et al. 2009; Li et al. 2012) (Figure 6.5c).

6.3.3 STOP-FLOW LC × LC

Stop-flow LC × LC can be considered as an automated version of offline LC × LC, with the important difference that the ^{2}D flow is stopped following direct transfer of a fraction to the ^{2}D column. The effect of this is to de-couple the sampling time from the ^{2}D analysis time, which makes this approach more similar to offline than online LC × LC. Because more time is then available for ^{2}D analysis, peak capacity in this dimension, and of stop-flow LC × LC, is similar to the offline approach. The total analysis time is also similar, and much longer than online LC × LC. An important aspect unique to stop-flow operation is the potential for additional ^{1}D band broadening during the stop-flow periods. However, this aspect has been shown to be of relatively little importance, at least in the case of moderate MW compounds (Bedani et al. 2006; Fairchild et al. 2009b; Kalili and de Villiers 2013a).

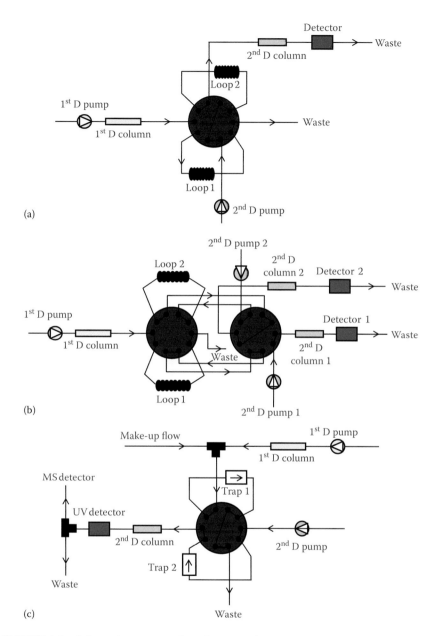

FIGURE 6.5 Schematic representation of common instrumental configurations for online LC × LC. (a) Basic configuration involving 2-position 10-port switching valve equipped with two trapping loops, (b) dual second dimension column configuration with two switching valves (From François et al., *J. Chromatogr. A*, 1178[1–2], 33–42, 2008. With permission.), and (c) configuration incorporating two trapping columns (From Li et al., *J. Chromatogr. A*, 1255, 237–243, 2012. With permission.). An optional splitter or make-up flow can be installed between the first dimension column and the valve(s) in each of the configurations, and additional detectors are optional.

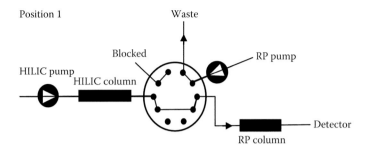

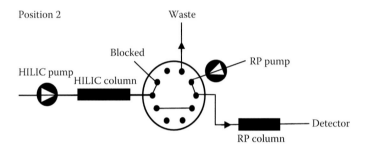

FIGURE 6.6 Schematic representation of a typical instrumental configuration for stop-flow LC × LC using a 10-port switching valve. An optional splitter can be installed between the first-dimensional column and the valve.

Instrumental requirements for stop-flow operation are identical to online LC × LC. A typical instrumental configuration for stop-flow LC × LC is shown in Figure 6.6 (Kalili and de Villiers 2013a). In valve position 1, a fraction from the ^{1}D column is transferred directly to the ^{2}D column, while the ^{2}D flow is directed to waste. In valve position 2, the ^{1}D flow is stopped while the column is connected to a blocked port; at the same time the previously transferred fraction is analyzed in the ^{2}D using the ^{2}D pump. This process is repeated until each ^{1}D fraction has been analyzed in ^{2}D. Special care should however be taken with the timing of valve switching and stopping the ^{1}D flow (preferably this should be done using the same instrument).

6.3.4 MOBILE PHASE CONSIDERATIONS IN HILIC × RP–LC
AND IMPLICATIONS FOR COLUMN DIMENSIONS

In HILIC × RP–LC analyses, one of the primary practical constraints stems from the fact that the HILIC mobile phase is effectively the sample solvent in ^{2}D. Since HILIC mobile phases as a rule contain high percentages of organic modifier, they also have high elution strength in RP–LC (and vice versa). As a consequence, injection band broadening is a significant concern in HILIC × RP–LC. The most common approach to minimize the effects of this constraint is to reduce the ^{1}D column internal diameter (i.d.) and flow rate. For example, considering common HILIC mobile phases and a standard 4.6 × 50 mm ^{2}D RP–LC column, the maximum permissible injection volume to avoid band broadening is somewhat less than 5 µL. For typical sampling times of less than

5 min, this implies that the ^{1}D flow rate should be in the order of less than 1 µL/min, which corresponds to the optimal mobile phase flow rate for columns of i.d. <0.2 mm.

One of the main advantages of offline HILIC × RP–LC is the flexibility that this mode offers in terms of allowing alteration of the sample solvent. The practitioner has three options to overcome injection band broadening issues: (1) inject only a small volume of each collected fraction, (2) diluting the collected fractions in a weak solvent to inject larger volumes, or (3) evaporation of the HILIC mobile phase followed by dissolution of the sample in a suitable solvent. Options 1 and 2 imply some loss of sensitivity, whereas option 3 can be time-consuming and increase the risk of sample alteration. A recent study reported the use of make-up counter-gradient in heart-cutting and offline HILIC × RP–LC analysis (Stevenson et al. 2014). This allowed normalization of the sample solvent composition to improve injection in ^{2}D, although in our view it has little utility in offline HILIC × RP–LC, since fraction solvents can be altered with much more flexibility through evaporation or dilution with aqueous solvent.

Because fractions are directly transferred to the ^{2}D column in online HILIC × RP–LC, injection band broadening is much more of a concern in this mode compared to offline operation. Again, the most common approach to minimize injection band broadening in online LC × LC is to reduce the column i.d. and therefore the flow rate in the ^{1}D. Operation of such narrow-bore columns does however place severe demands in terms of delay volume, pumping accuracy, and extra-column band broadening on commercial HPLC instrumentation (Lestremau et al. 2010). In cases where capillary LC instruments are not available, researchers often resort to using 1 or 2.1 mm i.d. columns at suboptimal flow rates. Increasing the i.d. of the ^{2}D column is of course also an option to reduce injection broadening, although this is less common due to increased dilution and very high flow rates in the ^{2}D which are incompatible with MS and lead to unacceptable levels of solvent consumption (Schoenmakers et al. 2006; Vivó-Truyols et al. 2010).

Other approaches to address this problem include incorporation of a flow splitter between the ^{1}D and ^{2}D columns (Fairchild et al. 2009b; Filgueira et al. 2011; Kalili and de Villiers 2013a, 2013b), which comes with the obvious limitation of reduced sensitivity. Note though that this methodology still meets the second criterion for a MD separation to be considered comprehensive (see Section 6.1), provided that the split ratio (SR) remains constant. An alternative which is gaining more traction is the use of trapping columns between the ^{2}Ds (Figure 6.5c) (Mihailova et al. 2008; Li et al. 2012; Wang et al. 2013). In the case of HILIC × RP–LC, RP-type traps are typically used, with an additional pump providing a make-up flow of an aqueous diluent prior to the trap(s) to facilitate trapping of analytes. Analytes are then released by backflushing the trap with the ^{2}D column downstream. Instrumental design and optimization of such systems is necessarily more complicated, although the benefits gained in terms of flexibility with regard to column dimensions and increased sensitivity are significant.

Since fractions are transferred directly to the ^{2}D column in stop-flow LC × LC, similar approaches to those used in online LC × LC may be utilized to reduce ^{2}D injection band broadening, namely, splitting the ^{1}D flow (Fairchild et al. 2009b; Kalili and de Villiers 2013a) and trapping analytes on a short (guard) column with an aqueous make-up flow (Wang et al. 2015).

Generally speaking, the i.d. of the ^{1}D column is smaller than that of the ^{2}D column in LC × LC (Schoenmakers et al. 2006). This is to minimize injection

band-broadening and artifacts, and is especially critical in the case of HILIC ×
RP–LC for the reasons outlined above. Also in offline HILIC × RP–LC analysis,
the use of a narrow-bore column is beneficial, as this coincides with a reduction in
the volumetric flow rate and therefore the volume of each collected fraction. As a
consequence, sensitivity loss due to dilution or injection of small volumes in ^{2}D is
reduced, and evaporation of fractions is also facilitated. Narrow bore columns also
reduce on-column dilution in the ^{1}D, although this is offset by the lower capacity
of these columns. On the other hand, where preparative isolation of compounds by
HILIC × RP–LC is required for example for nuclear magnetic resonance (NMR)
structural elucidation, (semi-) preparative columns can be used in both dimensions in
offline mode, with solvent evaporation performed before ^{2}D analysis (Fu et al. 2012).

In ^{2}D, column capacity is an important consideration to minimize injection band
broadening. Furthermore, because analysis speed in the ^{2}D is of paramount importance,
especially in online LC × LC, relatively short columns are preferred. As a consequence,
short, wide-bore (i.e., >2.1 mm i.d.) ^{2}D columns are typically used in HILIC × RP–LC
to address both speed and capacity constraints. To overcome the loss in efficiency associated with reduction in column length, a range of options are available. First, the use of
UHPLC instrumentation and columns packed with small particles is common due to
the improved performance of these systems for fast separations (De Villiers et al. 2006;
Schoenmakers et al. 2006; Vivó-Truyols et al. 2010; Sarrut et al. 2014). Recently, the
application of superficially porous phases has also found application in HILIC × RP–LC
due to similar advantages of these phases (Willemse et al. 2014). A complementary
approach is to use elevated temperature in ^{2}D (Yan et al. 2000; Schellinger et al. 2005;
Stoll et al. 2008): high temperature leads to an increase in the optimal mobile phase
velocity due to faster mass transfer and therefore is an effective means of increasing
analysis speed on the same column (Heinisch and Rocca 2009; Teutenberg 2009). While
the use of temperatures significantly above ambient does require dedicated instrumentation, significant speed gains can still be obtained at temperatures up to 80°C on commercial instruments, provided that careful optimization of experimental parameters is
performed. Finally, the use of monolithic columns has also found application in LC × LC
due to the high permeability of these phases (Tanaka et al. 2004; Bedani et al. 2006). The
very high flow rates that can be used on these columns significantly reduce instrumental
delay time and column re-equilibration times. However, it should be noted that for very
fast separations, packed columns generally provide much better separation performance
than monolithic columns (Desmet et al. 2006).

6.4 LC × LC METHOD DEVELOPMENT

Compared to 1D HPLC, method development in LC × LC is much more complex.
This is due to the complex interplay of a large number of experimental variables that
determine ultimate performance in LC × LC. Unlike the more mature technique of
comprehensive 2D GC (GC × GC), LC × LC is currently still largely a research tool
(the first commercial LC × LC instrument has only recently been released), and general method development strategies are therefore still lacking. Nevertheless, some
seminal papers on the topic provide some useful guidelines (Bedani et al. 2006;
Schoenmakers et al. 2006; Fairchild et al. 2009b; Vivó-Truyols et al. 2010; Bedani

et al. 2012; Gu et al. 2011). For an excellent recent overview of method development in online LC × LC, the reader is referred to Carr et al. (2012).

The first report on LC × LC method development comprehensively covered most experimental parameters of relevance in online LC × LC (Schoenmakers et al. 2006). In this work, input parameters required were the maximum pressures in both dimensions, the maximum ^{1}D analysis time (roughly equal to the total analysis time) and the minimum ^{1}D column diameter. The optimal kinetic performance in both dimensions was derived for different column packings using "Poppe" plots (Poppe 1997), and this information together with a sampling rate criterion was used to obtain the optimal column lengths and particle sizes for the maximum 2D peak capacity in the minimum time. The final step then involved selection of column diameters based on mobile phase flow rates to minimize injection band broadening and dilution. The applicability of this approach was demonstrated for RP–LC × SEC. Orthogonality of the two separation dimensions was not taken into account in Schoenmakers' work, although this can be considered a constant for a given combination of separation modes and is often neglected in LC × LC method development for this reason. Unfortunately, the authors used far too strict criteria for the sampling rate in this paper: $^1t_s = ^1\sigma$ was used, instead of the more common $^1t_s = 2$–$4^1\sigma$ based on the criteria proposed by Murphy and coworkers (1998). Therefore, the conclusions drawn in this work are qualitatively correct, but not necessarily quantitatively so.

Subsequently, Bedani et al. (2006) reported an alternative method development approach for the SEC × RP–LC analysis of peptides. Their starting point was an established ^{2}D analysis, and optimization involved the comprehensive hyphenation of a ^{1}D separation to provide the required 2D peak capacity. A general methodology to do so was derived based on van Deemter plate-height data. Initial findings indicated a very low mobile phase linear velocity in ^{1}D, which led the authors to the development of a stop-flow SEC × RP–LC methodology (this work included an elegant approach to determine additional ^{1}D band broadening due to stop-flow operation).

Guiochon and coworkers (Fairchild et al. 2009b) used another method based on a fixed ^{1}D separation to compare the performance of online, stop-flow, and offline LC × LC. These authors used accepted undersampling criteria and linear solvent strength (LSS) calculations to determine the effective performance as a function of ^{2}D column length, mobile phase composition, and ^{1}D analysis time. In a separate study, the same group reported an optimization study for offline LC × LC (Horváth et al. 2009a). A stepwise procedure was proposed, involving first of all estimation of the target 2D peak capacity and determination of the variation in peak capacity as a function of gradient time on the ^{2}D column selected. Next, a target peak capacity in ^{1}D is determined based on the total analysis time and target 2D peak capacity. This is then followed by selection and optimization of the ^{1}D separation and determination of the optimal sampling time. The final step then involves optimization of the ^{2}D separation to meet the established sampling criteria (Horváth et al. 2009a).

Vivó-Truyols et al. (2010) reported arguably the most comprehensive optimization scheme for online LC × LC. The methodology is based on the use of pareto-optimality (Massart et al. 1998) to obtain optimal performance descriptors—identified as 2D peak capacity, total analysis time, and detection limits—as a function of the important chromatographic parameters such as particle sizes, flow rates, column dimensions in both dimensions, and modulation period. Significantly, the approach also incorporated the effects of ^{1}D undersampling (treated as detection band broadening in this dimension) and injection band broadening due to large fraction volumes, which combined were shown to result in losses of up to 75% of the theoretical peak capacity under nonoptimal conditions. An example of the information obtained in this study is presented in Figure 6.7. Figure 6.7a illustrates the 2D peak capacity and dilution factors obtainable as a function of analysis time, whereas Figure 6.7b shows isodilution lines (i.e., lines of the peak capacity obtainable for a given analysis time and a constant total dilution factor) for the indicated dilution factors. These values were obtained assuming no focusing in ^{2}D. The authors also showed that when

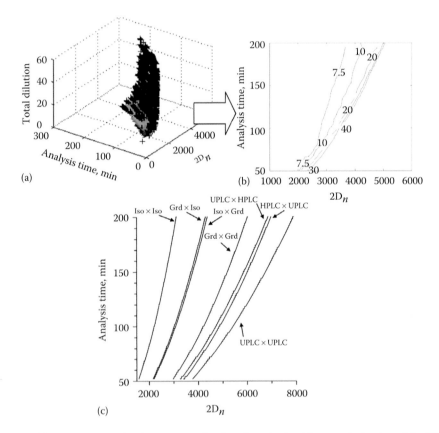

FIGURE 6.7 (a) Example of a pareto-optimal surface obtained from optimization of the 2D peak capacity, total analysis time, and total dilution for online LC × LC using gradient elution in both dimensions. (b) Shows the isodilution lines (dilution factors indicated by numbers) corresponding to the data in (a). (c) Pareto fronts illustrating the total peak capacity obtainable as a function of total analysis time for various online LC × LC systems. Abbreviations: Grd and Iso refer to gradient and isocratic separations, respectively; HPLC and UPLC indicate operation at pressures up to 400 and 1000 bar, respectively (using gradients in both dimensions). (Reproduced with permission from Vivó-Truyols, G. et al., *Anal. Chem.*, 82, 8525–8536, 2010.)

focusing is possible in ^{2}D, dilution can effectively be decreased without sacrificing peak capacity or analysis time. The pareto-optimality approach was also applied to study the effect of gradient and isocratic separations and elevated pressures in LC × LC. The results are summarized in Figure 6.7c, which clearly indicates that better performance is obtained in gradient mode and utilizing UHPLC instrumentation (Vivó-Truyols et al. 2010).

Carr and coworkers reported a practical approach to peak capacity optimization in online LC × LC (Gu et al. 2011). Two optimization schemes were evaluated for hypothetical samples comprising analytes with different retention properties. In the first case, the ^{1}D separation was optimized initially in terms of gradient time and profile, column length and flow rate using gradient Poppe plots (Wang et al. 2006b, 2006c). In ^{2}D, a fixed column length was used and the sampling time was then optimized using the experimentally determined relationship between the ^{2}D cycle time and peak capacity (see Equation 6.9) as well as the effect of sampling time on the degree of undersampling. The second approach involved single-step optimization involving five parameters for both dimensions simultaneously using the Solver function in Microsoft Excel. As input, a range of instrument, column, and analyte parameters were used to determine the performance in each dimension. This work did not consider dilution factors, mobile phase incompatibility, or SC for the optimization.

Heinisch and co-workers (D'Attoma et al. 2012; D'Attoma and Heinisch 2013) reported an alternative method to optimize and compare the online combination of various RP–LC and HILIC methods for the analysis of pharmaceuticals and peptides. The authors proposed a new metric for the measurement of orthogonality based on retention SC that employs LSS parameters (D'Attoma et al. 2012). Using this measure of orthogonality, peak capacities determined from measured isocratic efficiencies and LSS parameters in each dimension, as well as calculated dilution factors, the authors compared various LC × LC systems in terms of practical peak capacity and orthogonality. In this approach, sufficiently fast sampling times (and therefore sufficiently short ^{2}D analysis times) to neglect undersampling were used. Interestingly, the authors found that a high degree of orthogonality (as measured by the derived metric), while essential to obtain high practical peak capacity, did not guarantee high practical peak capacities for the systems under investigation.

From the above brief overview, it is evident that LC × LC method development is far from a straightforward task. In fact, in the majority of studies concerning method development, the task is simplified by pre-selecting certain experimental conditions in one of the dimensions. While previously reported method development strategies provide guidelines for parameter selection in LC × LC, the conclusions drawn are either relevant to specific column/sample combinations (such as RP–LC × SEC of polymers), or apply to LC × LC in general. HILIC × RP–LC separations do have unique constraints, primarily in terms of mobile phase compatibility between the two dimensions, which may somewhat alter the general conclusions drawn from these studies.

Therefore, in the following brief outline of method development, we adapt an alternative approach where practically feasible column dimensions are selected and evaluated in both dimensions as the first step in method optimization. This is done to accommodate the unique constraints placed on experimental conditions by the inherent nature of the mobile phases used in HILIC and RP–LC. Method development for each of the three hyphenation modes will be outlined, based loosely on our previous work on the development and optimization of HILIC × RP–LC methods for the analysis of phenolic compounds (Kalili and de Villiers 2013a, 2013b).

It is relevant to point out here the differences between the optimization scheme outlined below, termed a practical approach for HILIC × RP–LC method development, and a "true" optimization protocol such as derived for online LC × LC by Vivó-Truyols et al. (2010). In the latter case, the use of pareto-optimality ensures that simultaneous optimization of several parameters such as column lengths and diameters, particle sizes, and mobile phase velocities in both dimensions can be performed. Dilution factors and band broadening due to injection of large volumes in ^{2}D are also taken into account in the optimization scheme, the latter through incorporation of its effect on the ^{2}D performance. In contrast, in the optimization scheme outlined below, ^{2}D column dimensions are pre-selected based on practical considerations, and ^{1}D column dimensions are then chosen based on conditions where band broadening due to injection of HILIC solvents in ^{2}D "is entirely avoided." This rather strict condition is chosen because of the importance of injection band broadening in HILIC × RP–LC, and implies that the effect can essentially be ignored in method development (except for the consequences in terms of selection of other parameters to meet this criterion). In cases where injection effects are less dramatic, some loss in resolution due to this phenomenon can be incorporated in method development, as done by Vivó-Truyols et al. (2010). Furthermore, the fact that columns are pre-selected in our approach means that column length and particle sizes, beyond those included in the optimization, are not considered true variables—as they are in approaches based on pareto-optimality (or related methods such as Poppe plots and kinetic plots). While this has the effect of somewhat limiting the scope of the complete method optimization strategy, we feel that the findings are useful in the sense that they provide the best experimental conditions for the columns and instrumentation available.

6.5 A PRACTICAL APPROACH FOR HILIC × RP–LC METHOD DEVELOPMENT

Before addressing method development in detail, we note that the following discussion will focus on HILIC × RP–LC, that is, where HILIC is used in ^{1}D. However, most of the considerations outlined below apply equally in the case of RP–LC × HILIC. This is because of the relative mobile phase elution strengths in both dimensions: HILIC mobile phases are strong eluents in RP–LC and vice versa. The majority of LC × LC applications involving HILIC and RP–LC employ the former mode in the first dimension, with some notable exceptions (Liu et al. 2008; Feng et al. 2010; Wohlgemuth et al. 2010; D'Attoma et al. 2012; Fu et al. 2012; Xu et al. 2012; D'Attoma and Heinisch 2013; Guo et al. 2013). In fact, utilizing HILIC in the second dimension is potentially beneficial since this mode of separation is inherently compatible with high flow rates and fast analyses due to the low mobile phase viscosity (Appelblad et al. 2008; Song et al. 2014), the mobile phases used in HILIC are better suited to some detection methods than is the case for RP–LC (MS, evaporative light scattering detection [ELSD], charged aerosol detection [CAD], for example) and injection of aqueous-soluble samples is easier when RP–LC is used in the first dimension. On the other hand, the efficiency of HILIC for very fast separations is often much worse than is the case for RP–LC.

6.5.1 Selection of 1D Stationary and Mobile Phases

Arguably the most important choice in LC × LC involves selection of the column and mobile phase combinations to be used in each dimension. As a practical rule, it is worthwhile spending some time on the selection of individual 1D separations, as this process is much simpler in one dimension than in LC × LC, and can translate into significant performance gains in the latter. (In other words, one cannot rely solely on LC × LC operation to provide the resolving power, since comprehensive combination of two weak methods will still provide poor results.)

Separation conditions should be selected based on both efficiency and selectivity criteria. Of these, the latter is likely most important initially, as this determines the orthogonality of the overall LC × LC separation. In this context, it is worth revisiting the concept of "sample dimensionality (s)" originally defined by Giddings (1995) as "the number of independent variables (structural factors that collectively yield molecular identity) that must be specified to identify components in a sample." Ideally, the separation and sample dimensions should be equal—in these cases ordered chromatograms should be obtained. For example, for a sample where $s = 1$ (such as, e.g., a mixture of linear hydrocarbons, where chain length is the only independent variable), multidimensional separation will provide zero benefit. Most complex mixtures however have much higher dimensionality, through variation in MW, degree of saturation, substituent nature and position, and so on, and will therefore benefit from multidimensional separation. Unfortunately, most such samples also have a dimensionality larger than two, and higher-order separations are much harder to realize (Moore and Jorgenson 1995; Schoenmakers et al. 2006).

The concept of sample dimensionality is extremely relevant in the selection of 1D separations for LC × LC: the goal should be to address different "sample dimensions" in each of the separation dimensions. Many examples of such LC × LC methods exist, notably in polymer analysis, where structured contour plots are often obtained for mixtures varying, for example, in end-group chemistry and chain length. To provide an illustrative example from our own work, which will form the basis of much of the results presented in this section, the natural phenolic compounds proanthocyanidins will be used (Figure 6.8). Proanthocyanidins are a family of oligomeric compounds that vary in terms of MW (increasing with the degree of polymerization [DP]) and isomeric composition (procyanidins are composed of the isomeric monomers [+] catechin and [−] epicatechin, while esterification with gallic acid at 3–OH and trihydroxylation of the B-ring add additional variation for proanthocyanidins). The simplest subclass of proanthocyanidins, the procyanidins, can be considered a 2D sample, with DP and isomeric composition (i.e., number and position of catechin and epicatechin units) being considered the "independent variables" (the linkage pattern of individual units, Figure 6.8, can arguably be considered an additional dimension). HILIC separation on a diol phase provides separation according to DP (sample dimension one), whereas RP–LC separates isomeric compounds based on differences in hydrophobicity, that is, the second sample dimension. These two separation modes therefore access different sample properties, and their combination

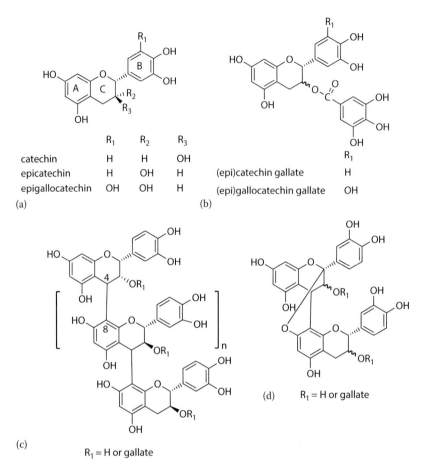

FIGURE 6.8 Structures of proanthocyanidins: (a) and (b) show the monomeric phenolic "building blocks" of the higher MW oligomers, the general linkage pattern of which is illustrated in (c) and (d).

in LC × LC provides more complete information about the sample than could be obtained using only one of them (Kalili and de Villiers 2009).

Looking specifically at column and mobile phase selection in each dimension, some general guidelines may be presented, although the ultimate choice will depend on the target analytes and the hyphenation mode (see further). Especially in the case of HILIC, the availability of a wide range of stationary phases nowadays, coupled to the significant effect that the stationary phase has on selectivity (Jandera 2011), makes the selection of a suitable phase challenging (Dejaegher et al. 2008). In our experience, it is preferable to evaluate a range of phases during initial method development. We have found, for example, that the same nominal type of stationary phase from different manufacturers often provides significantly different selectivity. The same applies to mobile phase selection, where of course sufficient retention and maximum resolution should be important criteria. High retention in HILIC is advantageous in that this allows the use of higher aqueous content in the mobile phase, which is beneficial from the perspective of hyphenation to RP–LC (this is also relevant for RP–LC × HILIC methods).

In the second dimension, speed and efficiency are important (even in offline and stop-flow HILIC × RP–LC because of the effect of the second dimension on the overall analysis time, see further). Also of relevance in terms of the stationary phase is to use columns with high retention properties to minimize injection band broadening issues. Depending on the ^{2}D column used, UHPLC instrumentation is often preferred to maximally exploit the column performance—in this dimension, the column length and flow rate are also typically chosen to allow operation close to the maximum system/column pressure, as this is where the maximum kinetic performance is attained (Desmet et al. 2006; Vivó-Truyols et al. 2010).

In the context of instrumentation, it is relevant to note that system volumes should be reduced as far as possible through reduction of connection tubing i.d. and length, as well as detector volumes. This is important in ^{1}D when a narrow-bore column is used to reduce extra-column band broadening, as well as in ^{2}D for the same reason and to reduce delay times.

6.5.1.1 Gradient versus Isocratic Separations

Both gradient and isocratic separations can be used in either of the two dimensions in LC × LC. For samples containing analytes with similar retention characteristics in the second dimension, isocratic separation in this dimension is an attractive option, especially in online LC × LC. This is beneficial in the sense that the need for column re-equilibration is avoided, thereby increasing the time available for analyte separation. However, peak capacities for short isocratic separations are typically very low, and elution of compounds beyond the sampling time will result in so-called "wrap-around," a phenomenon encountered more commonly in GC × GC where compounds elute in later modulation periods.

It has been shown that performance in online LC × LC increases in the sequence isocratic × isocratic < gradient × isocratic ~ isocratic × gradient < gradient × gradient (Vivó-Truyols et al. 2010) (Figure 6.7c). We will therefore limit our discussion to the case where gradient analyses are performed in both dimensions, as this is the most common practice. While the ^{1}D gradient is generally optimized in a similar manner as for 1D LC (i.e., to optimize resolution within a given analysis time), several different methodologies may be used in the second dimension. The most common approach is to use a fixed gradient for every ^{2}D analysis, with the beginning and end mobile phase compositions and gradient slope selected based on the analysis of the complete sample in this dimension. This approach, referred to as "full in fraction" (FIF) gradients (Jandera et al. 2010a), does suffer from the limitation that a large percentage of the time available in the second dimension is spent on re-equilibration of the column, thereby minimizing the time available for separation. In cases where the separation modes in both dimensions are partially correlated, such as RP–LC × RP–LC, it is often beneficial to vary the ^{2}D gradient throughout the first dimension separation (Cacciola et al. 2007; Bedani et al. 2009; Jandera 2012). Two different ways to do this can be distinguished: "segment in fraction" (SIF), which involves alteration of the start and end mobile phase compositions for each ^{2}D analysis according to the retention properties of the analytes eluting in that fraction, and "continuously shifting" (CS), where the second dimension gradient spans the entire ^{2}D analysis time. A graphical representation of each of these approaches is shown in Figure 6.9. The SIF approach often allows better utilization of the available second dimension time compared to FIF (at least for the case of LC × LC using correlated separation modes) and therefore enhanced peak capacities (Bedani et al. 2009; Jandera et al. 2010a; Leme et al. 2014). Also, the relatively steep ^{2}D gradient slopes used in this method provide sharp peaks, and allow some analyte focusing (depending on the mobile phases used). On the other hand, the CS approach uses a very flat gradient slope in the second dimension, resulting in broader peaks and a higher risk of injection band broadening. This is offset

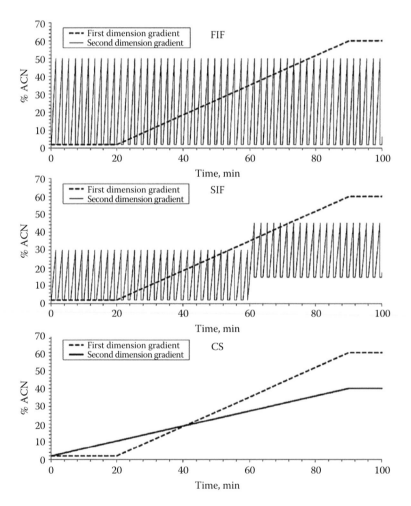

FIGURE 6.9 Schematic representation of gradient profiles used in the first (dotted lines) and second (solid lines) dimensions in LC × LC for full-in-fraction (FIF), segment-in-fraction (SIF), and continuously shifting (CS) modes. (Reproduced from Jandera, P. et al., *J. Chromatogr. A*, 1268, 91–101, 2012. With permission.)

by the fact that the available second dimension analysis time is more efficiently used since no time is required for column re-equilibration.

Note that in cases where ^{1}D peaks are sampled multiple times, and each fraction is analyzed in the second dimension using different conditions, the retention times will shift for each consecutive analyte "slice." A typical example is for overloaded peaks, where peak distortion in two dimensions reminiscent of GC × GC is observed (Stoll et al. 2015).

While the use of FIF, SIF, and CS gradients have been compared for RP–LC × RP–LC systems, to the best of our knowledge this has not been done for HILIC × RP–LC. In the latter case, the diametrically opposed retention orders would imply that reversed gradients would have to be used in the CS configuration, while the percentage organic modifier would systematically be reduced in the second dimension as a function of ^{1}D analysis time in SIF mode. Incorporation of the effects of shifted gradients on the performance in LC × LC is rather challenging, aside from the clear effect on the SC parameter. For this reason, and since SIF and CS methods are expected to provide minimal performance gains in HILIC × RP–LC, the following method optimization discussion will assume the use of the FIF gradient scheme only. In cases where the alternative gradient approaches are of interest, it is suggested that this be evaluated after method optimization according to the procedures outlined below.

6.5.2 SELECTION OF COLUMN DIMENSIONS

In HILIC × RP–LC, the choice of column dimensions should be governed first of all by the consideration that only small volumes of samples dissolved in HILIC mobile phases can be injected onto RP–LC columns if injection band broadening is to be minimized (the same applies if the two modes are swapped). While injection band broadening can be accounted for in method development based on chromatographic theory (Schoenmakers et al. 2006; Vivó-Truyols et al. 2010; Bedani et al. 2012), such an approach relies on parameters which are often hard to obtain for a wide range of analytes in real-life samples. Furthermore, this implies that injection band broadening in the second dimension is considered a variable in method optimization. We feel that this approach is susceptible to inaccuracies associated with assumptions made for a wide range of compounds. Since the contribution of this effect can indeed be very significant (reducing the peak capacity in the second dimension by up to 50% [Vivó-Truyols et al. 2010]), any potential inaccuracies will be amplified in the evaluation of LC × LC performance.

We therefore suggest that the ^{2}D column dimensions be selected first of all based on flow rate and analysis time considerations (Section 6.3.4). Note that these may vary between the different hyphenation modes: in general, longer ^{2}D columns can be accommodated in offline and stop-flow LC × LC compared to online operation. In all modes though, speed is an important criterion, and therefore a good starting point is to use sub-2 μm porous or superficially porous phases (Sarrut et al. 2014), preferably at temperatures above ambient to benefit from speed gains (Stoll and Carr 2005; Stoll et al. 2006). Elevated temperature also allows good separation at high flow rates, which significantly reduces the time required to re-equilibrate the column in gradient analysis (Yan et al. 2000; Thompson and Carr 2002; Stoll and Carr 2005; Schellinger et al. 2005; Stoll et al. 2006). Relatively wide-bore columns are also preferred in this dimension due to their higher capacity, although this should be bal-anced against the volumetric flow rate required, which may not be compatible with detectors such as MS. Furthermore, on-column dilution scales with the i.d. which negatively impacts on the overall sensitivity (Schure 1999; Horváth et al. 2009c). Typical column dimensions in the second dimension are ~30–150 × 2.1–4.6 mm (a notable exception is in proteomic research, where capillary columns are com-monly used in both dimensions hyphenated in offline mode). It is advisable to select several different column formats in the second dimension for method development.

Using the selected column(s), the maximum injection volume in the second dimension, $^2V_{max}$, should then be determined experimentally. This can be done by dissolving the sample in the HILIC mobile phases, and injecting various volumes onto the ^{2}D column(s) used in method optimization. $^2V_{max}$ is the maximum injection volume that does not lead to a reduction of chromatographic performance in the sec-ond dimension (note that this will vary with the HILIC mobile phases and analyte retention in the second dimension; the minimum value for $^2V_{max}$ is selected).

The next step is then selection of the ^{1}D column i.d.; this is done based on an ideal flow rate in this dimension. In LC × LC, the volume of each fraction (V_{frac}) is deter-mined by the sampling time and the ^{1}D volumetric flow rate (^1F):

$$V_{frac} = {}^1t_s \times {}^1F \qquad (6.7)$$

The perfect fraction volume will be equal to $^2V_{max}$, and from this value and a typical sampling time of 0.2–2 min, the ideal first dimension flow rate can be determined according to Equation 6.7. Knowing the range of ^1F, suitable first dimension col-umn diameters are selected where ^1F corresponds to the optimal mobile phase linear velocity. Obviously, such column dimensions are not always available or might be impractically small; in such cases the minimum practical i.d. should be selected.

To accommodate cases where $V_{frac} \neq {}^2V_{max}$, only a portion ($={}^2V_{max}$) of each fraction can be injected in the second dimension, and/or the flow rate in the first dimension can be reduced to operate below the optimal mobile phase linear velocity. The drawback of the former is a loss in sensitivity, and of the latter is a loss in ^{1}D resolution and an increase in analysis time; often a combination of both of these might provide the best compromise. Injection of only a portion of each fraction or dilution of each fraction is equivalent to using a flow splitter between the two dimensions, the SR of which is

$$SR = \frac{{}^1F \times {}^1t_s}{{}^2V_{max}} \qquad (6.8)$$

This equation emphasizes the importance of low ^{1}D flow rates in HILIC × RP–LC, where $^2V_{max}$ is generally small. (Note that SR can also be <1 in cases where fractions are evaporated and re-dissolved in smaller volumes in offline HILIC × RP–LC.) SR values are used in method optimization to account for the overall dilution of analyte peaks in HILIC × RP–LC, as will be outlined below.

The use of small particles is not essential in the first dimension, since speed is not an important criterion and narrow peaks will place severe restraints on sampling times. As a starting point therefore, conventional 3.5 or 5 μm phases in relatively long columns and using slow gradients may be used if short overall analysis times are not critical (Li et al. 2009). In fact, several studies have shown that increasing the gradient time in the first dimension generally results in higher practical performance of the overall 2D separation due to relaxation of sampling time constraints in especially online LC × LC (Horváth et al. 2009b; Huang et al. 2011; Carr et al. 2012).

Once the column dimensions have been selected, the performance in each dimension in terms of peak capacity (^1n_c and ^2n_c) can be determined using Equation 6.1 for all columns. Ideally, this should be done for several columns, flow rates, and gradient times in each dimension for a more complete optimization. In this case, experimental data of ^xn_c versus ^xt_g, that is, the peak capacity and gradient time in dimension x, can be fit to equations such as (Fairchild et al. 2009b)

$$^xn_{c,1D} = 1 + \frac{at_g}{b+t_g} \qquad (6.9)$$

where a and b are constants. From values of a and b determined from the fitting of Equation 6.9 to experimental data, ^xn_c can then be determined for any ^xt_g or ^xt_c without further experiments (Figure 6.2 presents an example of such experimental curves). Note though that such fitting should be performed for every column and flow rate under consideration (also in both dimensions if required), and that in some instances alternative equations to Equation 6.9 provide better fitting to experimental results (Li et al. 2009; Gu et al. 2011). It is also important to establish the relationship between the "cycle time" (t_c) and the gradient time in the second dimension. The relationship between these parameters is determined by the time required to sufficiently re-equilibrate the column before the next fraction is transferred ($t_{re\text{-}equil}$):

$$^xt_c = {}^xt_g + {}^xt_{re\text{-}equil} \qquad (6.10)$$

Consideration of Equations 6.2, 6.9, and 6.10 with Figure 6.2 clearly highlight the importance of minimizing the re-equilibration time, as this maximizes utilization of the time available for separation in the second dimension. In practice, the simplest way to reduce the re-equilibration time is to use a high flow rate, which is facilitated by elevated pressure (De Villiers et al. 2006) or temperature (Yan et al. 2000; Stoll and Carr 2005; Schellinger et al. 2005; Stoll et al. 2006) operation or, for example, the use of monolithic columns (Tanaka et al. 2004). The minimum re-equilibration time is determined practically as the minimum time required to provide reproducible results in the second dimension.

Equation 6.9 (or a similar function fit to experimental data of ^xt_g vs. ^xn_c) provides the relationship between performance and analysis time in each dimension which is essential in any method development strategy. Alternative approaches to obtain the same information include the use of Poppe plots or van Deemter coefficient data in isocratic (Schoenmakers et al. 2006) or gradient models (Gu et al. 2011), or calculated peak capacities based on gradient conditions used (Vivó-Truyols et al. 2010). We prefer the methodology outlined above, since this is based on experimentally measured peak widths for practically feasible column dimensions and therefore accurately reflects the performance in each 1D separation, also accounting for effects such as extra-column band broadening on a given instrument under actual experimental conditions.

6.5.3 SELECTION OF HYPHENATION MODE

The selection of the most suited hyphenation mode can be based on performance criteria such as overall peak capacity or analysis time, or on practical considerations such as instrumental availability. When insufficient instrumentation (i.e., lack of a second pump and valve) is available, offline HILIC × RP–LC is the only option, as this can be performed using a single HPLC instrument. Assuming availability of all required instrumentation, the choice of hyphenation mode will depend on the required resolving power (peak capacity) and/or available time (Fairchild et al. 2009b). It is important to note that in LC × LC, similar to 1D HPLC, significant performance gains can only be attained at the cost of longer analysis times. Therefore, in simple terms, if analysis time is important, online LC × LC should be used, as typical online LC × LC analysis times are in the same order of conventional HPLC methods. On the other hand, if maximum performance is required, offline or stop-flow methods are preferable. Figure 6.10 presents a typical example of the (practical) peak capacity obtainable as a function of analysis time for 1D RP–LC and online, offline, and stop-flow HILIC × RP–LC analyses. The choice between the latter two modes will depend on instrumental constraints (since stop-flow operation requires

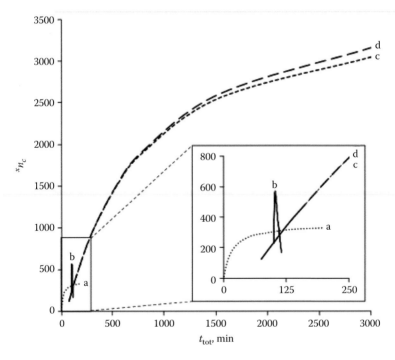

FIGURE 6.10 Comparison between peak capacities for the one-dimensional RP–LC (dotted line, curve a) and corrected two-dimensional peak capacities for online (solid line, curve b), stop-flow (dashed line, curve c), and offline (dashed line, curve d) HILIC × RP–LC analyses as a function of total analysis time. Data obtained for the HILIC × RP–LC analysis of procyanidins. (Reproduced with permission from Kalili, K. M. and de Villiers, A., *J. Chromatogr. A*, 1289, 69–79, 2013a.)

additional hardware and a more complex configuration) as well as additional criteria such as whether the first dimension mobile phase should be evaporated (offline would then be preferable) or if the target analytes are highly labile (in this case stop-flow operation might be the better option). The optimization of HILIC × RP–LC separations in each of these modes will be outlined below.

6.5.4   OFFLINE HILIC × RP–LC METHOD OPTIMIZATION

For offline HILIC × RP–LC analysis, the selection of column dimensions depends on whether the collected fractions are to be manipulated prior to their injection onto the ^{2}D column. If dilution with a weak solvent or evaporation of the HILIC mobile phase prior to dissolution in a weak RP–LC solvent is an option, the analyst is relatively flexible in the choice of column dimensions, and should be guided by the injection mass required in both dimensions. In cases where it is preferable to inject fractions directly without modification (such as where analyte stability is a concern and/or manipulation of fractions is unwanted), selection of column dimensions and evaluation of column performance should be performed as outlined in Section 6.5.2.

The starting point for method optimization is then a minimum of two sets of peak capacity data as a function of gradient time (and in the ^{2}D, cycle time). For each set of ^1n_c and ^2n_c values, ^1t_s is varied and the corrected 2D peak capacity calculated according to Equation 6.6 for each ^1t_s (refer to Section 6.2.2 for details on how to determine f_c and β). In most method development procedures, the finite orthogonality of the two separation modes is not taken into account, since any accurate quantitative measure of orthogonality should be roughly independent of the experimental parameters altered during method optimization. However, in cases where several methods differing in selectivity are evaluated, it is of course relevant to include this parameter in method optimization. Furthermore, doing so is beneficial in the sense that it provides a realistic measure of the performance of the LC × LC system as output, as limited orthogonality will have a significant effect on the practical performance (Huang et al. 2011).

To account for the dilution of compounds in LC × LC, the dilution factors in each dimension should be determined (Schure 1999; Schoenmakers et al. 2006; Horváth et al. 2009c; Vivó-Truyols et al. 2010; Bedani et al. 2012):

$$^1DF = \sqrt{2\pi}\, \frac{^1\sigma \times {}^1F}{^1V_{inj}} \tag{6.11}$$

$$^2DF = \sqrt{2\pi}\, \frac{^2F \times {}^2\sigma \times SR}{^1F \times {}^1t_s} \tag{6.12}$$

where:
^1DF and ^2DF are the dilution factors
$^1\sigma$ and $^2\sigma$ are the standard deviations in time units
^1F and ^2F the flow rates in dimensions one and two, respectively
$^1V_{inj}$ is the injection volume in ^{1}D
SR is the "split ratio" determined according to Equation 6.8 to account for the fact that only a portion of each fraction may be injected

The dilution of the overall LC × LC separation, ^{2D}DF, is then determined from the product of dilution factors in each dimension:

$$^{2D}DF = {}^1DF \times {}^2DF \tag{6.13}$$

Finally, the total analysis time for an offline LC × LC separation can be calculated according to (Kalili and de Villiers 2013b):

$$t_{tot,offline} = f \times {}^2t_c + f \times {}^1t_s = f \times {}^2t_c + {}^1t_g \tag{6.14}$$

where f is the number of fractions collected from the ^{1}D and is equal to $^1t_g/^1t_s$.

With all the information determined above in hand, it is possible to choose the optimal experimental conditions to provide the desired outcome in terms of practical peak capacity, total analysis time, and/or sample dilution. As an example, Figure 6.11 shows the variation in practical 2D peak capacity as a function of the ^{1}D sampling time and total analysis time for various ^{2}D cycle times for the offline HILIC × RP–LC

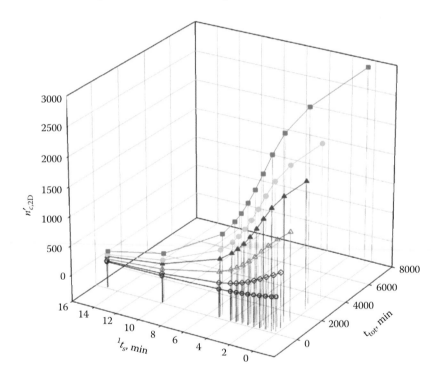

FIGURE 6.11 Illustration of the relationship of practical two-dimensional peak capacity ($n'_{c,2D}$, z-axis), first-dimensional sampling time (^1t_s, y-axis), and total analysis time (t_{tot}, x-axis) for the offline HILIC × RP–LC analysis of procyanidins. First dimension: Diol column (250 × 1 mm), 50 min gradient at 50 μL/min; second dimension: C18 column (50 × 4.6 mm), flow rate 1.5 mL/min. Data were obtained at various second-dimensional cycle times: 0.5 min (○), 2 min (◊), 5 min (Δ), 10 min (▲), 15 min (●), and 30 min (■). (Reproduced from Kalili, K. M. and de Villiers, A., *J. Chromatogr. A*, 1289, 69–79, 2013a. With permission.)

analysis of procyanidins (Kalili and de Villiers 2013b). The data were obtained for constant ^{1}D analysis conditions (${}^{1}t_{g}$ = 50 min, ${}^{1}F$ = 50 μL/min). While the values reported in this figure are applicable to the particular analytes/mobile phases/columns considered, some general conclusions that apply to offline HILIC × RP–LC operation may be drawn. First, the highest overall peak capacity will be obtained for the shortest sampling time and longest ^{2}D cycle time—this is because β is minimized and ${}^{2}n_{c}$ maximized under these conditions. The price to pay however is in terms of total analysis time, which would be exorbitantly long under these conditions (~5½ days for the highest $n'_{c,2D}$ values in Figure 6.11). It is also evident that, for the LC × LC system under consideration, the gain in 2D peak capacity upon reduction of the sampling time to less than 1 min is relatively limited, while the increase in total analysis time is significant under these conditions, especially when long ^{2}D cycle times are used. The data reported in Figure 6.11 were obtained using a 1 mm i.d. ^{1}D column operated at 50 μL/min, and injecting only 2 μL (=${}^{2}V_{max}$) in ^{2}D, where a 4.6 mm column was operated at 1.5 mL/min. Under these conditions, the total dilution of analyte peaks for a 2 min cycle time was an exorbitant 120 × (^{1}DF = 9.5 and ^{2}DF = 12.5). Nevertheless, by injection of a highly concentrated extract and the use of selective fluorescence (FL) detection, sufficient sensitivity was obtained for this application (see Figure 6.23).

It is relevant to note here that the peak production rate, defined as

$$\text{Peak production rate} = \frac{n'_{c,2D}}{t_{tot}} \tag{6.15}$$

is highest under conditions (${}^{1}t_{s}$ = 2.5 min, ${}^{2}t_{c}$ = 6.5 min) where relatively low performance ($n'_{c,2D} \approx 760$) is obtained. Since HILIC × RP–LC is generally performed in offline mode to obtain the best possible resolution, it is therefore questionable to use the peak production rate as criterion for method optimization in this mode.

6.5.5 ONLINE HILIC × RP–LC METHOD OPTIMIZATION

Online LC × LC is defined by the requirement that the ^{1}D sampling time is equal to the ^{2}D analysis (cycle) time, that is, ${}^{1}t_{s} = {}^{2}t_{c}$. Since the speed of the ^{2}D column is of primary importance, relatively short columns should be evaluated; otherwise, selection of column dimension in both dimensions can be performed according to the criteria outlined in Section 6.5.2, that is, based on the experimentally determined value of ${}^{2}V_{max}$ for the target analytes. This approach will in many cases deliver very low ^{1}D flow rates, and therefore impractically low values for the ^{1}D column i.d. In such cases, a practically relevant ^{1}D column diameter as close as possible to the ideal, that is, where the optimal volumetric flow rate and a typical sampling time correspond to V_{frac} volumes matching ${}^{2}V_{max}$ as close as possible according to Equation 6.7, should be selected.

In cases where $V_{frac} > {}^{2}V_{max}$, one of three approaches may be used to avoid injection band broadening in ^{2}D: (1) reducing ^{1}D flow rate (accepting a concomitant decrease in performance in this dimension and increase in total analysis time), (2) splitting the flow after ^{1}D (and sacrificing sensitivity), or (3) utilizing a trap with make-up flow between the two dimensions. Option (1) is the most attractive up to a certain point, since lower flow rates in ^{1}D result in broader peaks and therefore lessen restrictions placed on the sampling time by the requirement to avoid ^{1}D undersampling. Beyond a certain point, however, the reduction in ^{1}D peak capacity severely compromises the overall performance. The consequences of operating the ^{1}D column at a low flow rate are accounted for in method optimization by measuring ^1n_c under these conditions. Option (2) is accounted for by the incorporation of the SR in the determination of the total dilution according to Equations 6.8 and 6.11 through 6.13. Option (3), when properly implemented, has the effect of avoiding injection volume effects, and therefore SR is equal to one for all conditions. This is clearly beneficial from the perspective of sample dilution in ^{2}D (*cf.* Equation 6.12). However, design of such systems is relatively more complicated, as both efficient trapping and release of all analytes should be optimized.

Method optimization in online LC × LC is then relatively straightforward, since ^1t_s determines the ^{2}D cycle time (and corresponding gradient time), and therefore performance according to Equation 6.9, as well as the degree of ^{1}D undersampling. Accordingly, ^1t_s is used as primary variable, and the corresponding ^{2}D cycle time is used to determine $n'_{c,2D}$ according to Equation 6.6 to correct for ^{1}D undersampling and finite orthogonality. This can be done for a range of column formats and flow rates in both dimensions. A typical example of the relationship between $n'_{c,2D}$ and ^2t_c is shown in Figure 6.12, where it is clear that the 2D peak capacity passes through a maximum at a given cycle time. This observation is universal to online LC × LC (Horváth et al. 2009b; Fairchild et al. 2009b; Potts et al. 2010; Huang et al. 2011; Gu et al. 2011; Bedani et al. 2012; Carr et al. 2012; Kalili and de Villiers 2013b) and follows from the effect of the sampling time on ^{1}D undersampling and ^{2}D analysis time. For low ^2t_c times, undersampling is negligible, and any increase in the sampling time results in a significant increase in ^{2}D peak capacity. On the other hand, for longer ^1t_s times, increasing ^{1}D undersampling dominates the increase in ^{2}D peak capacity, resulting in a decrease of overall peak capacity. This compromise is inherent to online LC × LC, although the optimal sampling time required to deliver the maximum peak capacity will depend on the analytes, column dimensions and flow rates in each dimension. For the specific analytes and separations used to generate the data reported in Figure 6.12, a relatively long optimal sampling time (~3 min) is observed. This is a consequence of the comparatively broad ^{1}D peaks obtained for procyanidins in HILIC. In the majority of online LC × LC studies reported in literature, shorter optimal sampling times are the norm. It is also relevant to note that in the case where flow splitting is used between the two dimensions, the required SR (and therefore the total dilution, ^{2D}DF) also increases with the ^{2}D cycle time (Figure 6.12).

In general, a reduction in ^{1}D flow rate and a concomitant increase in gradient time will result in higher overall peak capacities (up to a point). This is because peak widths in this dimension become broader (although the peak capacity increases), and therefore the optimal sampling time becomes longer. This then provides for longer ^{2}D analyses and higher peak capacity values in this dimension (Horváth et al. 2009b;

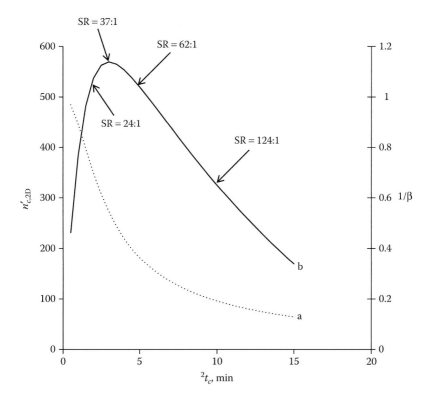

FIGURE 6.12 Relationship between the second dimension cycle time (^2t_c [x-axis], which is equal to the first dimension sampling time, ^1t_s), the corrected two-dimensional peak capacity ($n'_{c,2D}$, primary y-axis, curve b), and the inverse of the undersampling correction factor ($1/\beta$, secondary y-axis, curve a) for the online HILIC × RP–LC analysis of procyanidins. Data are plotted for a first dimension flow rate of 25 μL/min and gradient time of 100 min; other conditions as in Figure 6.11. The split ratios (SR) required to avoid second dimension band broadening are indicated for several second dimension cycle times. (Adapted from Kalili, K. M. and de Villiers, A., *J. Chromatogr. A*, 1289, 69–79, 2013a. With permission.)

Gu et al. 2011; Huang et al. 2011; Willemse et al. 2015). Another added benefit is that the required SR decreases with the ^{1}D flow (despite the fact that the optimal sampling time increases) (Willemse et al. 2015).

Finally, the peak production rate can be determined according to Equation 6.15, where the total analysis time in online LC × LC is determined by

$$t_{\text{tot,online}} = f \times {}^2t_c + {}^2t_c \tag{6.16}$$

As a rule, peak production rates are much higher in online compared to offline LC × LC. The highest peak production rate is obtained at the optimal sampling time corresponding to the maximum peak capacity.

6.5.6 STOP-FLOW HILIC × RP–LC METHOD OPTIMIZATION

Method optimization in stop-flow HILIC × RP–LC is essentially performed in exactly the same manner as outlined for offline approach (Section 6.5.4), with the notable exception that additional ^{1}D band broadening due to stop-flow operation should be accounted for. This follows from the fact that axial dispersion in the mobile phase during stop-flow periods may result in additional zone broadening in this dimension (Bedani et al. 2006; Fairchild et al. 2009b). This effect can be incorporated in method optimization by correcting the ^{1}D peak capacity. Two approaches can be distinguished, based on how the effective diffusion coefficients, D_{eff}, for the target analytes are obtained. In the first instance, these may be estimated theoretically (Fairchild et al. 2009b). A more accurate method is the experimental determination of D_{eff} using arrested elution methods (Knox and Scott 1983; Bedani et al. 2006; Kalili and de Villiers 2013b). In the latter case, the ^{1}D separation of the target analytes is performed with the flow interrupted for a range of time periods, followed by restarting the flow and measuring the peak variances for all analytes. A plot of measured peak variance versus stop-flow time can then be used to determine D_{eff} (Bedani et al. 2006; Kalili and de Villiers 2013b). This value might vary significantly for different analytes.

The effective stop-flow time, t_{stop}, experienced by each analyte during stop-flow LC × LC operation can be calculated according to

$$t_{stop} = t_{tot} - t_R \qquad (6.17)$$

where:

t_{tot} is the retention time for the specific analyte band in a stop-flow experiment
t_R is its retention time under continuous-flow conditions

Additional band broadening resulting from stop-flow operation, $\sigma_{stop\text{-}flow}$, can then be determined for each peak:

$$\sigma_{stop\text{-}flow} = D_{eff} \times t_{stop} \qquad (6.18)$$

The effective ^{1}D peak width under stop-flow conditions for the corresponding peak is then:

$$\sigma_{tot} = \sqrt{\sigma_{cont}^2 + \sigma_{stop\text{-}flow}^2} \qquad (6.19)$$

where:

σ_{tot} is the total standard deviation of the peak under stop-flow conditions
σ_{cont} the standard deviation for an uninterrupted analysis
$\sigma_{stop\text{-}flow}$ the additional standard deviation resulting from the stop-flow periods

This procedure should be followed for each (or the majority of) ^{1}D peaks to determine the average value of σ_{tot}. These values are then translated to the average peak widths at baseline ($=4\sigma_{tot}$) and used to obtain a value for the "corrected" ^{1}D peak capacity, $^1n'_c$, according to Equation 6.1 (for further details on the procedure, the reader is referred to Bedani et al. [2006] and Kalili and de Villiers [2013b]). Note that the ^{2}D cycle time directly determines the stop-flow time in stop-flow LC × LC. In order to use ^2t_c as a variable in method optimization, corrected ^{1}D peak capacities should therefore be determined for a range of stop-flow times. This is relatively easy once D_{eff} is known (Equation 6.18).

From this point on, method optimization can proceed as outlined for offline HILIC × RP–LC. The only notable difference is that fraction volumes larger than $^2V_{max}$ should be accommodated either by incorporating a suitable split of the flow prior to the ^{2}D column (Equation 6.8) or by using a trapping column between the two dimensions as outlined for online operation. In either case, dilution is accounted for by the dilution factor determined according to Equations 6.11 through 6.13. Total analysis times can be determined as for offline operation using Equation 6.14, and peak production rates according to Equation 6.15.

The results obtained from the procedure outlined above are at first glance somewhat surprising. An example of the relationship between the relative corrected ^{1}D peak capacity and the ^{2}D cycle time (which is equal to the stop-flow period) is presented for the HILIC analysis of procyanidins in Figure 6.13. As expected,

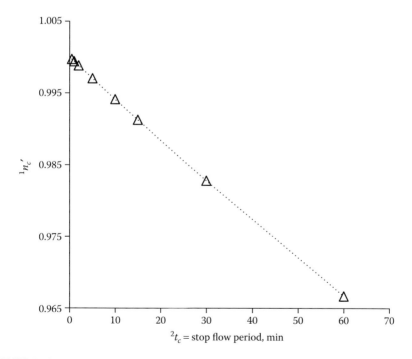

FIGURE 6.13 Illustration of the loss in first dimension peak capacity as a function of stop-flow time for the stop-flow HILIC × RP–LC analysis of procyanidins. The relative corrected first dimension peak capacity ($^1n'_c$) is plotted as a function of the second dimension cycle time (^2t_c, which is equal to the stop-flow time). Experimental conditions as in Figure 6.11. (Adapted from Kalili, K. M. and de Villiers, A., *J. Chromatogr. A*, 1289, 69–79, 2013a. With permission.)

the effective ^{1}D peak capacity decreases consistently with an increase in stop-flow time, since band broadening increases with the stop-flow period. However, it is also evident that the decrease in "corrected" ^{1}D peak capacity is relatively minor, even for stop-flow periods of up to 60 min. This observation may in part be ascribed to the relatively high MW analytes and the experimental conditions used: since procyanidin retention in HILIC increases with their DP, the higher DP molecules, which are characterized by lower effective diffusion coefficients, are exposed to the longest stop-flow periods (Kalili and de Villiers 2013b). This tends to negate the effect of stop-flow band broadening. However, it should be pointed out that very similar effective diffusion coefficients were obtained experimentally for dimeric and pentameric peptides (MW 198 and 343, respectively) (Bedani et al. 2006). These authors reported only slightly higher losses in ^{1}D peak capacity as a function of stop-flow time.

It can therefore be concluded that stop-flow operation has a relatively small detrimental effect on the overall performance of LC × LC separations. For modest stop-flow times, similar performance compared to offline operation can be expected. The choice between offline or stop-flow HILIC × RP–LC should therefore be governed not by performance criteria, but by considerations such as operational and optimization complexity (offline is simpler), automation and the risk of sample degradation (where stop-flow is beneficial).

6.6 COMBINING HILIC × RP–LC SEPARATION WITH DIFFERENT DETECTION STRATEGIES

When two or more complementary chromatographic methods are applied to a sample, a deeper insight into the complexity of the sample is provided by the combination of different retention mechanisms. However, the amount of information obtained will depend on the properties of the analytes and the detector(s) used. A few aspects are relevant when selecting a detector for LC × LC: (1) sensitivity, (2) quantitative performance, (3) compatibility with experimental conditions, (4) selectivity, and (5) structural elucidation properties. The first of these, sensitivity, is logical, although it should be stressed that experimental considerations such as column dimensions and hyphenation mode also directly determine the overall sensitivity of an LC × LC method as outlined in Section 6.5 (Stoll et al. 2015). Generally speaking, the relatively high dilution factors common to LC × LC operation places high demands on detector sensitivity.

For quantitative LC × LC analyses, similar to 1D LC, detectors such as UV and FL are often preferred to, for example, MS. For such applications, it may therefore be preferable to combine a quantitative detector with MS to confirm compound identity. Detectors can either be combined inline, where the non-destructive detector is positioned first in line, or by splitting the flow after the ^{2}D column.

In terms of compatibility, the obvious requirement will be compatibility with the mobile phase used in ^{2}D (in the case of HILIC × RP–LC and RP–LC × HILIC, this is generally not a concern). Second, the very fast ^{2}D separations required especially for online HILIC × RP–LC methods imply very narrow peak widths in this dimension. As a consequence, detectors with fast acquisition rates are required to accurately define these narrow peaks. For such applications, time-of-flight (TOF) MS detectors are more suitable than tandem MS (MS/MS) on triple quadrupole instruments, for example. Furthermore, the contribution of the detection process to band broadening should be considered: for very narrow, low-volume peaks, it is essential to minimize additional broadening due to for example large flow cells or extensive tubing between the column and detector.

The use of selective detectors such as FL and MS/MS or high resolution (HR)–MS is highly beneficial in the sense that this simplifies the demands placed on the separation performance (although matrix effects should always be kept in mind with ESI–MS detection). However, selective detectors will provide information only for specific analytes, and are therefore not the first choice for untargeted screening analyses.

When LC × LC is combined with one or more spectroscopic detector(s), information on structural features of analytes is provided, thereby assisting peak identification. This is critically important when complex samples of unknown composition are analyzed. From this perspective, MS is certainly the most important detector in LC × LC nowadays.

HILIC × RP–LC separations have been combined with several detection strategies, including ultraviolet–visible (UV–vis) or diode-array detection (DAD), FL, ELSD, CAD, and MS^n (including a range of mass analyzers). Of these, UV and MS are by far the most common (the exception is for polymer analysis, where ELSD is often used [Jandera et al. 2006; Abrar and Trathnigg 2010; Elsner et al. 2012]). Below, selected examples of the application of different detectors together with HILIC × RP–LC separation will be outlined, with the emphasis on non-standard configurations.

Schiesel et al. (2012) developed an offline RP–LC × HILIC method combined with both ion trap MS (IT–MS) and CAD detection for qualitative and quantitative profiling of polar impurities in a nutritional infusion solution comprising amino acids and dipeptides. The polar fraction was collected from the RP–LC analysis of the complete infusion sample, and was subsequently submitted to offline analysis utilizing two RP–LC columns in series in ^2D and a monolithic HILIC column in ^2D. Two identical instruments were used in ^2D, with one hyphenated to the CAD, and the other to the MS. The CAD is a relatively recent addition to the array of HPLC detectors (Gamache et al. 2005) which provides a universal (although mobile-phase dependent [Gorecki et al. 2006]) response for nonvolatile analytes and is well suited for quantification purposes involving compounds lacking chromophores, while the IT–MS allows selective ion trapping and fragmentation, thus serving as an excellent tool for structural elucidation. The combination of these complementary detectors allowed identification of unknown impurities based on IT–MS data, while CAD allowed their accurate quantification. Quantification by CAD was studied in two calibration modes, namely the unified calibration mode based on the average of compound-specific calibration functions, and the slope-corrected calibration mode, where response variation of the CAD as a function of mobile phase composition is accounted for. Both calibration methods afforded acceptable quantification accuracy (values ranging between 75% and 130% in comparison to authentic standards), and impurities were classified into those that need to be reported (above 0.05% of the active ingredients), identified (>0.1%), or quantified (>0.15%) based on the universal calibration data. RP–LC × HILIC–CAD results were further cross-validated with HPLC–MS/MS using synthesized authentic standards. The findings of this study illustrate that LC × LC–CAD is a viable technique for quantification of nonvolatiles in complex matrices, provided that the mobile phase-dependent response of the detector is sufficiently accounted for in the case where standards are not available.

As an example of the use of multiple detectors in series, Kalili et al. (2013) reported an online HILIC × RP–LC–FL–ESI–Q–TOF–MS method that proved effective in the characterization of grape seed proanthocyanidins. In this work, the ^{2}D flow was split ~1:1 after the FL detector to provide a flow of ~0.7 mL/min to the mass spectrometer. The benefits of using complementary separation and detection methods are well-exemplified in this study. For example, HILIC and MS cannot distinguish between proanthocyanidins of the same MW (these isomers are not separated based on polarity in HILIC), whereas isomeric separation is provided by RP–LC. RP–LC on the other hand cannot separate the large number of isomers of different MW, which were distinguished on the basis of polarity (HILIC) and mass spectral data. FL detection was useful in discriminating between non-galloylated and galloylated procyanidins (the latter do not fluoresce at the excitation and emission wavelengths used), while HR–MS allowed investigation of the distribution of individual classes of compounds by means of single ion extraction (Figure 6.14). Using this approach, identification of procyanidins comprising up to 12 monomeric units as well as monogalloylated—octagalloylated procyanidins of DP as high as 16 could be tentatively identified.

Montero et al. (2013a, 2013b) reported an alternative approach for the analysis of proanthocyanidins in grape seeds and apple, and phlorotannins in brown algae (Montero et al. 2014) based on online HILIC × RP–LC–DAD–ESI–MS/MS using an ion trap mass spectrometer. Compounds were tentatively identified based on UV, MS, and MS/MS spectra. It is relevant to note that since neither HR–MS nor MS/MS allow distinguishing between proanthocyanidin isomers, chromatographic separation is essential for these analyses. HILIC × RP–LC–MS (or MS/MS) approaches using a variety of mass analyzers have also found extensive use in the analysis of biomolecules such as lipids (Wang et al. 2013) and peptides (Lam et al. 2010; Wohlgemuth et al. 2010; Xu et al. 2012; Vanhoenacker et al. 2015) (see Section 6.7.1).

LC × LC can also be combined with post-column reaction assays, provided that certain practical constraints are sufficiently accommodated. In a recent study, we investigated the feasibility of coupling the 2,2′-azino-bis(3-ethylbenzothiazoline)-6 sulfonic acid (ABTS) radical scavenging assay to offline and online HILIC × RP–LC separation (Kalili et al. 2014). This assay is commonly used to identify antioxidant species by their reaction with the ABTS radical, which results in the formation of a non-colored product. This reaction can therefore be monitored by UV detection at 414 nm following post-column reaction in a reactor coil. Significantly, even for relatively fast (2 min) ^{2}D analyses, good sensitivity was achieved for all compounds for reaction times of 4.26 s (Figure 6.15) in a small-volume reactor coil optimized to minimize extra-column band broadening, indicating that the ABTS assay is sufficiently fast for hyphenation to online LC × LC. Visual comparison of the contour plots at 280 and 414 nm allows distinguishing radical scavengers from nonactive sample components.

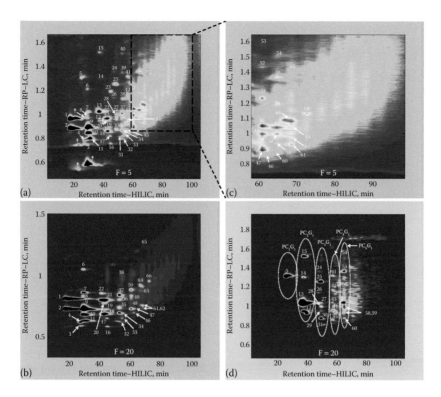

FIGURE 6.14   Total ion chromatogram (TIC) (a) and fluorescence (b) contour plots for the HILIC × RP–LC–FL–ESI–MS separation of a grape seed extract. (c) shows a detailed section of the TIC, and (d) presents the extracted ion contour plot for monogalloylated procyanidins obtained from the HILIC × RP–LC–MS data of the same analysis. First dimension: Diol column (250 × 1 mm), 100 min gradient at 25 μL/min; sampling time: 2 min (split 1 : 24); second dimension: C18 column (50 × 4.6 mm), flow rate 1.5 mL/min. Peak labels: PC_XG_Y refers to a procyanidin of degree of polymerisation X and galloylation Y. F denotes the z-axis scale. (Adapted from Kalili, K. M. et al., *Anal. Chem.*, 85, 9107–9115, 2013. With permission.)

This study demonstrated the compatibility of LC × LC with post-column reactors for the screening of bioactive molecules. Especially offline LC × LC in combination with such detection methods could serve as a viable alternative to the laborious classical assay-guided identification of bioactives in complex natural product extracts, provided that the reaction involved is sufficiently fast.

These selected examples clearly show that HILIC × RP–LC separation is amenable to hyphenation with a wide range of detection methods which may be used to enhance the performance of the method. Especially selective and sensitive detection methods such as MS are indispensable in the task of unraveling the complexity of real-world samples. It is also worth highlighting that the use of high resolution analytical techniques may remove the need for exhaustive sample preparation steps, which could result in loss of low level analytes, sample alteration or

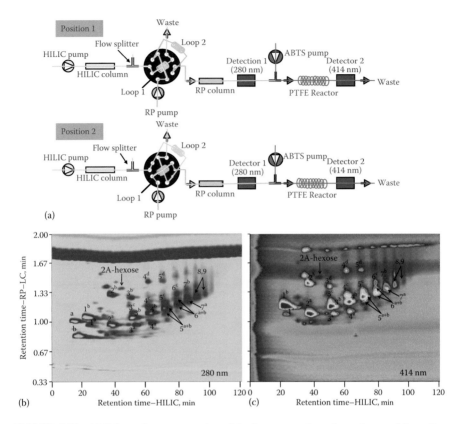

FIGURE 6.15 (a) Schematic representation of the instrumental configuration used for online HILIC × RP–LC–ABTS analysis of phenolics, (b) shows the contour plots obtained from data at 280 nm for detector 1 (post-column), and (c) shows the corresponding contour plot obtained at 414 nm for detector 2 (post reactor) for the analysis of cocoa procyanidins. First dimension: Diol column (250 × 1 mm), 100 min gradient at 25 μL/min; sampling time: 2 min (split 1: 32); second dimension: C18 column (50 × 4.6 mm), flow rate 1.5 mL/min; reactor coil: 2.8 m × 0.25 mm, ABTS solution flow: 0.5 mL/min. Peak labels: numbers correspond to the degree of polymerization (DP) of procyanidin molecules, superscripts distinguish between isomers of the same DP; 2A-hexose denotes an A-type glycosylated procyanidin dimer (Figure 8). (Adapted from Kalili, K. M. et al., *Anal. Bioanal. Chem.*, 406, 4233–4242, 2014. With permission.)

degradation of labile compounds, as simultaneous separation and selective detection of multiple, structurally diverse compounds is feasible in a single analysis.

6.7 APPLICATIONS OF HILIC × RP–LC

Since the first report of the comprehensive combination of HILIC and RP–LC, the technique has found increasing application in a range of fields. The most important of these are undoubtedly in the areas of proteomics, metabolomics, and natural product analysis, although HILIC × RP–LC has also been used for the analysis of polymers, pharmaceutical compounds, and so on. In the following sections, brief overviews of the application of HILIC × RP–LC in each of these fields will be presented. The aim here is not to present a comprehensive review of all application papers, but rather to provide insight into the most important recent developments in the application of HILIC × RP–LC separation for each of these sample classes.

6.7.1 HILIC × RP–LC in the Analysis of Biological Samples

One of the most important application areas of MD–LC is in the analysis of complex biological samples (Issaq et al. 2005). Especially in recent years, HILIC has increasingly been applied in bioanalysis (Boersema et al. 2008; Jian et al. 2010; Spagou et al. 2010; Ivanisevic et al. 2013; Heckendorf et al. 2014; Klavins et al. 2014) due to the complementary nature of the retention mechanisms in HILIC compared to RP–LC. It is therefore not surprising that the comprehensive combination of HILIC and RP–LC has also found utility in bioanalysis. Some of the main application areas of these methods will be outlined briefly below, with the relevant experimental details summarized in Table 6.1.

It is noted here that while the 2D combination of HILIC and RP–LC separations has found widespread application in the analysis of complex biological samples, many experimental approaches cannot be considered comprehensive (*cf.* Section 6.2.2). For example, the ^{1}D separation can effectively be used as a fractionation step (similar to what is done in strong cation exchange [SCX] separation of peptides). If large "steps" are used, the resultant combination of partially separated peaks means that the approach is not comprehensive in nature. This is not to say that such approaches are not useful—in fact, excellent performance has been demonstrated for such methods (Wu et al. 2011; Zhao et al. 2012; Cífková et al. 2013; Wang et al. 2013). However, such methodologies will not be covered in the current work, which is focused on comprehensive HILIC × RP–LC, unless of interest due to the coupling mode used. Similarly, heart-cutting 2D HILIC–RP–LC has been used with some success in the analysis of complex peptide (Liu et al. 2009; Simon et al. 2014) and lipid (Lísa et al. 2011; Ling et al. 2014) samples, but will not be addressed here. By far the majority of applications of HILIC × RP–LC pertain to proteomic and lipidomic studies, which will be addressed in more detail below.

6.7.1.1 Proteomics

In the field of proteomics, the so-called multidimensional protein identification technology (MudPIT) approach has found extensive use in peptide separation. In original form, MudPIT involves the use of a column containing sections of both SCX and RP–LC phases, where peptides are eluted stepwise from the SCX phase and separated by RP–LC prior to MS detection (Link et al. 1999; Washburn et al. 2001; Wolters et al. 2001). As alternative approaches, the offline hyphenation of SCX, SEC, and RP–LC with RP–LC (the latter typically using a different pH mobile phase) have also been exploited. HILIC has been shown to be a useful mode for peptide separations (Yoshida 2004), and the combination of HILIC and RP–LC has been shown to provide a very high degree of orthogonality for peptide separations (Gilar et al. 2005; Boersema et al. 2008). Consequently, the offline hyphenation of HILIC with RP–LC has also found application in this field (Boersema et al. 2007; Mihailova et al. 2008; Garbis et al. 2011; Di Palma et al. 2012).

TABLE 6.1
Summary of Recent Applications of HILIC × RP-LC (or RP-LC × HILIC) for the Analysis of Biological Samples

Sample(s)	Compound Class(es)	Hyphenation Mode (Sampling Time)	First-Dimension Conditions	Second-Dimension Conditions	Detection	Analysis Time/ Performance	Reference
Human serum (fractionated by SEC, trypsin digested)	Peptides	Offline (1 min)	HILIC Column: ZIC-HILIC 200 Å (160 × 0.2 mm, 3.5 µm; trapping column used in vented mode (50 × 0.2 mm)) Mobile phases: (A) 80% ACN[a] and buffer (B) 40% ACN and buffer	RP-LC Column: Zorbax C18 120 Å (254 × 0.05 mm, 3 µm), 40°C; trapping column (10 × 0.05 mm) Mobile phases: (A) 0.5% acetic acid and (B) 0.5% acetic acid in 80/20 ACN/H_2O	nESI-MS/MS[a] (Orbitrap)	~3000 min/1040/4973 or 1284/6625 proteins/peptides identified	Boersema et al. (2007)
Angiotensin 1-inhibiting peptides in milk hydrolysates	Peptides	Offline (1 min)	RP-LC Column: Inertsil 5 ODS3 (150 × 2.1 mm, 5 µm), 60°C Mobile phases: (A) 0.1% TFA[a] and (B) 0.1% TFA in ACN	HILIC Column: HILIC Atlantis (150 × 2.1 mm, 3 µm), 40°C Mobile phases: (A) 0.1% FA[a] in ACN and (B) 10 mM ammonium acetate+0.1% FA in H_2O	UV[a]-ESI-MS/ MS (Q-TOF[a])	~6480 min/not specified	Van Platerink et al. (2008)
HeLa cell lysate	Phospho-peptides	Offline (2 min)	HILIC Column: TSKgel Amide-80 (250 × 4.6 mm, 5 µm) Mobile phases: (A) 0.1% TFA in 98% ACN and (B) 0.1% TFA in 2% ACN	RP-LC Column: LC Packings C18 PepMap100 (15 × 0.075 mm, 5 µm) Mobile phases: (A) 0.1% FA and (B) 0.1% FA in ACN	ESI-MS/MS (ion trap)	~1400 min/814 phosphopeptides identified (>1000 phosphorylation sites)	McNulty and Annan (2008)
Digests of bovine α1-acid glycoprotein and IgG1	Glyco-peptides	Offline (2 min)	RP-LC Column: ProteoSpher RP18e (150 × 0.1 mm) Mobile phases: (A) 0.1% FA in 2% ACN and (B) 0.1% FA in 90% ACN	HILIC Column: ZIC-HILIC-modified monolithic silica (300 × 0.1 mm) Mobile phases: (A) 90% ACN and (B) 2% ACN	ESI-MS/MS (ion trap)	~4500 min/not specified	Wohlgemuth et al. (2010)

(Continued)

TABLE 6.1 (Continued)
Summary of Recent Applications of HILIC × RP–LC (or RP–LC × HILIC) for the Analysis of Biological Samples

Sample(s)	Compound Class(es)	Hyphenation Mode (Sampling Time)	First-Dimension Conditions	Second-Dimension Conditions	Detection	Analysis Time/ Performance	Reference
Human urine protein tryptic digests	Peptides	Offline (1 min)	HILIC Column: ZIC-HILIC 200 Å (150 × 2.1 mm, 5 μm) Mobile phases: (A) 20/80 20 mM ammonium acetate/ACN and (B) 60/40 20 mm ammonium acetate/ACN	RP-LC Trap: Kromasil C18 (5 × 1 mm, 5 μm), 20 mM FA as diluent Column: Kromasil C18 100 Å (100 × 0.32 mm, 5 μm), 40°C Mobile phases: (A) 95/5 20 mM FA/ACN and (B) 5/95 20 mM FA/ACN	UV-ESI–MS/MS (ion trap)	~1020 min/ ~438/1668 proteins/peptides identified	Loftheim et al. (2010)
HeLa cell lysate	Peptides	Offline (1 min)	HILIC Column: ZIC-HILIC 200 Å (250 × 0.075 mm, 3.5/5 μm); trapping column used in vented mode (20 × 0.1 mm, 3.5/5 μm) Mobile phases: (A) 0.5% acetic acid or 2% FA, 5 mM ammonium acetate in 95% ACN and (B) 0.07% FA, 5 mM ammonium acetate or 5 mM ammonium acetate	RP-LC Column: Reprosil (250 × 0.05 mm, 3 μm); trapping column (20 × 0.1 mm, 5 μm) Mobile phases: (A) 0.1 M acetic acid and (B) 0.1 M acetic acid in 80/20 ACN/H₂O	nESI–MS/MS (Orbitrap)	~3280 min/ ~3500/20000 proteins/peptides identified	Di Palma et al. (2011a)
FACSᵃ-sorted colon stem cells isolated from mouse intestine	Peptides	Offline (1 min)	HILIC Column: ZIC-HILIC 200 Å (270 × 0.075 mm, 5 μm); trapping column used in vented mode (20 × 0.1 mm, 5 μm) Mobile phases: (A) 0.5% acetic acid, 5 mM ammonium acetate in 95% ACN and (B) 5 mM ammonium acetate	RP-LC Column: Reprosil (400 × 0.05 mm, 3 μm); trapping column (20 × 0.1 mm, 5 μm) Mobile phases: (A) 0.1 M acetic acid and (B) 0.1 M acetic acid in 80/20 ACN/H₂O	nESI–MS/MS (Orbitrap)	~5600 min / ~3775/15775 proteins/peptides identified	Di Palma et al. (2011b)

(Continued)

TABLE 6.1 (Continued)
Summary of Recent Applications of HILIC × RP–LC (or RP–LC × HILIC) for the Analysis of Biological Samples

Sample(s)	Compound Class(es)	Hyphenation Mode (Sampling Time)	First-Dimension Conditions	Second-Dimension Conditions	Detection	Analysis Time/ Performance	Reference
Human serum (fractionated by SEC, trypsin digested)	Peptides	Offline (variable, peak signal dependent)[b]	HILIC Column: ZIC-HILIC (150 × 4.6 mm, 5 μm), 30°C Mobile phases: (A) 0.1% FA, 15 mM ammonium formate in ACN and (B) 0.1% FA, 15 mM ammonium formate in H_2O	RP-LC Column: Zorbax C18 300 Å (150 × 0.075 mm, 1.8 μm), 40°C Mobile phases: (A) 0.1% FA in 3/97 ACN/H_2O and (B) 0.1% FA in 97/3 ACN/H_2O	nESI–MS/MS (quadrupole-ion trap)	~4400 min/1955 proteins identified	Garbis et al. (2011)
Buthus martensi scorpion venom	Short-chain peptides	Offline (1 min)	RP-LC Column: XTerra MS C18 (100 × 19 mm, 5 μm) Mobile phases: (A) 0.1% TFA and (B) 0.1% TFA in ACN	HILIC Column: Click Maltose (150 × 4.6 mm, 5 μm), 30°C Mobile phases: (A) ACN, (B) H_2O, and (C) 100 mM triethylamine phosphate	Offline ESI–MS, ESI–MS/MS	~3000 min/18 peptides purified and sequenced	Xu et al. (2012)
Human cell lines (digested)	Peptides	Offline (1–2 min)	HILIC Column: ZIC-HILIC 200 Å (200 × 0.075 mm, 5 μm); trapping column used in vented mode (20 × 0.1 mm) Mobile phases: (A) 2% FA, 5 mM ammonium acetate in 95% ACN and (B) 0.07% FA, 5 mM ammonium acetate in H_2O	RP-LC Column: C18 (400 × 0.05 mm); trapping column (20 × 0.1 mm) Mobile phases: (A) 0.6% acetic acid and (B) 0.6% acetic acid in 80/20 ACN/H_2O	nESI–MS/MS (Orbitrap)	~2400 min/ ~3600/20000 proteins/peptides identified	Di Palma et al. (2012)
Rat muscle tissue dialysate	Peptides	Online (1.5 min)	HILIC Column: ZIC-HILIC (150 × 0.3 mm, 5 μm) Mobile phases: (A) 0.5 mM ammonium acetate in 95% ACN and (B) 9.5 mM ammonium acetate in 5% ACN	RP-LC Trap: Kromasil C18 (50 × 1 mm), 0.1% FA/ACN 95 / 5 as diluent Column: PLRP-S 300 Å (150 × 0.3 mm, 3 μm) Mobile phases: (A) 0.095% FA in 5% ACN and (B) 0.005% FA in 95/5 ACN/H_2O	UV/ESI–MS (TOF)	~960 min/not specified	Wilson et al. (2007)

(Continued)

TABLE 6.1 (Continued)
Summary of Recent Applications of HILIC × RP–LC (or RP–LC × HILIC) for the Analysis of Biological Samples

Sample(s)	Compound Class(es)	Hyphenation Mode (Sampling Time)	First-Dimension Conditions	Second-Dimension Conditions	Detection	Analysis Time/ Performance	Reference
Rat brain tissue	Neuro-peptides	Online (2/5 min)[b]	HILIC Column: ZIC–HILIC 200 Å (150 × 0.3 mm, 3.5 μm) Mobile phases: (A) 10 mM ammonium acetate in 95% ACN and (B) 10 mM ammonium acetate in 5% ACN	RP-LC Trap: Kromasil C18 (50 × 1 mm), 0.1% FA/ACN 95/5 as diluent Column: Kromasil C18 (100 × 0.32 mm, 3.5 μm) Mobile phases: (A) 0.6% acetic acid and (B) 0.6% acetic acid in 80/20 ACN/H$_2$O	μESI–MS/MS (ion trap)	855 min/2508–4595 peaks detected	Mihailova et al. (2008)
Lysates of rat PC12 cell lines and *Saccharomyces cerevisiae*	Peptides	Online (5 min)[b]	HILIC Column: TSKgel Amide 80 300 Å (170 × 0.25 mm, 3 μm) Mobile phases: (A) 90/10 ACN/solvent B/0.1% FA in ACN and (B) 2% ACN in 20 mM ammonium formate / 0.1% FA	RP-LC Trap: Poly-Sulfoethyl SCX 300 Å (50 × 0.15 mm, 5 μm); 0.5% FA in 2% ACN as diluent Column: Jupiter C18e 300 Å (170/150 × 0.15/0.075 mm, 3 μm) Mobile phases: (A) 0.5% FA in 2% ACN and (B) 0.5% FA in 98% ACN	ESI–MS/MS (Q-TOF)	~1050 min/2554/ 16916 proteins/ peptides identified	Zhao et al. (2012)
Tryptic digest of 3 proteins (lysozyme, BSA, and myoglobin)	Peptides	Online (0.77 min)	RP-LC Column: Hypersil Gold (50 × 1 mm, 3 μm), 30°C Mobile phases: (A) 10 mM ammonium acetate and (B) ACN	HILIC Column: Acquity BEH HILIC, (50 × 2.1 mm, 1.7 μm), 60°C Mobile phases: (A) ACN and (B) 10 mM ammonium acetate	UV	~200 min/$n'_{c,2D}$ (effective peak capacity): 2600	D'Attoma and Heinisch (2013)
Tryptic digest of the monoclonal antibody trastuzumab (Herceptin)	Peptides	Online (0.45 min)	HILIC Column: Zorbax RRHD 300-HILIC, (100 × 2.1 mm, 1.8 μm), 30°C Mobile phases: (A) 15 mM ammonium formate in 90% ACN and (B) 15 mM ammonium formate	RP-LC Column: Zorbax Eclipse Plus C18 (50 × 4.6 mm, 3.5 μm), 30°C Mobile phases: (A) 0.1% FA (MS)/0.1% phosphoric acid (DAD[a]) and (B) ACN	DAD/MS, MS/MS (Q-TOF)	~80 min/not specified	Vanhoenacker et al. (2015)

(Continued)

TABLE 6.1 (Continued)
Summary of Recent Applications of HILIC × RP–LC (or RP–LC × HILIC) for the Analysis of Biological Samples

Sample(s)	Compound Class(es)	Hyphenation Mode (Sampling Time)	First-Dimension Conditions	Second-Dimension Conditions	Detection	Analysis Time/ Performance	Reference
Porcine organs	Lipids	Offline (variable, peak signal dependent)[b]	HILIC Column: Spherisorb Silica (250 × 4.6 mm, 5 μm), 40°C Mobile phases: (A) ACN and (B) 5 mM ammonium acetate	RP-LC Column: Kinetex C18 (150 × 2.2 mm, 2.6 μm), 40°C Mobile phases: (A) 5 mM ammonium acetate and (B) ACN/IPA[a] (1:1)	ESI–MS/MS (ion trap)	~770 min/>160 species identified	Cifková et al. (2013)
Cow's milk and plasma	Phospho-lipids	Stop-flow (~1 min)	HILIC Column: Ascentis Express HILIC (150 × 2.1 mm, 2.7 μm) Mobile phases: (A) 10 mM ammonium formate in ACN (10/90), and (B) ACN/MeOH[a]/10 mM ammonium formate (55/35/10)	RP-LC Column: Ascentis Express C18 (150 × 4.6 mm, 2.7 μm), 65°C Mobile phases: (A)10 mm Ammonium formate /IPA/THF[a] (30/55/15) and (B) ACN	ESI–MS	~345 min/50/33 phospholipids identified	Dugo et al. (2013)
Human plasma	Lipids	Stop-flow (~1–7 min variable)[b]	HILIC Column: ACQUITY BEH HILIC (100 × 2.1 mm, 1.7 μm) Mobile phases: (A) 10 mM ammonium formate in IPA/ACN/H$_2$O (15/80/5), and (B) 10 mM ammonium formate in MeOH/H$_2$O (1/1)	RP-LC Trap: ACQUITY BEH C8 (5 × 2.1 mm, 1.7 μm); 10 mM ammonium formate as diluent Column: ACQUITY BEH C8 (100 × 2.1 mm, 1.7 μm), 50°C Mobile phases: (A) IPA/ACN/H$_2$O (20/48/32) and (B) IPA/ACN/H$_2$O (80/12/8)	ESI–MS/MS	~130 min/ $n_{c,2D}$ = 415 (372 lipids identified)	Wang et al. (2013)

(Continued)

TABLE 6.1 (Continued)

Summary of Recent Applications of HILIC × RP–LC (or RP–LC × HILIC) for the Analysis of Biological Samples

Sample(s)	Compound Class(es)	Hyphenation Mode (Sampling Time)	First-Dimension Conditions	Second-Dimension Conditions	Detection	Analysis Time/ Performance	Reference
Nutritional infusion solution	Amino acids and dipeptides	Offline (0.3–2.2 min, variable)[b]	RP-LC Columns: Gemini C18 (150 × 3.0 mm, 3 μm) + Synergi Fusion-RP (150 × 3.0 mm, 3 μm) Mobile phases: (A) 0.1% TFA and (B) 0.1% TFA in ACN	HILIC Column: Chromolith Performance Si monolith (100 × 4.6 mm) Mobile phases: (A) 1% H₂O, 1.5% 200 mM ammonium acetate buffer in ACN, and (B) 1.5% 200 mM ammonium acetate buffer	CAD[a], ESI–IT–MS	1200/1800 min/not specified	Schiesel et al. (2012)
Di- to deca-oligonucleotide standards	Oligo-nucleotides	Online (2 min)	HILIC Column: Ascentic Silica 100 Å (150 × 1 mm, 3 μm) Mobile phases: (A) 90/10 ACN/5 mM ammonium formate and (B) 5 mM ammonium formate	RP-LC Trap: ACQUITY BEH C8 (5 × 2.1 mm, 1.7 μm); 10 mM ammonium formate as diluent Column: XBridge C18 (50 × 4.6 mm, 3.5 μm), 35°C Mobile phases: (A) 0.1 M triethylamine acetate and (B) 20/80 0.1 M triethylamine acetate/ACN	UV, ESI–MS	~155 min/n_{c2D} ~ 500	Li et al. (2012)

[a] ACN = acetonitrile, CAD = charged aerosol detector, DAD = diode array detector ESI = electrospray ionization, FA = formic acid, FACS = fluorescence-activated cell sorting, IPA = isopropanol, MeOH = methanol, MS/MS = tandem mass spectrometry, Q-TOF = quadrupole-time-of-flight, TFA = trifluoroacetic acid, THF = tetrahydrofurane, UV = ultraviolet-visible detection.

[b] Analysis not truly comprehensive due to excessive sampling times.

Mohammed and coworkers evaluated the offline combination of zwitterionic HILIC (ZIC–HILIC) and RP–LC as an alternative MudPIT strategy in a series of papers (Boersema et al. 2007; Di Palma et al. 2011a, 2011b, 2012). The authors found that peptide separation on the ZIC–HILIC phase depended significantly on the mobile phase pH, but generally the ZIC–HILIC × RP–LC approach showed better performance than conventional SCX × RP–LC strategies. For the offline coupling of HILIC and RP–LC, narrow-bore HILIC columns were used, and the collected fractions were diluted with acidified aqueous solutions prior to their large-volume injection in ²D (a typical experimental protocol is illustrated graphically in Figure 6.16). These authors demonstrated how the approach could be used to identify more than 1000 proteins in a cellular nuclear lysate (Boersema et al. 2007). Subsequent work demonstrated the utility of this (slightly modified) approach for shotgun proteomic studies involving HeLa cell lysates (Di Palma et al. 2011a) and fluorescence-activated cell sorted (FACS) colon stem cells isolated from mouse intestine (Di Palma et al. 2011b).

Loftheim et al. (2010) reported an offline ZIC–HILIC × RP–LC method with high protein recovery for urinary proteomics following centrifugal filtration,

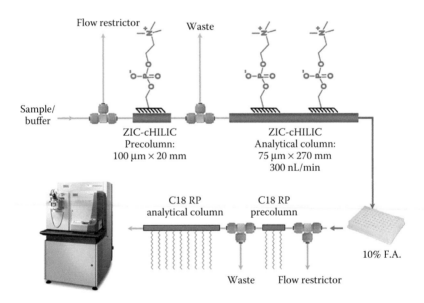

FIGURE 6.16 Schematic representation of the experimental protocol used for the offline ZIC–HILIC × RP–LC analysis of peptides. Trapping of desalted digests was achieved using a vented column set-up (15 µL/min, 10 min). The first dimension separation was performed at a flow rate of 300 nL/min, with one minute fractions collected in a well plate containing 40 µL 10% formic acid. In the second dimension, trapping was performed at 5 µL/min, and elution at 100 nL/min, prior to ESI–MS and MS/MS detection on an Orbitrap instrument. (Reproduced with permission from Di Palma, S. et al., *J. Proteome Res.*, 10, 3814–3819, 2011b.)

human serum albumin (HSA) removal and tryptic digestion. As another example of the power of the technique, Garbis et al. (2011) developed a novel MudPIT strategy combining several chromatographic modes for the characterization of serum proteins from patients with benign prostatic hyperplasia (BPH). The serum was first subjected to high-pressure SEC fractionation. Each of the fractions were dialyzed and subjected to trypsin proteolysis, followed by HILIC fractionation on the ZIC–HILIC phase. Subsequent nano-RP–LC–MS/MS of each of the HILIC fractions allowed identification of 1955 proteins spanning 12 orders of magnitude in serum samples.

Wohlgemuth et al. (2010) demonstrated that the offline combination of RP–LC and ZIC–HILIC (using monolithic phases in both dimensions) offers a powerful approach for profiling of glycopeptides through the combination of complementary separation modes: HILIC provided good separation according to glycan composition, whereas RP–LC separated compounds based on the peptide backbone.

Greibrokk and coworkers developed an interesting approach to combine HILIC and RP–LC separation of peptides in an online manner based on the use of several (18) C18 trapping columns. HILIC fractions were diluted with an aqueous-rich diluent prior to their trapping on these cartridges, facilitated by the use of a six-port two-position valve and two column selectors (Wilson et al. 2007). The instrumental configuration used is illustrated in Figure 6.17. Since the fractions could be stored on the traps until completion of the HILIC separation, this approach has the benefit of de-coupling the ^{2}D analysis time from the sampling time—although the sampling time is determined in this case by the number of traps available. The authors used a 53 min gradient in ^{2}D to obtain high peak capacities for the separation of tryptic digests of a rat muscle dialysate. In a subsequent study, Mihailova et al. (2008) reported that this approach employing a ZIC–HILIC phase was more powerful than online SCX × RP–LC for the analysis of neuropeptides in rat brain. Interestingly, these authors also found through the use of a downstream Hypercarb trap that breakthrough occurred for some peptides in the earlier HILIC fractions.

Van Platerink et al. (2008) used an offline RP–LC × HILIC approach to characterize hydrophilic peptides exhibiting inhibition of the angiotensin I-converting enzyme (ACE) involved in blood pressure regulation. The authors collected 1-min fractions which were neutralized, evaporated, and reconstituted before injection in ^{2}D. HILIC fractions were tested for activity using an at-line assay. This approach proved beneficial in that active hydrophilic peptides which exhibit poor retention on RP–LC columns could be identified for the first time following their improved separation in the HILIC dimension.

McNulty and Annan (2008) reported an offline HILIC × RP–LC method for the analysis of phosphopeptides. An amide HILIC column was used to fractionate cell lysates, with the increased retention of phosphorylated peptides exploited to elute nonphosphorylated peptides close to the column void. One-minute fractions were then collected and submitted to immobilized metal affinity chromatography (IMAC) enrichment; the selectivity of this step was significantly increased following HILIC pre-fractionation. This procedure improved the efficiency of

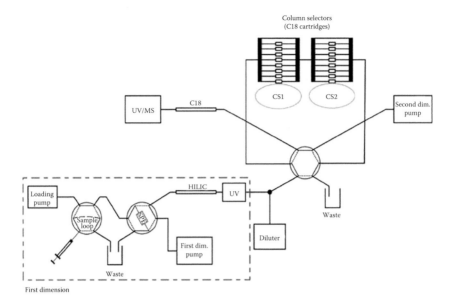

FIGURE 6.17 Schematic of the experimental configuration used for the SPE–HILIC–SPE–RP analysis of peptides. Samples were loaded onto a reversed-phase SPE cartridge (valve 2) and back-flushed onto the first dimension HILIC column. HILIC fractions were diluted (70 μL/min) prior to their trapping on several SPE cartridges using two column selectors (CS1 and 2). Trapped fractions were back-flushed onto the second dimension RP–LC column for separation (5 μL/min, 53 min gradient) prior to UV–MS detection. (Reproduced with permission from Wilson, S. R. et al., *Chromatographia*, 66, 469–474, 2007.)

RP–LC–MS/MS identification of phosphopeptides in the final step of the method (McNulty and Annan 2008).

The versatility of approaches combining HILIC and RP–LC is illustrated by a study by Zhao et al. (2012), although the methodology is not comprehensive by nature. In this work, the authors used a HILIC column in ^{1}D, and in an instrumental configuration employing four valves, transferred a nonretained (flow-through) and 10 different retained fractions to ^{2}D using a form of step gradient. Each of the fractions was diluted in a loop and collected on a SCX trap before being released to ^{2}D by injection of a high ionic strength buffer.

Heinisch and coworkers derived optimized online RP–LC × RP–LC and RP–LC × HILIC methods, the former based on the use of different pH mobile phases, for the analysis of peptides (D'Attoma et al. 2012). In accordance with a theoretical study comparing these systems (D'Attoma and Heinisch 2013), the authors found that RP–LC × RP–LC provided higher practical peak capacities, although the RP–LC × HILIC method displayed better coverage of the 2D space (Figure 6.18). Both methods provided much better performance than optimized 1D methods for peptide separation.

Vanhoenacker et al. (2015) compared online SCX × RP–LC, RP–LC × RP–LC at different pHs and HILIC × RP–LC for the analysis of tryptic digests of the monoclonal antibody trastuzumab (Herceptin) on a commercial LC × LC instrument. The authors found that for these samples, SCX × RP–LC and RP–LC × RP–LC provided better

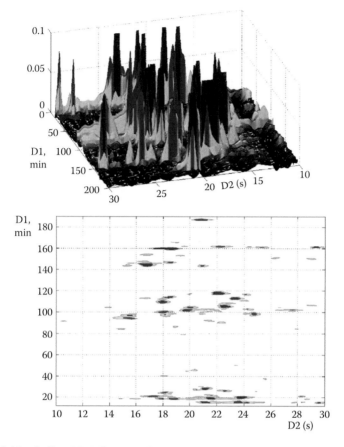

FIGURE 6.18 Online RP–LC × HILIC analysis of a tryptic digest of lysozyme, bovine serum albumin (BSA), and myglobin. The first dimension separation was performed on a 50 × 1 mm Hypersil Gold column (10 µL/min) and the second dimension on an Acquity BEH HILIC column (50 × 2.1 mm, 1.8 mL/min). Detection: UV at 210 nm. The sampling time was 0.77 min (injection volume in the second dimension 7.7 µL). (Reproduced with permission from D'Attoma, A. and Heinisch, S., *J. Chromatogr. A*, 1306, 27–36, 2013.)

separation than HILIC × RP–LC. Indeed, method optimization for HILIC × RP–LC was also more challenging than for the former two methods due to mobile phase considerations. To overcome these challenges, the authors used a lower flow rate (0.05 mL/min on a 2.1 mm HILIC column), split the flow 1:1 before the ^{2}D column and used a purely aqueous mobile phase at the start of the RP–LC gradient. Despite these adaptations, however, peptide breakthrough could not be entirely eliminated.

In a contribution demonstrating the possibility of upscaling LC × LC separations involving HILIC, Xu et al. (2012) used an offline RP–LC × HILIC method with a preparative column in ^{1}D to purify short-chain peptides from scorpion venom. Fractions were collected after the ^{2}D column and submitted to high resolution MS and collision-induced dissociation (CID)-MS/MS analysis to determine the peptide sequences.

6.7.1.2 Lipidomics

The utility of HILIC in lipidomic characterization is well illustrated by the report of Cífková et al. (2013). In this work, a non-targeted approach was developed for lipid analysis, involving HILIC–ESI–MS in positive and negative ionization modes for fractionation of lipid classes, followed by offline collection, evaporation, and RP–LC–ESI–MS analysis of specific compounds in each class. Fatty acids were determined by GC with flame ionization detection (FID), and more than 160 individual species were identified in this manner. The collection of fractions for each class of lipids as done in this study is an often-used methodology, although this is per definition an automated heart-cutting approach rather than a comprehensive one due to excessive sampling times.

Two interesting applications of stop-flow HILIC × RP–LC separation of lipids were recently reported. Dugo et al. used a stop-flow HILIC × RP–LC method for the analysis of phospholipids. HILIC provided class separation of six major phospholipid classes, whereas RP–LC afforded efficient separation of the molecular species in each class due to the relatively long ^{2}D separation allowed by stop-flow operation (Dugo et al. 2013) (Figure 6.19).

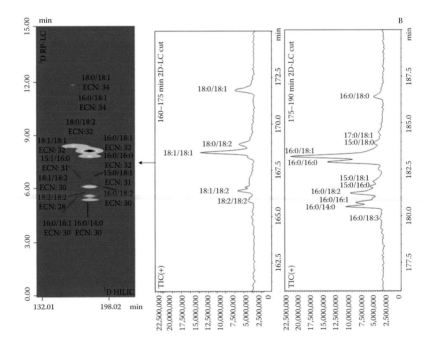

FIGURE 6.19 Example of the stop-flow HILIC × RP–LC–MS analysis of phospholipids in cow's milk. First dimension: Ascentis Express HILIC (150 × 2.1 mm), flow rate 0.1 mL/min, 35 min gradient. Second dimension: Ascentis Express C18 (150 × 4.6 mm), flow rate 0.9 mL/min isocratic (15 min), 65°C. Sampling time ~1 min, injection volume in the second dimension 100 μL. Peal labels: Equivalent carbon number (ECN); numbers correspond to fatty acid chain length and number of double bonds. (Reproduced with permission from Dugo, P. et al., *J. Chromatogr. A*, 1278, 46–53, 2013.)

Wang et al. (2013) reported an alternative methodology for the stop-flow HILIC × RP–LC analysis of lipids where a RP trap was used together with a water-rich diluent to trap analytes after the ^{1}D HILIC separation. In this manner dilution effects associated with conventional stop-flow HILIC × RP–LC were avoided. A group-type separation was employed for the different lipid classes in ^{1}D, and since the corresponding sampling times were long, this method is not comprehensive according to the criteria outlined in Section 2.2. Nevertheless, a total of 372 lipids comprising 13 classes were identified in human plasma samples using this approach.

6.7.1.3   Miscellaneous

An online HILIC × RP–LC method utilizing ion pair reversed-phase liquid chromatography (IP-RP–LC) in ^{2}D for the analysis of oligonucleotides was developed by Li et al. (2012). The HILIC method employed a silica column, and the elution order observed for the different classes of oligonucleotides (oligodeoxythymidines < oligodeoxycytidines < oligodeoxyadenosines) was found to be the opposite of that observed in RP–LC. Within each class, the retention behavior of the corresponding homologues were similar. Initial online coupling of the two methods showed peak splitting due to the injection of large volumes of HILIC mobile phase onto the ^{2}D column. By reducing the ^{1}D column diameter and flow rate, this effect could be minimized, but not avoided. Therefore the authors designed a system incorporating two trapping columns in the valve (Figure 6.20a); ^{1}D fractions were trapped on the C18 traps with a make-up flow of water, and eluted in the reverse directions to the ^{2}D column. This approach demonstrated much better separation than could be obtained by either one-dimensional method, and by using a triethylamine acetate buffer in ^{2}D, hyphenation to ESI–MS was enabled to identify oligonucleotides in negative mode (Figure 6.20b).

Lämmerhofer and coworkers reported an offline RP–LC × HILIC method for the analysis of the polar fraction of a nutritional infusion containing amino acids and dipeptides (Schiesel et al. 2012). This approach was prompted by insufficient separation of these compounds by RP–LC; the first 8-min fraction of the RP–LC method was therefore collected and submitted to offline separation. By evaporation of each fraction, low-level impurities could be identified using HILIC–ESI–IT–MS and quantified by HILIC–CAD (for further details on the detection strategies, refer to Section 6.6).

6.7.2   HILIC × RP–LC in the Analysis of Natural Products, Foods, and Beverages

Natural products, due to their complexity, have been the subject of an increasing number of LC × LC studies in recent years. Interest in natural products stems from the broad spectrum of bioactive roles that their constituents are known to possess. Chemical studies of natural products may involve characterization of chemical constituents and/or biologically active compounds. In the former case, the goal is to identify all constituents, whereas the latter focuses on compounds displaying specific reactivity. Traditionally, identification of bioactives in whole natural product extracts relied on bioassay-guided fractionation, which is a tedious process (Van Beek et al. 2009; Malherbe et al. 2012). Furthermore, these offline isolation procedures could lead to sample degradation, artifact formation, or even loss of

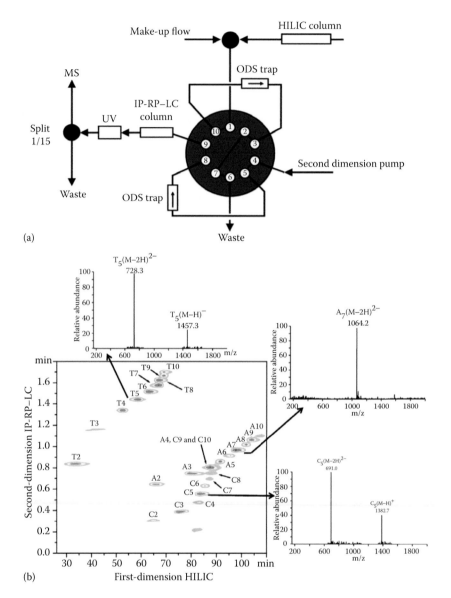

FIGURE 6.20 (a) Schematic representation of the configuration used for the online HILIC × RP–LC–MS analysis of oligonucleotides utilizing two reversed-phase (ODS) traps. First dimension: Ascentis Silica (150 × 1 mm), 155 min gradient at 5 μL/min; dilution: 150 μL/min, sampling time 2 min; second dimension: Xbridge C18 (50 × 2.1 mm), 1.55 min gradient at 3.5 mL/min. (b) Contour plot and example MS spectra obtained for the analysis of oligonucleotide standards. Peak labels: A, C, and T correspond to homologous series of oligodeoxyadenosines, oligodeoxycytidines, and oligodeoxythymidines, respectively. (Reproduced with permission from Li, Q. et al., *J. Chromatogr. A*, 1255, 237–243, 2012.)

analytes (Malherbe et al. 2012). As a means of speeding up this process and to provide more reliable and accurate information on individual compounds, high-resolution screening (HRS) approaches have been developed. In HRS, chromatographic methods are combined with single or multiple detection strategies such as UV–vis or DAD, FL, electrochemical detection (ECD), biochemical detection (BCD), MS, and NMR.

RP–LC, and recently also HILIC, have demonstrated their suitability in the analysis of organic compounds in natural product extracts and foods (Bernal et al. 2011) and as such have found widespread application. A multitude of papers on the application of HILIC and RP–LC, either as single methods or as comprehensive or heart-cutting 2D LC in online, offline or stop-flow modes, may be found in literature. A brief overview of some of the recent applications of HILIC × RP–LC to natural product analysis is presented below (Table 6.2 summarizes the relevant experimental details).

Because of its simplicity and very high resolving power, the offline approach is the most widely used mode for interfacing HILIC and RP–LC separations in natural product analysis. For example, Liang and coworkers investigated the suitability of an in-house made Click β-cyclodextrin (CD) column in ^{2}D for the offline RP–LC × HILIC analysis of traditional Chinese medicines (TCMs) (Liu et al. 2008). The new column was found to provide good separation of both polar and medium-polar components and a high degree of orthogonality relative to RP–LC separation on a C18 column. The same column combination was used to assess the purity of fractions isolated by preparative RP–LC (Wang et al. 2009). In a subsequent study, these authors used the same CD column in both HILIC and RP–LC modes using acidified aqueous acetonitrile mobile phases for the offline RP–LC × HILIC separation of a *Herba Hedyotis Diffusae* extract (Feng et al. 2010). The column was found to exhibit dual retention mechanisms depending on the ratio of acetonitrile and water in the mobile phase, affording a reversed-phase separation at low and HILIC separation at high acetonitrile contents, respectively. (Such behavior has been noted for various HILIC columns [Dos Santos Pereira et al. 2009; Jandera et al. 2010b].) Furthermore, complementary separations were obtained between the two separation modes, and good performance was achieved under both HILIC and RP–LC conditions. Extending this work further, Liang et al. (2012) fractionated a water extract of TCM into polar and medium-polar fractions by solid-phase extraction (SPE) and subjected the polar fraction to offline HILIC × HILIC separation (using a HILIC silica column in ^{1}D and an XAmide column in ^{2}D) and the medium-polar fraction to offline HILIC × RP–LC analysis (using an XAmide column in ^{1}D and a C18 column in ^{2}D). A total of 749 individual components were detected using this approach (206 compounds in the polar fraction and 543 in the medium-polar fraction), and practical peak capacities in excess of 2600 were measured for both systems. Offline HILIC × RP–LC using a combination of CD, C18, and oligo-ethylene glycol (OEG) columns was shown to be more effective than parallel 1D separations for the identification of flavonoids in complex mixtures (Zeng et al. 2012).

Liang and coworkers also devised an offline comprehensive HILIC × RP–LC–MS/MS method for characterization of saponins in a crude extract of *Panax notoginseng* (Xing et al. 2012). One-dimensional fractions were collected and evaporated to dryness under a nitrogen stream and re-dissolved in a solvent which is more

TABLE 6.2
Summary of Recent Applications of HILIC × RP-LC (or RP-LC × HILIC) in Natural Product and Food Analysis

Sample(s)	Compound Class(es)	Hyphenation Mode (Sampling Time)	First-Dimension Conditions	Second-Dimension Conditions	Detection	Analysis Time/Performance	Reference
Carthamus tinctorius Linn.	Polar and medium-polar fractions	Offline (1 min)	RP-LC Column: Inertsil C18 (250 × 4.6 mm, 5 μm)[b] Mobile phases: (A) 0.1% FA[a] in H$_2$O and (B) 0.1% FA in ACN[a]	HILIC Column: Click β-Cyclodextrin (150 × 4.6 mm, 5 μm), 30°C Mobile phases: (A) 0.1% FA in ACN and (B) 0.1% FA in water	UV[a]	~600 min/$n'_{c,2D}$: 1487 (not corrected for β)	Liu et al. (2008)
Cocoa beans Apple	Procyanidins Flavonols Dihydrochalcones Phenolic acids	Offline (1 min)	HILIC Column: Develosil Diol (250 × 1 mm, 5 μm), ambient Mobile phases: (A) 1% acetic acid in ACN and (B) 1% acetic acid in MeOH/H$_2$O (94.05:4.95)	RP-LC Column: Zorbax SB-C18 (50 × 4.6 mm, 1.8 μm), 50°C Mobile phases: (A) 0.1% FA in water and (B) ACN	UV FL[a] ESI–MS[a]	1590 min/$n'_{c,2D}$: 2334–3512	Kalili and de Villiers (2009)
Green tea	Proanthocyanidins Flavonols Phenolic acids	Offline (1 min)	HILIC Column: Develosil Diol (250 × 1 mm, 5 μm), ambient Mobile phases: (A) 1% acetic acid in ACN and (B) 1% acetic acid in MeOH/H$_2$O (94.05:4.95)	RP-LC Column: Zorbax SB-C18 (50 × 4.6 mm, 1.8 μm), 50°C Mobile phases: (A) 0.1% FA in water and (B) ACN	UV FL ESI–MS	1350 min/$n'_{c,2D}$: 2186–2725	Kalili and de Villiers (2010)
Herba Hedyotis Diffusae	Not specified	Offline (1 min)	RP-LC Column: Click β-Cyclodextrin (150 × 4.6 mm, 5 μm), 30°C Mobile phases: (A) 0.1% FA and (B) ACN	HILIC Column: Click β-Cyclodextrin (150 × 4.6 mm, 5 μm), 30°C Mobile phases: (A) ACN and (B) 0.1% FA	UV	~980 min/not specified	Feng et al. (2010)

(Continued)

TABLE 6.2 (Continued)

Summary of Recent Applications of HILIC × RP–LC (or RP–LC × HILIC) in Natural Product and Food Analysis

Sample(s)	Compound Class(es)	Hyphenation Mode (Sampling Time)	First-Dimension Conditions	Second-Dimension Conditions	Detection	Analysis Time/ Performance	Reference
Stevia rebaudiana	Steviol glycosides	Offline (1 min)	RP-LC (preparative) Column: XCharge C18 (150 × 20 mm, 10 μm)[b] Mobile phases: (A) H₂O and (B) ACN	HILIC Column: XAmide (150 × 4.6 mm, 5 μm)[b] Mobile phases: (A) ACN and (B) H₂O	UV ESI-MS[n] NMR[a] (¹H and ¹³C)	~750 min/not specified	Fu et al. (2012)
Scutellaria barbata D. Don	Flavonoids Medium-polar fraction	Offline (1 min)	HILIC Column: XAmide (150 × 4.6 mm, 5 μm)[b] Mobile phases: (A) 0.1% FA in ACN and (B) 0.1% FA	RP-LC Column: XUnion C18 (150 × 2.1 mm, 5 μm)[b] Mobile phases: (A) 0.1% FA and (B) 0.1% FA in ACN	DAD[a] ESI-MS/MS	~540 min/$n'_{c,2D}$: 2698–2879 (not corrected for orthogonality)	Liang et al. (2012)
Panax notoginseng	Saponins	Offline (1 min)	HILIC Column: XAmide (150 × 4.6 mm, 5 μm), 30°C Mobile phases: (A) ACN and (B) H₂O	RP-LC Column: ACQUITY BEH C18 (100 × 2.1 mm, 1.7 μm), 30°C Mobile phases: (A) H₂O and (B) 0.1% FA in ACN	DAD ESI-MS/MS	~770 min/$n_{c,2D}$: 10200 (theoretical), (224 saponins identified)	Xing et al. (2012)
A mixture of 6 traditional Chinese medicines	Flavonoids	Offline (2 min)[c]	HILIC Column: Click CD (150 × 4.6 mm, 5 μm)[b] Mobile phases: (A) H₂O, (B) ACN, and (C)100 mM ammonium formate	RP-LC Columns: XTerra MS C18 (150 × 2.1 mm, 5 μm) and Click oligo (ethylene glycol) (OEG) (150× 4.6 mm, 5 μm)[b] Mobile phases: (A) 0.2% FA and (B) 0.2% FA in ACN	DAD ESI-MS	~450 min/not specified (25 compounds identified)	Zeng et al. (2012)
Panax notoginseng	Saponins	Offline (1 min)	RP-LC (preparative) Column: XUnion C18 column (220 × 80 mm, 10 μm)[b] Mobile phases: (A) H₂O and (B) ACN	HILIC Column: XAmide (150 × 4.6 mm, 5 μm), 30°C Mobile phases: (A) ACN and (B) H₂O	UV ESI-MS[n] NMR (¹H and ¹³C)	~1925 min/not specified	Guo et al. (2013)

(Continued)

TABLE 6.2 (Continued)
Summary of Recent Applications of HILIC × RP–LC (or RP–LC × HILIC) in Natural Product and Food Analysis

Sample(s)	Compound Class(es)	Hyphenation Mode (Sampling Time)	First-Dimension Conditions	Second-Dimension Conditions	Detection	Analysis Time/ Performance	Reference
Blueberries Grape skins Black beans Red radish Red cabbage	Anthocyanins	Offline (0.5 min)	HILIC Column: XBridge BEH Amide (150 × 4.6 mm, 2.5 μm), 50°C Mobile phases: (A) 0.4% TFA[a] in ACN and (B) 0.4% TFA in water	RP-LC Column: Kinetex C18 (50 × 4.6 mm, 2.6 μm), 50°C Mobile phases: (A) 7.5% FA in water and (B) 7.5% FA in ACN	DAD ESI–MS ESI–MS[E]	2310–3300 min/$n'_{c,2D}$: 2255–6953	Willemse et al. (2014)
Rooibos tea (Aspalathus linearis)	Flavonoids	Offline and online (1 min)	HILIC Column: Develosil Diol (250 × 1 mm, 5 μm), ambient Mobile phases: (A) 2% acetic acid in ACN and (B) 2% acetic acid in MeOH/H₂O (93.05/4.95)	RP-LC Column: Zorbax SB-C18 (50 × 4.6 mm, 1.8 μm), 50°C Mobile phases: (A) 0.1% acetic acid and (B) ACN	DAD ESI–MS ESI–MS/MS	132 min (online), 2095 min (offline)/ $n'_{c,2D}$: 556 (online) and 2096 (offline)	Beelders et al. (2012)
Standard solutions	Phenolic and flavonoid standards	Online (1.5 min)	Columns: Lichrospher DIOL (125 × 4.0 mm, 5 μm), ZIC-HILIC sulfobetaine (250 × 2.1 mm, 3.5 μm), Discovery PEG (150 × 2.1 mm, 3 μm), DioEDMA polymethacrylate sulfobetaine monolith (170 × 0.53 mm); all columns at 40 °C Mobile phases: (A) ACN and (B) 10 mM ammonium acetate	Columns: Chromolith Flash RP18-e (25 × 4.6 mm), Poroshell 120 SB-C18 (30 × 3.0 mm, 2.7 μm), Ascentis Express C18 (30 × 3.0 mm, 2.7 μm), Kinetex C18 (50 × 3.0 mm, 2.6 μm), Kinetex XB-C18 (50 × 3.0 mm, 2.6 μm), Kinetex XB-C18 (30 × 3.0 mm, 2.6 μm), Ascentis Express C8 (30 × 3.0 mm, 2.7 μm), Kinetex PFP[a] (30 × 3.0 mm, 2.7 μm), Kinetex PFP (50 × 3.0 mm, 2.6 μm), Kinetex PFP (30 × 3.0 mm, 2.6 μm), Ascentis Express Phenyl-Hexyl (30 × 3.0 mm, 2.7 μm), Ascentis Express RP-Amide (30 × 3.0 mm, 2.7 μm); all columns at 50°C Mobile phases: (A) 10 mM ammonium acetate and (B) ACN	DAD	~60 min/$n'_{c,2D}$: 551–1064	Jandera et al. (2012)

(Continued)

TABLE 6.2 (Continued)
Summary of Recent Applications of HILIC × RP–LC (or RP–LC × HILIC) in Natural Product and Food Analysis

Sample(s)	Compound Class(es)	Hyphenation Mode (Sampling Time)	First-Dimension Conditions	Second-Dimension Conditions	Detection	Analysis Time/ Performance	Reference
Cocoa beans	Procyanidins	Offline (1 min) Online (2 min) Stop flow (1 min)	HILIC Column: Develosil Diol (250 × 1 mm, 5 μm), ambient Mobile phases: (A) ACN and acetic acid (99:1) and (B) MeOH, H_2O and acetic acid (94.05:4.95:1)	RP-LC Column: Zorbax SB-C18 (50 × 4.6 mm, 1.8 μm), 50°C Mobile phases: (A) 0.1% FA and (B) ACN	UV FL ESI-MS	800 min/n'_{c2D}: 1760 (offline); 102 min/ n'_{c2D}: 462 (online); 800 min/n'_{c2D}: 1387 (stop-flow)	Kalili and de Villiers (2013a)
Grape seeds	Condensed tannins	Online (2 min)	HILIC Column: Develosil Diol (250 × 1 mm, 5 μm), ambient Mobile phases: (A) 1% acetic acid in ACN and (B) 1% acetic acid in MeOH/H_2O (94.05/4.95)	RP-LC Column: Zorbax SB-C18 (50 × 4.6 mm, 1.8 μm), 50°C Mobile phases: (A) 0.1% FA and (B) ACN	UV FL ESI-MS	102 min/78 compounds identified	Kalili et al. (2013)
Grape seeds	Procyanidins	Online (1.3 min)	HILIC Column: Lichrospher diol-5 (150 × 1.0 mm, 5 μm)[b] Mobile phases: (A) 2% acetic acid in ACN and (B) 2% acetic acid in MeOH/H_2O (95/3)	RP-LC Column: Ascentis Express C18 (50 × 4.6 mm, 2.7 μm) and monolithic C18 (100 × 4.6 mm)[b] Mobile phases: (A) 0.1% FA and (B) 50/50 ACN/MeOH	DAD ESI-MS/MS	~85 min/n'_{c2D}: 875 (not corrected for orthogonality, 46 compounds identified)	Montero et al. (2013b)
Apples	Procyanidins, Flavonols Dihydro-chalcones Phenolic acids	Online (1.3 min)	HILIC Column: Lichrospher diol-5 (150 × 1.0 mm, 5 μm)[b] Mobile phases: (A) 2% acetic acid in ACN and (B) 2% acetic acid in MeOH/H_2O (95/3)	RP-LC Column: Ascentis Express C18 (50 × 4.6 mm, 2.7 μm)[b] Mobile phase 1: (A) 0.1% FA and (B) ACN; Mobile phase 2: (A) 0.1% FA and (B) 50/50 ACN/MeOH	DAD ESI-MS/ MS	~50 min/not specified (65 compounds identified)	Montero et al. (2013a)

(Continued)

TABLE 6.2 (Continued)
Summary of Recent Applications of HILIC × RP–LC (or RP–LC × HILIC) in Natural Product and Food Analysis

Sample(s)	Compound Class(es)	Hyphenation Mode (Sampling Time)	First-Dimension Conditions	Second-Dimension Conditions	Detection	Analysis Time/ Performance	Reference
Cystoseira abies-marina (brown algae)	Phlorotannins	Online (1.3 min)	HILIC Column: Lichrospher diol-5 (150 × 1.0 mm, 5 μm)[b] Mobile phases: (A) 2% acetic acid in ACN and (B) 2% acetic acid in MeOH/H$_2$O (95/3)	RP-LC Column: Ascentis Express C18 (50 × 4.6 mm, 2.7 μm) and Kinetex PFP (50 × 4.6 mm, 2.6 μm)[b] Mobile phases: (A) 0.1% FA and (B) ACN	DAD ESI–MS/MS	~85 min/$n'_{c,2D}$: 739 and 992 (not corrected for orthogonality, 52 compounds identified)	Montero et al. (2014)
Panax ginseng roots	Triterpenoid saponins	Stop-flow (variable)[c]	HILIC Column: TSKGEL Amide-80 (150 × 2.0 mm, 3.0 μm)[b] Mobile phases: (A) 0.05% acetic acid in ACN and (B) 0.05% acetic acid	RP-LC Column: Acquity BEH C18 (100 × 2.1 mm, 1.7 μm), 35°C Mobile phases: (A) 0.05% acetic acid and (B) 0.05% acetic acid in ACN	ESI–MS/MS	~120 min/$n_{c,2D}$: 747 (theoretical, 94 compounds identified)	Wang et al. (2015)

a ACN = acetonitrile, DAD = diode array detection, ESI = electrospray ionization, FA = formic acid, FL = fluorescence, MeOH = methanol, MS/MS = tandem mass spectrometry, NMR = nuclear magnetic resonance spectroscopy, PFP = pentafluorophenyl, TFA = trifluoroacetic acid, UV = ultraviolet-visible detection.

b Analysis temperature not specified.

c Analysis not truly comprehensive due to excessive sampling times.

favorable for injecting onto the ^{2}D column. Tentative identification of 224 saponins was possible based on MS and MS/MS data. While high theoretical peak capacity values were reported for this study, the actual performance is much lower due to the effects of undersampling (excessive fraction collection times were used) and finite orthogonality. This work was extended to the offline preparative RP–LC × HILIC isolation of saponins from *Panax notoginseng* (Guo et al. 2013) and steviol glycosides from *Stevia rebaudiana* (Fu et al. 2012). Using this approach, 13 steviol glycosides and 9 saponins were identified on the basis of chromatographic retention time, accurate MWs, diagnostic fragmentation ions, and NMR data.

In our group, we have utilized offline HILIC × RP–LC for the analysis of diverse classes of phenolic compounds, including proanthocyanidins, anthocyanins, flavonols, flavones, and phenolic acids, in various natural product extracts (Kalili and de Villiers 2009, 2010; Beelders et al. 2012; Willemse et al. 2014). To avoid the need to evaporate the mobile phase between the two dimensions, concentrated extracts were injected onto a microbore (1 mm i.d.) diol HILIC column in ^{1}D. This ensured minimal sample dilution in ^{1}D, with fraction volumes for 1-min sampling times corresponding to ~50 µL. Using this approach, good sensitivity was achieved in ^{2}D when ≤2 µL was injected (2 µL was the maximum permissible injection volume of HILIC solvents onto a 4.6 mm i.d. RP–LC column to avoid peak distortion). The complementarity of HILIC and RP–LC separation for phenolic analysis was clearly illustrated for a range of natural products, with contour plots displaying well-structured elution patterns where phenolics were separated according to MW or polarity in the HILIC dimension and on the basis of hydrophobicity in the RP–LC dimension. These systems provided exceptionally high practical peak capacities (as high as 7000) for long analysis times. As an example, Figure 6.21 shows contour plots obtained for the offline HILIC × RP–LC analysis of anthocyanins in red grape skins and red cabbage. By employing optimized separation conditions in both dimensions to accommodate the unique chromatographic properties of anthocyanins (De Villiers et al. 2009a, 2011), practical peak capacities of 2800 and 4550, respectively, were attained for these analyses. Also of interest is the fact that different anthocyanin classes were grouped together in the ^{2}D retention space based on their degree of glycosylation and acylation (Willemse et al. 2014).

In spite of increased operational complexity and challenges associated with the online coupling of HILIC and RP–LC methods, this hyphenation mode has also been applied in natural product analysis with notable success.

In our group, we adopted a flow-splitting approach as a means of overcoming the solvent issues in online HILIC × RP–LC systems (Beelders et al. 2012; Kalili and de Villiers 2013a; Kalili et al. 2013, 2014; Willemse et al. 2015). The use of highly concentrated extracts and minimal dilution on the ^{1}D column provided sufficient sensitivity in ^{2}D using UV, FL, and MS detection (Figure 6.14). For example, this approach allowed separation of 78 proanthocyanidins, including 33 procyanidins, 25 monogalloylated-, 11 digalloylated-, 6 trigalloylated-, and 1 tetragalloylated procyanidin, as well as 2 (epi)catechin monoglycosides in grape seed (Kalili et al. 2013). In addition, procyanidins of DP up to 16 were detected in the grape seed extract. Practical peak capacities in the region of 550 were determined for these analyses.

Herrero and coworkers developed an alternative online HILIC × RP–LC method for the analysis of grape seed procyanidins (Montero et al. 2013b). Combining this

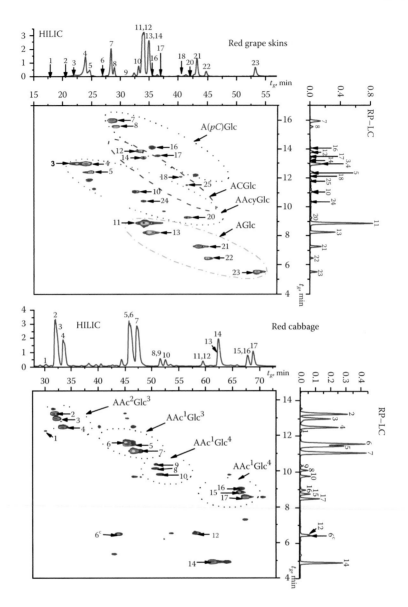

FIGURE 6.21 Contour plots obtained for the offline HILIC × RP–LC analysis of anthocyanins in grape skins (top) and red cabbage (bottom). First dimension: XBridge Amide (150 × 4.6 mm), 100/113 min gradients at 0.2 mL/min, 50°C; fraction collection: 0.5 min, 2 μL injected; second dimension: Kinetex C18 (50 × 4.6 mm), 25 min gradient at 0.5 mL/min, 50°C. Peak labels: AGlc, anthocyanidin-glucosides; AAcyGlc, anthocyanidin-acetyl-glucosides; ACGlc, anthocyanidin-caffeoyl-glucosides; A(pC)Glc, anthocyanidin-(p-coumaroyl)glucosides; AAc²Glc³, anthocyanidin-di-acylated-triglucosides; AAc¹Glc³, anthocyanidin-mono-acylated-triglucosides; AAc²Glc⁴, anthocyanidin-di-acylated-tetra-glucosides; AAc¹Glc⁴, anthocyanidin-mono-acylated-tetra-glucosides. (Reproduced with permission from Willemse, C. M. et al., *J. Chromatogr. A*, 1359, 189–201, 2014.)

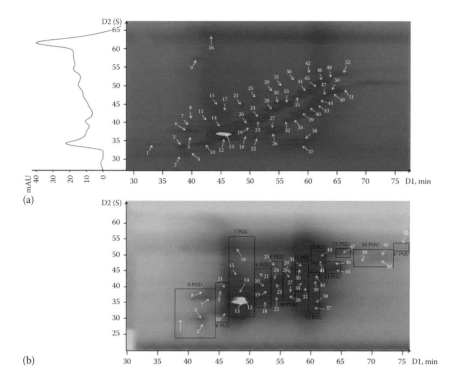

FIGURE 6.22 Contour plots obtained for the online HILIC × RP–LC analysis of phlorotannins in brown algae. First dimension: diol (150 × 1 mm), 85 min gradient at 15 µL/min; modulation period 1.3 min, injection volume 19.5 µL (30 µL loops); Second dimension: (a) Ascentis Express C18 (50 × 4.6 mm) and (b) Kinetex PFP (50 × 4.6 mm), 1.3 min gradient at 3 mL/min. Peak labels: PGU refers to the number of phloroglucinol units of phlorotannins tentatively identified by MS/MS. (Reproduced with permission from Montero, L. et al., *Electrophoresis*, 35, 1644–1651, 2014.)

method with DAD and MS/MS detection, these researchers successfully characterized 46 compounds in a total analysis time of ~85 min. In subsequent studies, a slightly modified methodology was employed in the analysis of different classes of phenolics in apples (Montero et al. 2013a) and phlorotannins in brown algae (Montero et al. 2014) allowing identification of 65 components in the apple extract and 52 compounds in the brown algae extract (Figure 6.22). Practical peak capacities ranging between 793 and 992 were reported for these analyses, although these values were not corrected for lack of orthogonality. To minimize injection band broadening due to solvent effects, these authors used relatively large sampling loops (30 µL for 19.5 µL fractions) to dilute the ¹D eluent before injection onto the ²D column.

In an effort to overcome the mobile phase compatibility issues associated with HILIC × RP–LC systems, Jandera and coworkers (2012) synthesized a polar monolithic sulfobetaine polymethacrylate capillary column (0.5 mm i.d.) for use in the online HILIC × RP–LC separation of phenolic and flavonoid compounds. When used in the first (HILIC) dimension together with various 3.0 mm i.d. superficially porous C18 columns in the ²D, peak distortion was significantly suppressed owing to the small fraction volumes provided by the capillary column. The monolithic column provided high orthogonality, good separation performance and sensitivity in comparison to commercial silica-based diol, polyethylene glycol (PEG) or ZIC columns.

Stop-flow HILIC × RP–LC has also been successfully applied in the characterization of natural products, despite concerns regarding the relative operational complexity of this approach and the risk of additional band broadening during stop-flow periods (Shalliker and Gray 2006). To assess the effect of stopping the ^{1}D flow on band broadening, we have quantified the loss in peak capacity in the case of stop-flow HILIC × RP–LC analysis of procyanidins (Kalili and de Villiers 2013a, 2013b). First, the effective diffusion coefficients of procyanidins of different DP were determined from experiments where band broadening was measured for various stop-flow times. From these values, the effect of stopping the flow on the ^{1}D peak capacity was determined using the effective stop-flow times for each peak (*cf.* Section 6.5.6). These calculations showed that additional band broadening due to the stop-flow operation was minimal, reaching a maximum of only 3.6% loss in ^{1}D peak capacity for stop-flow periods (corresponding to ^{2}D cycle times) of up to 60 min (Kalili and de Villiers 2013b). These predictions were confirmed experimentally, as comparable elution profiles were obtained when cocoa procyanidins were analyzed using an analysis time of 15 min in ^{2}D under offline and stop-flow HILIC × RP–LC conditions (Kalili and de Villiers 2013a) (Figure 6.23).

Wang et al. (2015) reported a stop-flow 2D HILIC–RP–LC method for the characterization of triterpenoid saponins from a ginseng root extract. Note that this method is not truly comprehensive because relatively long sampling times were used, resulting is recombination of peaks separated in ^{1}D (Schoenmakers et al. 2003; Marriot et al. 2012). Nevertheless, the approach was suitable for its intended purpose and is interesting in that a trapping column was used to facilitate fraction transfer between the two dimensions. The ^{1}D and ^{2}D columns were connected via an 8-port 2-position switching valve. Essentially, the sample was injected on the ^{1}D column and a fraction was transferred to a mixer where it was diluted with a make-up flow of acidified water to weaken the solvent strength of the HILIC fraction before transfer onto a trapping column. After trapping, the valve was switched and the ^{1}D flow stopped to allow the fraction to be analyzed in ^{2}D. Once ^{2}D analysis was completed, the valve was switched back to the trap position for transfer of the next fraction, and this procedure was repeated throughout the ^{1}D separation. The method demonstrated good linearity and repeatability, although the separation performance was relatively low for a stop-flow system (a theoretical peak capacity of 747 was determined). The method was successfully applied for the classification of ginseng extracts differing in age using statistical analysis. A total of 94 triterpenoid saponins were identified based on mass spectral data and 19 triterpenoid saponins were found to differ in concentration depending on the age of the ginseng analyzed.

6.7.3 HILIC × RP–LC in Polymer Analysis

HILIC × RP–LC has also received some application in polymer analysis, although to date much less than is the case for biological samples and natural products. HILIC offers some potential benefits compared to the more conventional forms of

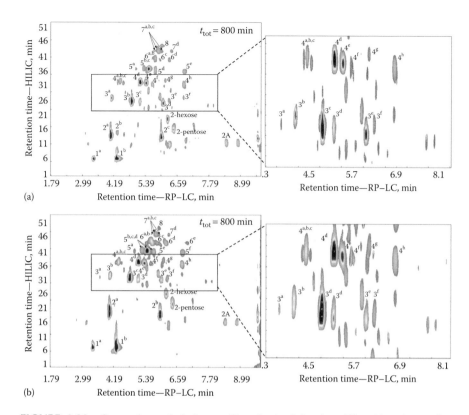

FIGURE 6.23 Comparison of elution profiles obtained for the offline (a) and stop-flow (b) HILIC × RP–LC–FL analysis of cocoa procyanidins. First dimension: Diol column (250 × 1 mm), 50 min gradient at 50 µL/min; sampling time: 1 min (1 : 32 split for stop-flow); second dimension: C18 column (50 × 4.6 mm), flow rate 1.5 mL/min. Peak labels: numbers correspond to the degree of polymerization (DP) of procyanidin molecules, superscripts distinguish between isomers of the same DP. (Reproduced from Kalili, K. M. and de Villiers, A., *J. Chromatogr. A,* 1289, 58–68, 2013b. With permission.)

HPLC used in polymer analysis such as SEC, NP–LC, RP–LC, and LCCC. In the latter mode, the mobile phase composition and temperature are chosen such that enthalpic and entropic contributions to retention are balanced, and retention is governed by a given structural unit independent of the molar mass or other structural features (Malik and Pasch 2014). This is referred to as the critical point of adsorption (CAP) for the target structural unit, often a particular end group.

Jandera and coworkers explored the use of HILIC as opposed to NP–LC in ^{2}D for the online LC × LC separation of ethylene oxide–propylene oxide (EO–PO) co-oligomers (Jandera et al. 2006). The reasoning behind this approach was that HILIC is more compatible with RP–LC used in ^{1}D, since injection of aqueous fractions lead to stationary phase deactivation and poor reproducibility in classical NP–LC. A 1 mm i.d. aminopropyl column was used under isocratic conditions with a mobile phase consisting of 1.2% ethanol and 1.5% water in dichloromethane in ^{2}D to separate the co-oligomers based on the number of EO units (it is therefore questionable whether this method can truly be considered HILIC). This method provided good orthogonality since RP–LC separation in ^{1}D separated co-oligomers according to the number of PO units. Injection of 10 µL fractions did not cause problems in ^{2}D separation.

Abrar and Trathnigg (2010) reported an online RP–LC × HILIC system for the analysis of polysorbates and fatty esters of PEG. In the RP–LC dimension, conditions suitable for LCCC were used, whereas HILIC provided separation according to the number of terminal hydroxyl groups. Two configurations (HILIC × RP–LC and RP–LC × HILIC) were investigated using Luna HILIC and Jupiter C18 columns in the HILIC and RP–LC dimensions, respectively. A constant mobile phase composition was used throughout the analysis, at 0.1 and 2 or 2.5 mL/min in ^{1}D and ^{2}D, respectively. The sampling time was kept constant at 2 min. It was found that higher ester oligomers were too strongly retained on the RP column, requiring analysis times greater than 2.5 min when a flow rate of 2.5 mL/min was used on a 15 cm column. The use of higher flow rates was not possible due to pressure limitations, and shorter columns did not provide satisfactory resolution. On the other hand, HILIC separation could be completed with sufficient resolution within 2 min on a 25 cm column operated at a flow rate of 2.0 mL/min. Therefore, the authors opted for the RP–LC × HILIC configuration, allowing satisfactory resolution of different polysorbates according to functionality within 90 min.

Wu and Marriott (2011) evaluated NP–LC, HILIC, and RP–LC for the LC × LC analysis of alkylphenol polyethoxylates. RP–LC offered good separation for octyl and nonyl alkylphenol polyethoxylates under isocratic conditions and was selected for use in ^{2}D. NP–LC afforded better resolution compared to HILIC, although the comprehensive combination of NP–LC and RP–LC was not attempted due to mobile phase immiscibility. However, HILIC × RP–LC delivered good separation performance, allowing separation of analytes on the basis of both alkyl end groups (RP–LC) and ethoxymer chain lengths (HILIC).

Elsner et al. (2012) explored the combination of ZIC–HILIC in ^{1}D and reversed phase in ^{2}D for the simultaneous analysis of anionic (fatty alcohol sulfates, fatty alcohol ether sulfates), nonionic (alkyl polyglucosides, fatty alcohol ethoxylates), and amphoteric (cocamidopropyl betaines) surfactants. In ^{1}D, an acetonitrile/ammonium acetate gradient was used at a flow rate of 25 µL/min, whereas an ammonium acetate/methanol gradient was used in ^{2}D on a C8 column at a flow rate of 3 mL/min. Detection was performed by ESI–Q–TOF–MS. This system provided baseline separation and a well-structured representation of all surfactants according to the degree of ethoxylation in the ZIC–HILIC dimension and with respect to alkyl chain lengths in the RP–LC dimension (Figure 6.24). This method has been shown to provide good reproducibility in terms of retention time and peak area on a commercial LC × LC instrument (Elsner et al. 2015).

6.8   CONCLUDING REMARKS

The field of LC × LC has undergone rapid expansion in the last decade. This has been made possible first of all by important advances in our understanding of the fundamental aspects of LC × LC, especially with regard to sampling times and the "practical"

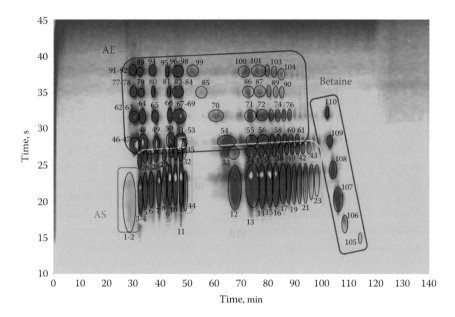

FIGURE 6.24 Contour plot obtained for the online HILIC × RP–LC separation of anionic, nonionic, and amphoteric surfactants. First dimension: ZIC–HILIC column (250 × 2.1 mm), 110 min gradient at 25 µL/min; sampling time: 1 min; second dimension: C8 column (30 × 4.6 mm), 3 mL/min; detection: +ESI-(Q-TOF)-MS. Peak labels: AE; fatty alcohol ethoxylates, AS; fatty alcohol sulfates, AG; alkyl polyglycosides, AES; fatty alcohol ether sulfates. (Reproduced with permission from Elsner, V. et al., *J. Chromatogr. A*, 1268, 22–28, 2012.)

performance attainable and how to measure this accurately. The second major contributing factor to the growth of LC × LC has been instrumental developments, with especially the availability of high-speed columns and instruments optimized to provide maximum performance on these columns proving highly beneficial in ²D of online LC × LC systems. Other important developments include columns and instruments enabling high-temperature operation, and of course the availability of more powerful MS detectors. Some aspects that do however require further research include the need for an accurate and universally accepted means of estimating orthogonality, and method development in LC × LC, which currently still represents a significant challenge.

From inspection of Figure 6.1 in this chapter, it is clear that HILIC × RP–LC has also found increasing application during this period, although the technique currently still represents only a small percentage of all LC × LC applications. Certainly, techniques such as IEC × RP–LC, SEC × RP–LC, and even RP–LC × RP–LC are much more common due to their proven performance for specific analytes (peptides, polymers) and the relative ease with which these modes may be coupled. However, several recent developments provide evidence that HILIC × RP–LC will find increasing application in the coming years. First of all, important findings regarding the complimentary nature of HILIC for the analysis of biological samples (especially peptides and lipids) and natural products (especially phenolic compounds) is expected to drive increasing use in the fields of proteomics, metabolomics, and natural product analysis.

In fact, the development of improved 1D HILIC–MS methods in these fields underpins much of the HILIC × RP–LC work reported to date, and this trend can be expected to continue in the near future. Secondly, the recent availability of the first commercial LC × LC instrumentation is sure to accelerate the more widespread use of the technique.

On the other hand, the relative elution strengths of the mobile phases used in HILIC and RP–LC do make the hyphenation of these techniques uniquely challenging. As a consequence, method development in HILIC × RP–LC is arguably much more challenging than is the case for techniques such as IEC × RP–LC. This is most likely a contributing factor to the observation (Figure 6.1) that the majority of applications involving HILIC × RP–LC are performed in the offline mode (53% compared to 40% for online operation), which is the simplest and most flexible way of doing LC × LC. Another reason for this lies in the complexity of the samples encountered in especially proteomics: for these samples, maximum separation performance is the most important criterion, and offline HILIC × RP–LC will therefore remain the preferred mode for such applications. For relatively simpler samples, the utilization of online HILIC × RP–LC can realistically be expected to increase due to the associated benefits such as automation and speed.

Finally, it is important to stress that optimal performance in LC × LC, that is, performance much better than is attainable by 1D LC, is only attained through careful optimization of several related experimental parameters—for this, a thorough understanding of the theory of LC × LC is essential. To date, too many of the reports utilizing HILIC × RP–LC are performed using suboptimal experimental conditions, resulting in separations not demonstrably better than optimized 1D HPLC methods. Nevertheless, as more researchers are exposed to the field and instrumentation is improved, more widespread and efficient application of HILIC × RP–LC can be expected in the near future. The benefits provided by optimized HILIC × RP–LC methods in combination with suitable detection strategies can then be more fully exploited by the research community.

ACKNOWLEDGMENTS

The authors gratefully acknowledge financial support from Stellenbosch University, Sasol, the National Research Foundation (NRF, South Africa), and the University of Namibia.

SYMBOLS AND ABBREVIATIONS

SYMBOLS

α	selectivity factor
β	undersampling correction factor
$^{x}\sigma$	standard deviation of a chromatographic peak in the xth dimension
$\sigma_{stop\text{-}flow}$	band broadening (standard deviation) resulting from stop-flow operation
σ_{cont}	standard deviation for a continuous flow analysis
σ_{tot}	total standard deviation, applicable in stop-flow analysis

^xD	xth dimension
D_{eff}	effective diffusion coefficient
^xDF	dilution factor in the xth dimension
^xF	volumetric flow rate in the xth dimension
f	number of fractions collected in the first dimension
f_c	fractional surface coverage
N	number of theoretical plates
$n'_{c,\text{2D}}$	(practical) peak capacity of a two-dimensional separation
$n_{c,\text{1D}}$	peak capacity of a one-dimensional separation
$n_{c,\text{2D}}$	(theoretical) peak capacity of a two-dimensional separation
^xn_c	peak capacity of separation in the xth dimension
$^1n'_c$	corrected first-dimension peak capacity
R_s	(chromatographic) resolution
s	sample dimensionality
SR	*split* ratio
^xt_c	cycle time in the xth dimension (sum of ^xt_g and $^xt_{\text{re-equil}}$)
^xt_g	gradient time in the xth dimension
t_R	retention time
$^xt_{\text{re-equil}}$	re-equilibration time in the xth dimension
^1t_s	first dimension sampling time
t_{stop}	effective stop-flow time
t_{tot}	total analysis time
V_{frac}	fraction volume
$^xV_{\text{inj}}$	injection volume in the xth dimension
$^2V_{\text{max}}$	maximum injection volume in the second dimension
w_b	peak width at baseline ($= 4\sigma$)

ABBREVIATIONS

1D	one-dimensional
2D	two-dimensional
ABTS	2,2′-azino-bis(3-ethylbenzothiazoline)-6 sulfonic acid
ACE	angiotensin I-converting enzyme
AE	fatty alcohol ethoxylates
AES	fatty alcohol ether sulfates
AG	alkyl polyglycosides
AS	fatty alcohol sulfates
BCD	biochemical detection
BPH	benign prostatic hyperplasia
BSA	bovine serum albumin
CAD	charged aerosol detector
CAP	critical point of adsorption
CD	cyclodextrin
CH	convex hull method to estimate orthogonality
CID	collision-induced dissociation
CS	continuously shifting gradients

DAD	diode array detection
DP	degree of polymerization
ECD	electrochemical detection
ECN	equivalent carbon number
ELSD	evaporative light scattering detector
EO	ethylene oxide
ESI	electrospray ionization
FACS	fluorescence-activated cell sorting
FID	flame ionization detector
FIF	full in fraction gradients
FL	fluorescence
GC	gas chromatography
GC × GC	comprehensive two-dimensional gas chromatography
HILIC	hydrophilic interaction chromatography
HPLC	high-performance liquid chromatography
HR–MS	high-resolution mass spectrometry
HRS	high-resolution screening
HSA	human serum albumin
i.d.	internal diameter
IEC	ion exchange chromatography
IMAC	immobilized metal affinity chromatography
IP	ion pair
IT	ion trap
LC	liquid chromatography
LCCC	liquid chromatography under critical conditions
LC × LC	comprehensive two-dimensional liquid chromatography
LC–LC	heart-cutting two-dimensional liquid chromatography, or serial LC
LSS	linear solvent strength
MD–LC	multidimensional liquid chromatography
MS	mass spectrometry
MS/MS	tandem mass spectrometry
MS^n	multistage mass spectrometry (where $n \geq 2$)
MudPIT	multidimensional protein identification technology
MW	molecular weight
NMR	nuclear magnetic resonance
NP–LC	normal phase liquid chromatography
ODS	octadecyl-silica
OEG	oligo-ethylene glycol
PEG	polyethylene glycol
PFP	pentafluorophenyl
PGU	phloroglucinol unit(s) of phlorotannins
PO	propylene oxide
Q	quadrupole
RP–LC	reversed-phase liquid chromatography
SC	surface coverage

SCX	strong cation exchange
SEC	size exclusion chromatography
SIF	segment in fraction gradients
SPE	solid-phase extraction
SR	split ratio
TCM	traditional Chinese medicine
TOF	time-of-flight
UHPLC	ultra high-pressure liquid chromatography
UV–vis	ultraviolet–visible
ZIC-HILIC	zwitterionic-HILIC

REFERENCES

Abrar, S. and B. Trathnigg. 2010. Separation of Nonionic Surfactants according to Functionality by Hydrophilic Interaction Chromatography and Comprehensive Two-Dimensional Liquid Chromatography. *Journal of Chromatography A* 1217(52): 8222–8229. doi:10.1016/j.chroma.2010.10.118.

Al Bakain, R., I. Rivals, P. Sassiat, D. Thiébaut, M. C. Hennion, G. Euvrard, and J. Vial. 2011. Comparison of Different Statistical Approaches to Evaluate the Orthogonality of Chromatographic Separations: Application to Reverse Phase Systems. *Journal of Chromatography A* 1218(20):2963–2975. doi:10.1016/j.chroma.2011.03.031.

Alpert, A. J. 1990. Hydrophilic-Interaction Chromatography for the Separation of Peptides, Nucleic Acids and Other Polar Compounds. *Journal of Chromatography* 499:177–196.

Anderson, N. L. and N. G Anderson. 2002. The Human Plasma Proteome: History, Character, and Diagnostic Prospects. *Molecular & Cellular Proteomics* 1:845–867. doi:10.1074/mcp. R200007-MCP200.

Appelblad, P., T. Jonsson, W. Jiang, and K. Irgum. 2008. Fast Hydrophilic Interaction Liquid Chromatographic Separations on Bonded Zwitterionic Stationary Phase. *Journal of Separation Science* 31(9):1529–1536. doi:10.1002/jssc.200800080.

Balke, S. T. and R. D. Patel. 1983. Orthogonal Chromatography: Polymer Cross-Fractionation by Coupled Gel Permeation Chromatographs. In *Advances in Chemistry Series*, vol. 203, pp. 281–310. Washington, DC: American Chemical Society.

Bedani, F., W. T. Kok, and H.-G. Janssen. 2006. A Theoretical Basis for Parameter Selection and Instrument Design in Comprehensive Size-Exclusion Chromatography X Liquid Chromatography. *Journal of Chromatography A* 1133(1–2):126–134. doi:10.1016/j.chroma.2006.08.048.

Bedani, F., W. T. Kok, and H.-G. Janssen. 2009. Optimal Gradient Operation in Comprehensive Liquid Chromatography X Liquid Chromatography Systems with Limited Orthogonality. *Analytica Chimica Acta* 654(1):77–84. doi:10.1016/j.aca.2009.06.042.

Bedani, F., P. J. Schoenmakers, and H.-G. Janssen. 2012. Theories to Support Method Development in Comprehensive Two-Dimensional Liquid Chromatography—A Review. *Journal of Separation Science* 35(14):1697–1711. doi:10.1002/jssc.201200070.

Beelders, T., K. M. Kalili, E. Joubert, D. de Beer, and A. de Villiers. 2012. Comprehensive Two-Dimensional Liquid Chromatographic Analysis of Rooibos (Aspalathus Linearis) Phenolics. *Journal of Separation Science* 35(14):1808–1820. doi:10.1002/jssc.201200060.

Bernal, J., A. M. Ares, J. Pól, and S. K. Wiedmer. 2011. Hydrophilic Interaction Liquid Chromatography in Food Analysis. *Journal of Chromatography A* 1218(42):7438–7452. doi:10.1016/j.chroma.2011.05.004.

Boersema, P. J., N. Divecha, A. J. R. Heck, and S. Mohammed. 2007. Evaluation and Optimization of ZIC-HILIC-RP as an Alternative MudPIT Strategy. *Journal of Proteome Research* 6(3):937–946. doi:10.1021/pr060589m.

Boersema, P. J., S. Mohammed, and A. J. R. Heck. 2008. Hydrophilic Interaction Liquid Chromatography (HILIC) in Proteomics. *Analytical and Bioanalytical Chemistry* 391(1):151–159. doi:10.1007/s00216-008-1865-7.

Burgman, M. A. and J. C. Fox. 2003. Bias in Species Range Estimates from Minimum Convex Polygons: Implications for Conservation and Options for Improved Planning. *Animal Conservation* 6(1):19–28. doi:10.1017/S1367943003003044.

Cacciola, F., P. Jandera, Z. Hajdú, P. Česla, and L. Mondello. 2007. Comprehensive Two-Dimensional Liquid Chromatography with Parallel Gradients for Separation of Phenolic and Flavone Antioxidants. *Journal of Chromatography A* 1149:73–87. doi:10.1016/j. chroma.2007.01.119.

Carr, P. W., J. M. Davis, S. C. Rutan, and D. R. Stoll. 2012. Principles of Online Comprehensive Multidimensional Liquid Chromatography. In *Advances in Chromatography*, vol. 50, E. Grushka and N. Grinberg (eds.), 50th edn., pp. 139–236. Boca Raton, FL: CRC Press.

Cífková, E., M. Holčapek, and M. Lísa. 2013. Nontargeted Lipidomic Characterization of Porcine Organs Using Hydrophilic Interaction Liquid Chromatography and Off-Line Two-Dimensional Liquid Chromatography-Electrospray Ionization Mass Spectrometry. *Lipids* 48(9):915–928. doi:10.1007/s11745-013-3820-4.

D'Attoma, A., C. Grivel, and S. Heinisch. 2012. On-Line Comprehensive Two-Dimensional Separations of Charged Compounds Using Reversed-Phase High Performance Liquid Chromatography and Hydrophilic Interaction Chromatography. Part I: Orthogonality and Practical Peak Capacity Considerations. *Journal of Chromatography A* 1262:148–159. doi:10.1016/j.chroma.2012.09.028.

D'Attoma, A. and S. Heinisch. 2013. On-Line Comprehensive Two Dimensional Separations of Charged Compounds Using Reversed-Phase High Performance Liquid Chromatography and Hydrophilic Interaction Chromatography. Part II: Application to the Separation of Peptides. *Journal of Chromatography A* 1306:27–36. doi:10.1016/j.chroma.2013.07.048.

Davis, J. M. and J. C. Giddings. 1983. Statistical-Theory of Component Overlap in Multicomponent Chromatograms. *Analytical Chemistry* 55(3):418–424. doi:10.1021/ac00254a003.

Davis, J. M., D. R. Stoll, and P. W. Carr. 2008a. Dependence of Effective Peak Capacity in Comprehensive Two-Dimensional Separations on the Distribution of Peak Capacity between the Two Dimensions. *Analytical Chemistry* 80(21):8122–8134. doi:10.1021/ ac800933z.

Davis, J. M., D. R. Stoll, and P. W. Carr. 2008b. Effect of First-Dimension Undersampling on Effective Peak Capacity in Comprehensive Two-Dimensional Separations. *Analytical Chemistry* 80:461–473. doi:10.1021/ac800933z.

De Villiers, A., D. Cabooter, F. Lynen, G. Desmet, and P. Sandra. 2009a. High Performance Liquid Chromatography Analysis of Wine Anthocyanins Revisited: Effect of Particle Size and Temperature. *Journal of Chromatography A* 1216(15):3270–3279. doi:10.1016/j. chroma.2009.02.038.

De Villiers, A., D. Cabooter, F. Lynen, G. Desmet, and P. Sandra. 2011. High-Efficiency High Performance Liquid Chromatographic Analysis of Red Wine Anthocyanins. *Journal of Chromatography. A* 1218(29):4660–4670. doi:10.1016/j.chroma.2011.05.042.

De Villiers, A., F. Lestremau, R. Szucs, S. Gélébart, F. David, and P. Sandra. 2006. Evaluation of Ultra Performance Liquid Chromatography. Part I. Possibilities and Limitations. *Journal of Chromatography A* 1127(1–2):60–69.

De Villiers, A., F. Lynen, and P. Sandra. 2009b. Effect of Analyte Properties on the Kinetic Performance of Liquid Chromatographic Separations. *Journal of Chromatography A* 1216(16):3431–3442. doi:10.1016/j.chroma.2008.11.101.

Dejaegher, B., D. Mangelings, and Y. V. Heyden. 2008. Method Development for HILIC Assays. *Journal of Separation Science* 31(9):1438–1448. doi:10.1002/jssc.200700680.

Desmet, G., D. Cabooter, P. Gzil, H. Verelst, D. Mangelings, Y. V. Heyden, and D. Clicq. 2006. Future of High Pressure Liquid Chromatography: Do We Need Porosity or Do We Need Pressure? *Journal of Chromatography A* 1130(1):158–166. doi:10.1016/j.chroma.2006.05.082.

Di Palma, S., P. J. Boersema, A. J. R. Heck, and S. Mohammed. 2011a. Zwitterionic Hydrophilic Interaction Liquid Chromatography (ZIC-HILIC and ZIC-cHILIC) Provide High Resolution Separation and Increase Sensitivity in Proteome Analysis. *Analytical Chemistry* 83:3440–3447.

Di Palma, S., S. Mohammed, and A. J. R. Heck. 2012. ZIC-cHILIC as a Fractionation Method for Sensitive and Powerful Shotgun Proteomics. *Nature Protocols* 7(11):2041–2055.

Di Palma, S., D. Stange, M. Van De Wetering, H. Clevers, A. J. R. Heck, and S. Mohammed. 2011b. Highly Sensitive Proteome Analysis of FACS-Sorted Adult Colon Stem Cells. *Journal of Proteome Research* 10:3814–3819. doi:10.1021/pr200367p.

Dos Santos Pereira, A., F. David, G. Vanhoenacker, and P. Sandra. 2009. The Acetonitrile Shortage: Is Reversed HILIC with Water an Alternative for the Analysis of Highly Polar Ionizable Solutes? *Journal of Separation Science* 32(12):2001–2007. doi:10.1002/jssc.200900272.

Dück, R., H. Sonderfeld, and O. J. Schmitz. 2012. A Simple Method for the Determination of Peak Distribution in Comprehensive Two-Dimensional Liquid Chromatography. *Journal of Chromatography A* 1246:69–75. doi:10.1016/j.chroma.2012.02.038.

Dugo, P., F. Cacciola, T. Kumm, G. Dugo, and L. Mondello. 2008. Comprehensive Multidimensional Liquid Chromatography: Theory and Applications. *Journal of Chromatography A* 1184(1–2):353–368. doi:10.1016/j.chroma.2007.06.074.

Dugo, P., O. Favoino, R. Luppino, G. Dugo, and L. Mondello. 2004. Comprehensive Two-Dimensional Normal-Phase (Adsorption)-Reversed-Phase Liquid Chromatography. *Analytical Chemistry* 76(9):2525–2530.

Dugo, P., N. Fawzy, F. Cichello, F. Cacciola, P. Donato, and L. Mondello. 2013. Stop-Flow Comprehensive Two-Dimensional Liquid Chromatography Combined with Mass Spectrometric Detection for Phospholipid Analysis. *Journal of Chromatography A* 1278:46–53. doi:10.1016/j.chroma.2012.12.042.

Elsner, V., S. Laun, D. Melchior, M. Köhler, and O. J. Schmitz. 2012. Analysis of Fatty Alcohol Derivatives with Comprehensive Two-Dimensional Liquid Chromatography Coupled with Mass Spectrometry. *Journal of Chromatography A* 1268:22–28. doi:10.1016/j.chroma.2012.09.072.

Elsner, V., V. Wulf, M. Wirtz, and O. J. Schmitz. 2015. Reproducibility of Retention Time and Peak Area in Comprehensive Two-Dimensional Liquid Chromatography. *Analytical and Bioanalytical Chemistry* 407(1):279–284. doi:10.1007/s00216-014-8090-3.

Fairchild, J. N., K. Horváth, and G. Guiochon. 2009a. Theoretical Advantages and Drawbacks of On-Line, Multidimensional Liquid Chromatography Using Multiple Columns Operated in Parallel. *Journal of Chromatography A* 1216(34):6210–6217. doi:10.1016/j.chroma.2009.06.085.

Fairchild, J. N., K. Horváth, and G. Guiochon. 2009b. Approaches to Comprehensive Multidimensional Liquid Chromatography Systems. *Journal of Chromatography A* 1216(9):1363–1371. doi:10.1016/j.chroma.2008.12.073.

Feng, J.-T., Z.-M. Guo, H. Shi, J.-P. Gu, Y. Jin, and X.-M. Liang. 2010. Orthogonal Separation on One Beta-Cyclodextrin Column by Switching Reversed-Phase Liquid Chromatography and Hydrophilic Interaction Chromatography. *Talanta* 81(4–5): 1870–1876. doi:10.1016/j.talanta.2010.03.007.

Filgueira, M. R., Y. Huang, K. Witt, C. Castells, and P. W. Carr. 2011. Improving Peak Capacity in Fast On-Line Comprehensive Two-Dimensional Liquid Chromatography with Post First Dimension Flow-Splitting. *Analytical Chemistry* 83:9531–9539.

François, I., A. de Villiers, and P. Sandra. 2006. Considerations on the Possibilities and Limitations of Comprehensive Normal Phase-Reversed Phase Liquid Chromatography (NPLC × RPLC). *Journal of Separation Science* 29(4):492–498.

François, I., A. de Villiers, B. Tienpont, F. David, and P. Sandra. 2008. Comprehensive Two-Dimensional Liquid Chromatography Applying Two Parallel Columns in the Second Dimension. *Journal of Chromatography A* 1178(1–2):33–42. doi:10.1016/j.chroma.2007.11.032.

François, I., K. Sandra, and P. Sandra. 2009. Comprehensive Liquid Chromatography: Fundamental Aspects and Practical Considerations-A Review. *Analytica Chimica Acta* 641(1–2):14–31. doi:10.1016/j.aca.2009.03.041.

Fu, Q., Z. Guo, X. Zhang, Y. Liu, and X. Liang. 2012. Comprehensive Characterization of Stevia Rebaudiana Using Two-Dimensional Reversed-Phase Liquid Chromatography/ Hydrophilic Interaction Liquid Chromatography. *Journal of Separation Science* 35(14):1821–1827. doi:10.1002/jssc.201101103.

Gamache, P. H., R. S. McCarthy, S. M. Freeto, D. J. Asa, M. J. Woodcock, K. Laws, and R. O. Cole. 2005. HPLC Analysis of Nonvolatile Analytes Using Charged Aerosol Detection. *LCGC North America* 23(6SUPPL):101–106.

Garbis, S. D., T. I. Roumeliotis, S. I. Tyritzis, K. M. Zorpas, K. Pavlakis, and C. A. Constantinides. 2011. Analysis of the Serum Proteome and Phosphoproteome: Application to Clinical Sera Derived from Humans with Benign Prostate Hyperplasia. *Analytical Chemistry* 83:708–718.

Giddings, J. C. 1967. Maximum Number of Components Resolvable by Gel Filtration and Other Elution Chromatographic Methods. *Analytical Chemistry* 39(8):1027–1028.

Giddings, J. C. 1984. Two-Dimensional Separations: Concept and Promise. *Analytical Chemistry* 56(12):1258A–1270A.

Giddings, J. C. 1991. *Unified Separation Science*. New York: Wiley Interscience.

Giddings, J. C. 1995. Sample Dimensionality: A Predictor of Order-Disorder in Component Peak Distribution in Multidimensional Separation. *Journal of Chromatography A* 703(1–2):3–15. doi:10.1016/0021-9673(95)00249-M.

Gilar, M., J. Fridrich, M. R. Schure, and A. Jaworski. 2012. Comparison of Orthogonality Estimation Methods for the Two-Dimensional Separations of Peptides. *Analytical Chemistry* 84(20):8722–8732. doi:10.1021/ac3020214.

Gilar, M., P. Olivova, A. E. Daly, and J. C. Gebler. 2005. Orthogonality of Separation in Two-Dimensional Liquid Chromatography. *Analytical Chemistry* 77(19):6426–6434. doi:10.1021/ac050923i.

Gorecki, T., F. Lynen, R. Szucs, and P. Sandra. 2006. Universal Response in Liquid Chromatography Using Charged Aerosol Detection. *Analytical Chemistry* 78(9):3186–3192.

Gritti, F., A. dos Santos Pereira, P. Sandra, and G. Guiochon. 2009. Comparison of the Adsorption Mechanisms of Pyridine in Hydrophilic Interaction Chromatography and in Reversed-Phase Aqueous Liquid Chromatography. *Journal of Chromatography A* 1216(48):8496–8504. doi:10.1016/j.chroma.2009.10.009.

Gritti, F., A. dos Santos Pereira, P. Sandra, and G. Guiochon. 2010. Efficiency of the Same Neat Silica Column in Hydrophilic Interaction Chromatography and per Aqueous Liquid Chromatography. *Journal of Chromatography A* 1217(5):683–688. doi:10.1016/j.chroma.2009.12.004.

Gritti, F. and G. Guiochon. 2013a. Comparison between the Intra-Particle Diffusivity in the Hydrophilic Interaction Chromatography and Reversed Phase Liquid Chromatography Modes. Impact on the Column Efficiency. *Journal of Chromatography A* 1297:85–95. doi:10.1016/j.chroma.2013.04.055.

Gritti, F. and G. Guiochon. 2013b. Mass Transfer Mechanism in Hydrophilic Interaction Chromatography. *Journal of Chromatography A* 1302:55–64. doi:10.1016/j.chroma.2013.06.001.

Gu, H., Y. Huang, and P. W. Carr. 2011. Peak Capacity Optimization in Comprehensive Two Dimensional Liquid Chromatography: A Practical Approach. *Journal of Chromatography A* 1218(1):64–73. doi:10.1016/j.chroma.2010.10.096.

Guiochon, G., L. A. Beaver, M. F. Gonnord, A. M. Siouffi, and M. Zakaria. 1983. Theoretical Investigation of the Potentialities of the Use of a Multidimensional Column in Chromatography. *Journal of Chromatography* 255:415–437.

Guiochon, G., M. F. Gonnord, A. Siouffi, and M. Zakaria. 1982. Study of the Performances of Thin-Layer Chromatography VII. Spot Capacity in Two-Dimensional Thin-Layer Chromatography. *Journal of Chromatography* 250:1–20.

Guo, X., X. Zhang, J. Feng, Z. Guo, Y. Xiao, and X. Liang. 2013. Purification of Saponins from Leaves of Panax Notoginseng Using Preparative Two-Dimensional Reversed-Phase Liquid Chromatography/hydrophilic Interaction Chromatography. *Analytical and Bioanalytical Chemistry* 405(10):3413–3421. doi:10.1007/s00216-013-6721-8.

Haefliger, O. P. 2003. Universal Two-Dimensional HPLC Technique for the Chemical Analysis of Complex Surfactant Mixtures. *Analytical Chemistry* 75(3):371–378. doi:10.1021/ac020534d.

Hayes, R., A. Ahmed, T. Edge, and H. Zhang. 2014. Core-Shell Particles: Preparation, Fundamentals and Applications in High Performance Liquid Chromatography. *Journal of Chromatography A* 1357:36–52. doi:10.1016/j.chroma.2014.05.010.

Heckendorf, A., M. Gilar, I. S. Krull, and A. Rathore. 2014. Biotechnology Today. *LCGC North America* 32(1):38–53.

Heinisch, S. and J. L. Rocca. 2009. Sense and Nonsense of High-Temperature Liquid Chromatography. *Journal of Chromatography A* 1216:642–658. doi:10.1016/j.chroma.2008.11.079.

Horie, K., H. Kimura, T. Ikegami, A. Iwatsuka, N. Saad, O. Fiehn, and N. Tanaka. 2007. Calculating Optimal Modulation Periods to Maximize the Peak Capacity in Two-Dimensional HPLC. *Analytical Chemistry* 79(10):3764–3770. doi:10.1021/ac062002t.

Horváth, K., J. Fairchild, and G. Guiochon. 2009a. Optimization Strategies for Off-Line Two-Dimensional Liquid Chromatography. *Journal of Chromatography A* 1216(12):2511–2518. doi:10.1016/j.chroma.2009.01.064.

Horváth, K., J. N. Fairchild, and G. Guiochon. 2009b. Generation and Limitations of Peak Capacity in Online Two-Dimensional Liquid Chromatography. *Analytical Chemistry* 81(10):3879–3888. doi:10.1021/ac802694c.

Horváth, K., J. N. Fairchild, and G. Guiochon. 2009c. Detection Issues in Two-Dimensional On-Line Chromatography. *Journal of Chromatography A* 1216(45):7785–7792. doi:10.1016/j.chroma.2009.09.016.

Huang, Y., H. Gu, M. Filgueira, and P. W. Carr. 2011. An Experimental Study of Sampling Time Effects on the Resolving Power of On-Line Two-Dimensional High Performance Liquid Chromatography. *Journal of Chromatography A* 1218(20):2984–2994. doi:10.1016/j.chroma.2011.03.032.

Ikegami, T., K. Tomomatsu, H. Takubo, K. Horie, and N. Tanaka. 2008. Separation Efficiencies in Hydrophilic Interaction Chromatography. *Journal of Chromatography A* 1184(1–2):474–503. doi:10.1016/j.chroma.2008.01.075.

Issaq, H. J., K. C. Chan, G. M. Janini, T. P. Conrads, and T. D. Veenstra. 2005. Multidimensional Separation of Peptides for Effective Proteomic Analysis. *Journal of Chromatography B: Analytical Technologies in the Biomedical and Life Sciences* 817(1):35–47. doi:10.1016/j.jchromb.2004.07.042.

Ivanisevic, J., Z.-J. Zhu, L. Plate, R. Tautenhahn, S. Chen, P. J. O. Brien, C. H. Johnson, M. A. Marletta, G. J. Patti, and G. Siuzdak. 2013. Toward 'Omic Scale Metabolite Pro Fi Ling: A Dual Separation−Mass Spectrometry Approach for Coverage of Lipid and Central Carbon Metabolism. *Analytical Chemistry* 85:6876–6884.

Jandera, P. 2011. Stationary and Mobile Phases in Hydrophilic Interaction Chromatography: A Review. *Analytica Chimica Acta* 692(1–2):1–25. doi:10.1016/j.aca.2011.02.047.

Jandera, P. 2012. Programmed Elution in Comprehensive Two-Dimensional Liquid Chromatography. *Journal of Chromatography A* 1255:112–129. doi:10.1016/j.chroma. 2012.02.071.

Jandera, P., J. Fischer, H. Lahovská, K. Novotná, P. Česla, and L. Kolářová. 2006. Two-Dimensional Liquid Chromatography Normal-Phase and Reversed-Phase Separation of (Co)oligomers. *Journal of Chromatography A* 1119(1–2):3–10. doi:10.1016/j. chroma.2005.10.081.

Jandera, P., T. Hájek, and P. Česla. 2010a. Comparison of Various Second-Dimension Gradient Types in Comprehensive Two-Dimensional Liquid Chromatography. *Journal of Separation Science* 33:1382–1397. doi:10.1002/jssc.200900808.

Jandera, P., T. Hájek, V. Skeríková, and J. Soukup. 2010b. Dual Hydrophilic Interaction-RP Retention Mechanism on Polar Columns: Structural Correlations and Implementation for 2-D Separations on a Single Column. *Journal of Separation Science* 33(6–7):841–852. doi:10.1002/jssc.200900678.

Jandera, P., T. Hájek, M. Staňková, K. Vyňuchalová, and P. Česla. 2012. Optimization of Comprehensive Two-Dimensional Gradient Chromatography Coupling in-Line Hydrophilic Interaction and Reversed Phase Liquid Chromatography. *Journal of Chromatography A* 1268:91–101. doi:10.1016/j.chroma.2012.10.041.

Jian, W., R. W. Edom, Y. Xu, and N. Weng. 2010. Recent Advances in Application of Hydrophilic Interaction Chromatography for Quantitative Bioanalysis. *Journal of Separation Science* 33(6–7):681–697. doi:10.1002/jssc.200900692.

Kalili, K. M., S. De Smet, T. van Hoeylandt, F. Lynen, and A. de Villiers. 2014. Comprehensive Two-Dimensional Liquid Chromatography Coupled to the ABTS Radical Scavenging Assay: A Powerful Method for the Analysis of Phenolic Antioxidants. *Analytical and Bioanalytical Chemistry* 406(17):4233–4242. doi:10.1007/s00216-014-7847-z.

Kalili, K. M. and A. de Villiers. 2010. Off-Line Comprehensive Two-Dimensional Hydrophilic Interaction X Reversed Phase Liquid Chromatographic Analysis of Green Tea Phenolics. *Journal of Separation Science* 33(6–7):853–863. doi:10.1002/ jssc.200900673.

Kalili, K. M., J. Vestner, M. A. Stander, and A. de Villiers. 2013. Toward Unraveling Grape Tannin Composition: Application of Online Hydrophilic Interaction Chromatography × Reversed-Phase Liquid Chromatography-Time-of-Flight Mass Spectrometry for Grape Seed Analysis. *Analytical Chemistry* 85(19):9107–9115. doi:10.1021/ac401896r.

Kalili, K. M. and A. de Villiers. 2009. Off-Line Comprehensive 2-Dimensional Hydrophilic Interaction X Reversed Phase Liquid Chromatography Analysis of Procyanidins. *Journal of Chromatography A* 1216(35):6274–6284. doi:10.1016/j. chroma.2009.06.071.

Kalili, K. M. and A. de Villiers. 2013a. Systematic Optimisation and Evaluation of On-Line, Off-Line and Stop-Flow Comprehensive Hydrophilic Interaction Chromatography × Reversed Phase Liquid Chromatographic Analysis of Procyanidins. Part II : Application to Cocoa Proc. *Journal of Chromatography A* 1289:69–79. doi:10.1016/j.chroma.2013.03.008.

Kalili, K. M. and A. de Villiers. 2013b. Systematic Optimisation and Evaluation of On-Line, Off-Line and Stop-Flow Comprehensive Hydrophilic Interaction Chromatography × Reversed Phase Liquid Chromatographic Analysis of Procyanidins, Part I : Theoretical Considerations. *Journal of Chromatography A* 1289:58–68. doi:10.1016/j.chroma.2013.03.008.

Karger, B. L., L. R. Snyder, and C. Horvath. 1973. *An Introduction to Separation Science.* New York: John Wiley & Sons.

Klavins, K., H. Drexler, S. Hann, and G. Koellensperger. 2014. Quantitative Metabolite Pro Fi Ling Utilizing Parallel Column Analysis for Simultaneous Reversed-Phase and Hydrophilic Interaction Liquid Chromatography Separations Combined with Tandem Mass Spectrometry. *Analytical Chemistry* 86:4145–4150.

Knox, J. H. and H. P. Scott. 1983. B and C Terms in the Van Deemter Equation for Liquid Chromatography. *Journal of Chromatography A* 282:297–313. doi:10.1016/S0021-9673(00)91609-1.

Lam, M. P. Y., S. O. Siu, E. Lau, X. Mao, H. Z. Sun, P. C. N. Chiu, W. S. B. Yeung, D. M. Cox, and I. K. Chu. 2010. Online Coupling of Reverse-Phase and Hydrophilic Interaction Liquid Chromatography for Protein and Glycoprotein Characterization. *Analytical and Bioanalytical Chemistry* 398(2):791–804. doi:10.1007/s00216-010-3991-2.

Leme, G. M., F. Cacciola, P. Donato, A. J. Cavalheiro, P. Dugo, and L. Mondello. 2014. Continuous vs. Segmented Second-Dimension System Gradients for Comprehensive Two-Dimensional Liquid Chromatography of Sugarcane (Saccharum Spp.). *Analytical and Bioanalytical Chemistry* 406:4315–4324. doi:10.1007/s00216-014-7786-8.

Lestremau, F., D. Wu, and R. Szücs. 2010. Evaluation of 1.0 mm i.d. Column Performances on Ultra High Pressure Liquid Chromatography Instrumentation. *Journal of Chromatography A* 1217(30):4925–4933. doi:10.1016/j.chroma.2010.05.044.

Li, Q., F. Lynen, J. Wang, H. Li, G. Xu, and P. Sandra. 2012. Comprehensive Hydrophilic Interaction and Ion-Pair Reversed-Phase Liquid Chromatography for Analysis of Di- to Deca-Oligonucleotides. *Journal of Chromatography A* 1255:237–243. doi:10.1016/j.chroma.2011.11.062.

Li, X., D. R. Stoll, and P. W. Carr. 2009. Equation for Peak Capacity Estimation in Two-Dimensional Liquid Chromatography. *Analytical Chemistry* 81(2):845–850. doi:10.1021/ac801772u.

Liang, Z., K. Li, X. Wang, Y. Ke, Y. Jin, and X. Liang. 2012. Combination of off-Line Two-Dimensional Hydrophilic Interaction Liquid Chromatography for Polar Fraction and Two-Dimensional Hydrophilic Interaction Liquid Chromatography×Reversed-Phase Liquid Chromatography for Medium-Polar Fraction in a Traditional Chin. *Journal of Chromatography A* 1224:61–69. doi:10.1016/j.chroma.2011.12.046.

Linden, J. C. and C. L. Lawhead. 1975. Liquid Chromatography of Saccharides. *Journal of Chromatography A* 105:125–133.

Ling, Y. S., H.-J. Liang, M.-H. Lin, C.-H. Tang, K.-Y. Wu, M.-L. Kuo, and C. Y. Lin. 2014. Two-Dimensional LC-MS/MS to Enhance Ceramide and Phosphatidylcholine Species Profiling in Mouse Liver. *Biomedical Chromatography* 28(9):1284–1293. doi:10.1002/bmc.3162.

Link, A. J., J. Eng, D. M. Schieltz, E. Carmack, G. J. Mize, D. R. Morris, B. M. Garvik, and J. R. Yates III. 1999. Direct Analysis of Protein Complexes Using Mass Spectrometry. *Nature Biotechnology* 17(7):676–682. doi:10.1038/10890.

Lísa, M., E. Cífková, and M. Holčapek. 2011. Lipidomic Profiling of Biological Tissues Using Off-Line Two-Dimensional High-Performance Liquid Chromatography-Mass Spectrometry. *Journal of Chromatography A* 1218(31):5146–5156. doi:10.1016/j.chroma.2011.05.081.

Liu, A., J. Tweed, and C. E. Wujcik. 2009. Investigation of an On-Line Two-Dimensional Chromatographic Approach for Peptide Analysis in Plasma by LC-MS-MS. *Journal of Chromatography B, Analytical Technologies in the Biomedical and Life Sciences* 877(20–21):1873–1881. doi:10.1016/j.jchromb.2009.05.012.

Liu, Y., X. Xue, Z. Guo, Q. Xu, F. Zhang, and X. Liang. 2008. "Novel Two-Dimensional Reversed-Phase Liquid Chromatography/Hydrophilic Interaction Chromatography, An Excellent Orthogonal System for Practical Analysis. *Journal of Chromatography A* 1208(1–2):133–140. doi:10.1016/j.chroma.2008.08.079.

Liu, Z. and D. Patterson Jr. 1995. Geometric Approach to Factor Analysis for the Estimation of Orthogonality and Practical Peak Capacity in Comprehensive Two Dimensional Separations. *Analytical Chemistry* 67(7):3840–3845. doi:10.1021/ac00117a004.

Loftheim, H., T. D. Nguyen, H. Malerød, E. Lundanes, A. Asberg, and L. Reubsaet. 2010. 2-D Hydrophilic Interaction Liquid Chromatography-RP Separation in Urinary Proteomics-Minimizing Variability through Improved Downstream Workflow Compatibility. *Journal of Separation Science* 33:864–872. doi:10.1002/jssc.200900554.

MacNair, J. E., K. C. Lewis, and J. W. Jorgenson. 1997. Ultrahigh-Pressure Reversed-Phase Liquid Chromatography in Packed Capillary Columns. *Analytical Chemistry* 69(6):983–989. doi:10.1021/ac961094r.

Makarov, A. 2000. Electrostatic Axially Harmonic Orbital Trapping: A High-Performance Technique of Mass Analysis. *Analytical Chemistry* 72(6):1156–1162. doi:10.1021/ac991131p.

Makarov, A. and M. Scigelova. 2010. Coupling Liquid Chromatography to Orbitrap Mass Spectrometry. *Journal of Chromatography A* 1217(25):3938–3945. doi:10.1016/j.chroma.2010.02.022.

Malherbe, C. J., D. de Beer, and E. Joubert. 2012. Development of On-Line High Performance Liquid Chromatography (HPLC)-Biochemical Detection Methods as Tools in the Identification of Bioactives. *International Journal of Molecular Science* 13(3):3101–3133. doi:10.3390/ijms13033101.

Malik, M. I. and H. Pasch. 2014. Novel Developments in the Multidimensional Characterization of Segmented Copolymers. *Progress in Polymer Science* 39(1):87–123. doi:10.1016/j.progpolymsci.2013.10.005.

Marriot, P. J., P. Schoenmakers, and Z.-Y. Wu. 2012. Nomenclature and Conventions in Comprehensive Multidimensional Chromatography. *LCGC Europe* 25(5):266–275.

Massart, L., B. G. M. Vandeginste, L. M. C. Buydens, S. De Jong, P. J. Lewi, and J. Smeyers-Verbeke. 1998. Handbook of Chemometrics and Qualimetrics: Part A. *Data Handling in Science and Technology* 20:771–804 (Elsevier). doi:10.1016/S0922-3487(97)80056-1.

McCalley, D. V. and U. D. Neue. 2008. Estimation of the Extent of the Water-Rich Layer Associated with the Silica Surface in Hydrophilic Interaction Chromatography. *Journal of Chromatography A* 1192(2):225–229. doi:10.1016/j.chroma.2008.03.049.

McNulty, D. E. and R. S. Annan. 2008. Hydrophilic Interaction Chromatography Reduces the Complexity of the Phosphoproteome and Improves Global Phosphopeptide Isolation and Detection. *Molecular & Cellular Proteomics* 7(5):971–980. doi:10.1074/mcp.M700543-MCP200.

Mihailova, A., H. Malerød, S. R. Wilson, B. Karaszewski, R. Hauser, E. Lundanes, and T. Greibrokk. 2008. Improving the Resolution of Neuropeptides in Rat Brain with On-Line HILIC-RP Compared to On-Line SCX-RP. *Journal of Separation Science* 31(3):459–467. doi:10.1002/jssc.200700257.

Montero, L., M. Herrero, E. Ibáñez, and A. Cifuentes. 2013a. Profiling of Phenolic Compounds from Different Apple Varieties Using Comprehensive Two-Dimensional Liquid Chromatography. *Journal of Chromatography A* 1313:275–283. doi:10.1016/j.chroma.2013.06.015.

Montero, L., M. Herrero, E. Ibáñez, and A. Cifuentes. 2014. Separation and Characterization of Phlorotannins from Brown Algae Cystoseira Abies-Marina by Comprehensive Two-Dimensional Liquid Chromatography. *Electrophoresis* 35(11):1644–1651. doi:10.1002/elps.201400133.

Montero, L., M. Herrero, M. Prodanov, E. Ibáñez, and A. Cifuentes. 2013b. Characterization of Grape Seed Procyanidins by Comprehensive Two-Dimensional Hydrophilic Interaction × Reversed Phase Liquid Chromatography Coupled to Diode Array Detection and Tandem Mass Spectrometry. *Analytical and Bioanalytical Chemistry* 405(13):4627–4638. doi:10.1007/s00216-012-6567-5.

Moore, A. W. and J. W. Jorgenson. 1995. Comprehensive Three-Dimensional Separation of Peptides Using Size Exclusion Chromatography/Reversed Phase Liquid Chromatography/Optically Gated Capillary Zone Electrophoresis. *Analytical Chemistry* 67(19):3456–3463.

Murphy, R. E., M. R. Schure, and J. P. Foley. 1998. Effect of Sampling Rate on Resolution in Comprehensive Two-Dimensional Liquid Chromatography. *Analytical Chemistry* 70(8):1585–1594. doi:10.1021/ac971184b.

Neue, U. D. 1997. *HPLC Columns, Theory, Technology and Practice*, U. D. Neue (ed.). New York: Wiley-VCH.

Neue, U. D. 2005. Theory of Peak Capacity in Gradient Elution. *Journal of Chromatography A* 1079(1–2):153–161. doi:10.1016/j.chroma.2005.03.008.

Opiteck, G. J. and J. W. Jorgenson. 1997. Two-Dimensional SEC/RPLC Coupled to Mass Spectrometry for the Analysis of Peptides. *Analytical Chemistry* 69(13):2283–2291.

Opiteck, G. J., S. M. Ramirez, J. W. Jorgenson, and M. A. Moseley. 1998. Comprehensive Two-Dimensional High-Performance Liquid Chromatography for the Isolation of Overexpressed Proteins and Proteome Mapping. *Analytical Biochemistry* 258(2):349–361. doi:10.1006/abio.1998.2588.

Patel, K. D., A. D. Jerkovich, J. C. Link, and J. W. Jorgenson. 2004. In-Depth Characterization of Slurry Packed Capillary Columns with 1.0-Mm Nonporous Particles Using Reversed-Phase Isocratic Ultrahigh-Pressure Liquid Chromatography. *Analytical Chemistry* 76(19):5777–5786. doi:10.1021/ac049756x.

Periat, A., J. Boccard, J.-L. Veuthey, S. Rudaz, and D. Guillarme. 2013. Systematic Comparison of Sensitivity between Hydrophilic Interaction Liquid Chromatography and Reversed Phase Liquid Chromatography Coupled with Mass Spectrometry. *Journal of Chromatography A* 1312:49–57. doi:10.1016/j.chroma.2013.08.097.

Poppe, H. 1997. Some Reflections on Speed and Efficiency of Modem Chromatographic Methods/Review. *Journal of Chromatography A* 778:3–21. doi:10.1016/S0021-9673(97)00376-2.

Potts, L. W., D. R. Stoll, X. Li, and P. W. Carr. 2010. The Impact of Sampling Time on Peak Capacity and Analysis Speed in On-Line Comprehensive Two-Dimensional Liquid Chromatography. *Journal of Chromatography A* 1217(36):5700–5709. doi:10.1016/j.chroma.2010.07.009.

Regnier, F., A. Amini, A. Chakraborty, M. Geng, J. Ji, L. Riggs, C. Sioma, S. Wang, and X. Zhang. 2001. Multidimensional Chromatography and the Signature Peptide Approach to Proteomics. *LCGC* 19(2):200–213.

Rutan, S. C., J. M. Davis, and P. W. Carr. 2012. Fractional Coverage Metrics Based on Ecological Home Range for Calculation of the Effective Peak Capacity in Comprehensive Two-Dimensional Separations. *Journal of Chromatography A* 1255:267–276. doi:10.1016/j.chroma.2011.12.061.

Sarrut, M., G. Crétier, and S. Heinisch. 2014. Theoretical and Practical Interest in UHPLC Technology for 2D-LC. *Trends in Analytical Chemistry* 63:104–112. doi:10.1016/j.trac.2014.08.005.

Schellinger, A. P, D. R. Stoll, and P. W. Carr. 2005. High Speed Gradient Elution Reversed-Phase Liquid Chromatography. *Journal of Chromatography A* 1064(2):143–156. doi:10.1016/j.chroma.2004.12.017.

Schiesel, S., M. Lämmerhofer, and W. Lindner. 2012. Comprehensive Impurity Profiling of Nutritional Infusion Solutions by Multidimensional off-Line Reversed-Phase Liquid Chromatography × Hydrophilic Interaction Chromatography-Ion Trap Mass-Spectrometry and Charged Aerosol Detection with Universal Calibrati. *Journal of Chromatography A* 1259:100–110. doi:10.1016/j.chroma.2012.01.009.

Schoenmakers, P. J., P. Marriot, and J. Beens. 2003. Nomenclature and Conventions in Comprehensive Multidimensional Chromatography. *LCGC Europe* 16(6):335–339.

Schoenmakers, P. J., G. Vivó-Truyols, and W. M. C. Decrop. 2006. A Protocol for Designing Comprehensive Two-Dimensional Liquid Chromatography Separation Systems. *Journal of Chromatography A* 1120(1–2):282–290. doi:10.1016/j.chroma.2005.11.039.

Schure, M. R. 1999. Limit of Detection, Dilution Factors, and Technique Compatibility in Multidimensional Separations Utilizing Chromatography, Capillary Electrophoresis, and Field-Flow Fractionation. *Analytical Chemistry* 71(8):1645–1657. doi:10.1021/ac981128q.

Schure, M. R. 2011. The Dimensionality of Chromatographic Separations. *Journal of Chromatography A* 1218(2):293–302. doi:10.1016/j.chroma.2010.11.016.

Seeley, J. V. 2002. Theoretical Study of Incomplete Sampling of the First Dimension in Comprehensive Two-Dimensional Chromatography. *Journal of Chromatography A* 962(1–2):21–27. doi:10.1016/S0021-9673(02)00461-2.

Semard, G., V. Peulon-Agasse, A. Bruchet, J.-P. Bouillon, and P. Cardinaël. 2010. Convex Hull: A New Method to Determine the Separation Space Used and to Optimize Operating Conditions for Comprehensive Two-Dimensional Gas Chromatography. *Journal of Chromatography A* 1217(33):5449–5454. doi:10.1016/j.chroma.2010.06.048.

Shalliker, R. A. and M. J. Gray. 2006. Concepts and Practice of Multidimensional High-Performance Liquid Chromatography. In *Advances in Chromatography,* Vol. 44, E. Grushka and N. Grinberg (eds.), 44th ed., 45:177–236. New York: CRC Press.

Shen, Y., R. Zhang, R. J. Moore, J. Kim, T. O. Metz, K. K. Hixson, R. Zhao, E. A. Livesay, H. R. Udseth, and R. D. Smith. 2005. Automated 20 Kpsi RPLC-MS and MS/MS with Chromatographic Peak Capacities of 1000–1500 and Capabilities in Proteomics and Metabolomics. *Analytical Chemistry* 77(10):3090–3100. doi:10.1021/ac0483062.

Simon, R., S. Passeron, J. Lemoine, and A. Salvador. 2014. Hydrophilic Interaction Liquid Chromatography as Second Dimension in Multidimensional Chromatography with an Anionic Trapping Strategy: Application to Prostate-Specific Antigen Quantification. *Journal of Chromatography A* 1354:75–84. doi:10.1016/j.chroma.2014.05.063.

Slonecker, P. J., X. Li, T. H. Ridgway, and J. G. Dorsey. 1996. Informational Orthogonality of Two-Dimensional Chromatographic Separations. *Analytical Chemistry* 68(4):682–689. doi:10.1021/ac950852v.

Sneekes, E. J., B. Dolman, and R. Swart. 2007. *Novel Off-Line Multidimensional LC Method for Separation and Tandem MS Detection of Tryptic Peptides.* Dionex application note, Dionex Corporation, Amsterdam, the Netherlands.

Song, H., E. Adams, G. Desmet, and D. Cabooter. 2014. Evaluation and Comparison of the Kinetic Performance of Ultra-High Performance Liquid Chromatography and High-Performance Liquid Chromatography Columns in Hydrophilic Interaction and Reversed-Phase Liquid Chromatography Conditions. *Journal of Chromatography A* 1369:83–91. doi:10.1016/j.chroma.2014.10.002.

Spagou, K., H. Tsoukali, N. Raikos, H. Gika, I. D. Wilson, and G. Theodoridis. 2010. Hydrophilic Interaction Chromatography Coupled to MS for Metabonomic/metabolomic Studies. *Journal of Separation Science* 33(6–7):716–727. doi:10.1002/jssc.200900803.

Stevenson, P. G., D. N. Bassanese, X. A. Conlan, and N. W. Barnett. 2014. Improving Peak Shapes with Counter Gradients in Two-Dimensional High Performance Liquid Chromatography. *Journal of Chromatography A* 1337:147–154. doi:10.1016/j.chroma.2014.02.051.

Stoll, D. R. and P. W. Carr. 2005. Fast, Comprehensive Two-Dimensional HPLC Separation of Tryptic Peptides Based on High-Temperature HPLC. *Journal of the American Chemical Society* 127(14):5034–5035. doi:10.1021/ja050145b.

Stoll, D. R., J. D Cohen, and P. W. Carr. 2006. Fast, Comprehensive Online Two-Dimensional High Performance Liquid Chromatography through the Use of High Temperature Ultra-Fast Gradient Elution Reversed-Phase Liquid Chromatography. *Journal of Chromatography A* 1122:123–137. doi:10.1016/j.chroma.2006.04.058.

Stoll, D. R., E. S. Talus, D. C. Harmes, and K. Zhang. 2015. Evaluation of Detection Sensitivity in Comprehensive Two-Dimensional Liquid Chromatography Separations of an Active Pharmaceutical Ingredient and Its Degradants. *Analytical and Bioanalytical Chemistry* 407:265–277. doi:10.1007/s00216-014-8036-9.

Stoll, D. R., X. Wang, and P. W. Carr. 2008. Comparison of the Practical Resolving Power of One- and Two-Dimensional High-Performance Liquid Chromatography Analysis of Metabolomic Samples. *Analytical Chemistry* 80(1):268–278. doi:10.1016/j.chroma.2007.06.074.

Stoll, D. R. 2010. Recent Progress in Online, Comprehensive Two-Dimensional High-Performance Liquid Chromatography for Non-Proteomic Applications. *Analytical and Bioanalytical Chemistry* 397(3):979–986. doi:10.1007/s00216-010-3659-y.

Tanaka, N. 2002. Monolithic Silica Columns for High-Efficiency Chromatographic Separations. *Journal of Chromatography A* 965:35–49.

Tanaka, N., H. Kimura, D. Tokuda, K. Hosoya, T. Ikegami, N. Ishizuka, H. Minakuchi et al. 2004. Simple and Comprehensive Two-Dimensional Reversed-Phase HPLC Using Monolithic Silica Columns. *Analytical Chemistry* 76(5):1273–1281. doi:10.1021/ac034925j.

Teutenberg, T. 2009. Potential of High Temperature Liquid Chromatography for the Improvement of Separation Efficiency-A Review. *Analytica Chimica Acta* 643:1–12. doi:10.1016/j.aca.2009.04.008.

Thompson, J. D. and P. W. Carr. 2002. High-Speed Liquid Chromatography by Simultaneous Optimization of Temperature and Eluent Composition. *Analytical Chemistry* 74(16):4150–4159. doi:10.1021/ac0112622.

Unger, K. K., K. Racaityte, K. Wagner, T. Miliotis, L. E. Edholm, R. Bischoff, and G. Marko-Varga. 2000. Is Multidimensional High Performance Liquid Chromatography (HPLC) an Alternative in Protein Analysis to 2D Gel Electrophoresis? *Journal of High Resolution Chromatography* 23(3):259–265. doi:10.1002/(SICI)1521-4168(20000301)23:3<259::AID-JHRC259>3.0.CO;2-V.

Van Beek, T. A., K. K. R. Tetala, I. I. Koleva, A. Dapkevicius, V. Exarchou, S. M. F. Jeurissen, F. W. Claassen, and E. J. C. Van Der Klift. 2009. Recent Developments in the Rapid Analysis of Plants and Tracking Their Bioactive Constituents. *Phytochemistry Reviews* 8(2):387–399. doi:10.1007/s11101-009-9125-5.

Van Platerink, C. J., H.-G. M. Janssen, and J. Haverkamp. 2008. Application of at-Line Two-Dimensional Liquid Chromatography-Mass Spectrometry for Identification of Small Hydrophilic Angiotensin I-Inhibiting Peptides in Milk Hydrolysates. *Analytical and Bioanalytical Chemistry* 391:299–307. doi:10.1007/s00216-008-1990-3.

Vanhoenacker, G., I. Vandenheede, F. David, P. Sandra, and K. Sandra. 2015. Comprehensive Two-Dimensional Liquid Chromatography of Therapeutic Monoclonal Antibody Digests. *Analytical and Bioanalytical Chemistry* 407(1):355–366. doi:10.1007/s00216-014-8299-1.

Vivó-Truyols, G., S. van der Wal, and P. J. Schoenmakers. 2010. Comprehensive Study on the Optimization of Online Two-Dimensional Liquid Chromatographic Systems Considering Losses in Theoretical Peak Capacity in First- and Second-Dimensions: A Pareto-Optimality Approach. *Analytical Chemistry* 82(20):8525–8536.

Wagner, K., K. Racaityte, K. K. Unger, T. Miliotis, L. E. Edholm, R. Bischoff, and G. Marko-Varga. 2000. Protein Mapping by Two-Dimensional High Performance Liquid Chromatography. *Journal of Chromatography A* 893(2):293–305.

Wagner, K., T. Miliotis, R. Bischoff, and K. K. Unger. 2002. An Automated On-Line Multidimensional HPLC System for Protein and Peptide Mapping with. *Analytical Chemistry* 74(4):809–820.

Wang, S., J. Li, X. Shi, L. Qiao, X. Lu, and G. Xu. 2013. A Novel Stop-Flow Two-Dimensional Liquid Chromatography-Mass Spectrometry Method for Lipid Analysis. *Journal of Chromatography A* 1321:65–72. doi:10.1016/j.chroma.2013.10.069.

Wang, S., L. Qiao, X. Shi, C. Hu, H. Kong, and G. Xu. 2015. On-Line Stop-Flow Two-Dimensional Liquid Chromatography-Mass Spectrometry Method for the Separation and Identification of Triterpenoid Saponins from Ginseng Extract. *Analytical and Bioanalytical Chemistry* 407(1):331–341. doi:10.1007/s00216-014-8219-4.

Wang, X., W. E. Barber, and P. W. Carr. 2006a. A Practical Approach to Maximizing Peak Capacity by Using Long Columns Packed with Pellicular Stationary Phases for Proteomic Research. *Journal of Chromatography A* 1107:139–151. doi:10.1016/j.chroma.2005.12.050.

Wang, X., D. R. Stoll, P. W. Carr, and P. J. Schoenmakers. 2006b. A Graphical Method for Understanding the Kinetics of Peak Capacity Production in Gradient Elution Liquid Chromatography. *Journal of Chromatography A* 1125:177–181. doi:10.1016/j.chroma.2006.05.048.

Wang, X., D. R. Stoll, A. P. Schellinger, and P. W. Carr. 2006c. Peak Capacity Optimization of Peptide Separations in Reversed-Phase Gradient Elution Chromatography: Fixed Column Format. *Analytical Chemistry* 78(10):3406–3416. doi:10.1021/ac0600149.

Wang, Y., F. Zhang, X. Xue, Y. Xiao, J. Feng, and X. Liang. 2009. Use of an Orthogonal System, RP-LC–HILIC, for Evaluation of Purity. *Chromatographia* 69(11–12):1379–1384. doi:10.1365/s10337-009-1077-0.

Washburn, M. P., D. A. Wolters, and J. R. III Yates. 2001. Large-Scale Analysis of the Yeast Proteome by Multidimensional Protein Identification Technology. *Nature Biotechnology* 19(3):242–247. doi:10.1038/85686.

Willemse, C. M., M. A. Stander, A. G. J. Tredoux, and A. de Villiers. 2014. Comprehensive Two-Dimensional Liquid Chromatographic Analysis of Anthocyanins. *Journal of Chromatography A* 1359:189–201. doi:10.1016/j.chroma.2014.07.044.

Willemse, C. M., M. A. Stander, J. Vestner, A. G. J. Tredoux, and A. de Villiers. 2015. Comprehensive Two-Dimensional HILIC× Reversed-Phase Liquid Chromatography Coupled to High-Resolution Mass Spectrometry (RP-LC-UV-MS) Analysis of Anthocyanins and Derived Pigments in Red Wine. *Analytical Chemistry* 87: 12006–12015. doi:10.1021/acs.analchem.5b03615.

Wilson, S. R., M. Jankowski, M. Pepaj, A. Mihailova, F. Boix, G. Vivo Truyols, E. Lundanes, and T. Greibrokk. 2007. 2D LC Separation and Determination of Bradykinin in Rat Muscle Tissue Dialysate with On-Line SPE-HILIC-SPE-RP-MS. *Chromatographia* 66(7–8):469–474. doi:10.1365/s10337-007-0341-4.

Wohlgemuth, J., M. Karas, W. Jiang, R. Hendriks, and S. Andrecht. 2010. Enhanced Glyco-Profiling by Specific Glycopeptide Enrichment and Complementary Monolithic Nano-LC (ZIC-HILIC/RP18e)/ESI-MS Analysis. *Journal of Separation Science* 33(6–7):880–890. doi:10.1002/jssc.200900771.

Wolters, D. A., M. P. Washburn, and J. R. III Yates. 2001. An Automated Multidimensional Protein Identification Technology for Shotgun Proteomics. *Analytical Chemistry* 73(23):5683–5690. doi:10.1021/ac010617e.

Wu, C. J., Y. W. Chen, J. H. Tai, and S.-H. Chen. 2011. Quantitative Phosphoproteomics Studies Using Stable Isotope Dimethyl Labeling Coupled with IMAC-HILIC-nanoLC-MS/MS for Estrogen-Induced Transcriptional Regulation. *Journal of Proteome Research* 10:1088–1097. doi:10.1021/pr100864b.

Wu, Z.-Y. and P. J. Marriott. 2011. One- and Comprehensive Two-Dimensional High-Performance Liquid Chromatography Analysis of Alkylphenol Polyethoxylates. *Journal of Separation Science* 34(23):3322–3329. doi:10.1002/jssc.201100701.

Xing, Q., T. Liang, G. Shen, X. Wang, Y. Jin, and X. Liang. 2012. Comprehensive HILIC × RPLC with Mass Spectrometry Detection for the Analysis of Saponins in Panax Notoginseng. *The Analyst* 137(9):2239–2249. doi:10.1039/c2an16078a.

Xu, J., X. Zhang, Z. Guo, J. Yan, L. Yu, X. Li, X. Xue, and X. Liang. 2012. Short-Chain Peptides Identification of Scorpion Buthus Martensi Karsch Venom by Employing High Orthogonal 2D-HPLC System and Tandem Mass Spectrometry. *Proteomics* 12(19–20): 3076–3084. doi:10.1002/pmic.201200224.

Yan, B., J. Zhao, J. S. Brown, J. Blackwell, and P. W. Carr. 2000. High-Temperature Ultrafast Liquid Chromatography. *Analytical Chemistry* 72(6):1253–1262. doi:10.1021/ac991008y.

Yoshida, T. 2004. Peptide Separation by Hydrophilic-Interaction Chromatography: A Review. *Journal of Biochemical and Biophysical Methods* 60(3):265–280. doi:10.1016/j.jbbm.2004.01.006.

Zeng, J., X. Zhang, Z. Guo, J. Feng, J. Zeng, X. Xue, and X. Liang. 2012. Separation and Identification of Flavonoids from Complex Samples Using Off-Line Two-Dimensional Liquid Chromatography Tandem Mass Spectrometry. *Journal of Chromatography A* 1220:50–56. doi:10.1016/j.chroma.2011.11.043.

Zhao, Y., R. P. W. Kong, G. Li, M. P. Y. Lam, C. H. Law, S. M. Y. Lee, H. C. Lam, and I. K Chu. 2012. Fully Automatable Two-Dimensional Hydrophilic Interaction Liquid Chromatography-Reversed Phase Liquid Chromatography with Online Tandem Mass Spectrometry for Shotgun Proteomics. *Journal of Separation Science* 35(14):1755–1763. doi:10.1002/jssc.201200054.

7 Sample Preparation for Thin Layer Chromatography

Mieczysław Sajewicz, Teresa Kowalska, and Joseph Sherma

CONTENTS

7.1 INTRODUCTION

Prior to analyte determination by an instrumental analytical method, sample preparation is a crucial preliminary step that decisively affects the final analytical result. However, its importance in the total analytical method is often underestimated relative to the instrumental measurement steps. Sample preparation is usually given limited or no coverage in books and review articles on thin layer chromatography (TLC), so this chapter is devoted filling the gap of comprehensive information on techniques that have been used to prepare the sample applied to the plate prior to its development with the mobile phase in order to achieve high-quality analytical results. Laboratory samples must be representative of the original material to which the analytical method is to be applied, and they must be transported and stored so as to not change their composition prior to analysis. Use of both traditional and newer sample preparation technologies is discussed in the following sections for selected applications of TLC and high-performance TLC (HPTLC) to a wide range of analyte and sample types reported in the literature. Throughout this chapter, TLC is often used as a term that covers analysis on both HP and non-HP plates.

A primary demand imposed on samples that can effectively be analyzed by means of TLC is that they be liquid. Thus, with most samples the first step in sample preparation is extraction. The optimum case would be a selective extraction, but quite often the extraction process is not selective enough and unwanted compounds pass from the matrix to the extraction solvent. With the limited resolution power of TLC, coextractives may interfere in the analytical method. These problems can be solved by a second step known broadly as sample cleanup, that is, isolation of the compounds of interest from the rest of the extract, often carried out on an adsorbent column such as silica gel or Florisil. Once we have obtained an extract (step 1), and, if necessary, separated the compounds of interest from the rest (i.e., purified the extract in step 2), the concentration of these compounds may be too low and, hence, incompatible with the limit of detection (LOD) and limit of quantification (LOQ) of the employed TLC method. We then have to proceed to step 3, which is concentration of the extract by evaporating surplus solvent.

The majority of analytical applications of TLC target botanical samples (mostly used in traditional medicines of different geographical regions and analyzed for the composition of curative components and authentication), drugs (quality control), food samples (quality control and authentication), and environmental samples (mostly in order to detect contamination by pollutants). The sample preparation techniques discussed in the forthcoming sections of this chapter are used with all of these types of samples and others preceding TLC analysis.

7.2 TRADITIONAL SOLVENT EXTRACTION

Traditional solvent extraction by heating the sample in a solvent under the reflux is still very popular, especially in laboratories where the chemical composition of biological samples is investigated. These are mainly phytochemistry and pharmacognosy laboratories of biological and pharmaceutical facilities, as well as the learned institutions dedicated to the studies of traditional medicines (mainly in China and India, but also in the other geographical regions worldwide).

This simplest extraction approach usually is a two-step procedure, which in the preliminary step involves removal of the low polar chlorophyll and/or waxes from the dried plant matrix (the compounds that are considered as a ballast material that affects chromatographic analyses). This removal is obtained by heating a plant sample under reflux in a low polar solvent (usually *n*-hexane or petroleum ether). This extract, which contains ballast compounds, is usually discarded, while the pretreated botanical matrix is first dried and then extracted under the reflux for the second time with a solvent of choice. This main extraction step can either be total or target-specific classes of secondary metabolites (e.g., phenolics, coumarins, and alkaloids).

Total extraction is often carried out with methanol, which very efficiently isolates most secondary metabolites that belong to different classes and represent a wide spectrum of polarity. An additional advantage of using methanol is its strong antiseptic activity that preserves a plant or other biological extract from microorganism growth, which otherwise would rapidly destroy the extracted components. Recently, basically due to the health and environmental concerns, methanol is being replaced by ethanol, which provides similar extraction power compared to methanol but has lower toxicity. Olennikov and Partilkhayev [1] provided an example in their investigation of the flavonoid composition of the shrub *Caragana spinosa*, which grows in Siberia, Transbaikalia, and the far eastern regions of the Russian Federation and is recognized in Buryat and Tibetan folk medicine for its potent anti-inflammatory action. The authors carried out repeated extractions of the powdered herb material, in order to enhance the extraction yields, with environment-friendly ethanol, and, eventually, all ethanol extracts were combined for analysis using silica gel HPTLC plates, double development with ethyl acetate-1,2-dichloroethane-acetic acid-85% formic acid-water (10:2.51:1:0.8) mobile phase, Neu detection reagent (2-aminoethyl diphenylborinate), and scanning with a Sorbfil Videodensitometer 2.0.

Sajewicz et al. [2] carried out a multistep solvent extraction under reflux according to an elaborate protocol with different solvents in order to obtain six different extract fractions. The purpose of this study was to compare the TLC fingerprints of the phenolics present in three different sage (*Salvia* L.) species. Thus, the six extracts were obtained for each individual plant (containing free phenolic acids, two types of phenolic acids glycosides, flavonoid aglycons, and two types of flavonoid glycosides), which were then analyzed by means of silica gel 60 F_{254} TLC to obtain the respective densitograms serving as fingerprints. In that way, each individual sage species was characterized by a set of six fingerprints, which furnished an extensive platform for a comparison of the curative potential of each individual herb. The same group of authors presented analogous results of a multistep extraction with which six medicinal herbs belonging to the three different genera of the Lamiaceae family (i.e., *Salvia*, *Thymus*, and *Dracocephala*) were compared using silica gel 60 F_{254} plates [3]. It should be noted that use of silica gel 60 F_{254} TLC and HPTLC plates, almost always from Merck KGaA (EMD Millipore Corp. in the USA), is most widely reported in the literature by a wide margin; these plates contain layers prepared from 6 nm (60 angstrom) average pore diameter silica gel particles with a fluorescent phosphor incorporated to allow detection of compounds that quench the fluorescence produced by irradiation with 254 nm UV light.

7.3   SOXHLET EXTRACTION

This is a reflux-type extraction technique in which a solid sample is put in a thick, stiff paper thimble through which the solvent repeatedly fills and siphons. The analyte is exhaustively extracted, usually over a period of hours, without operator intervention. Currently, Soxhlet extraction can be regarded as a widespread and standard approach to isolation of active components from botanical, food, environmental, and other samples. This is a relatively cheap and well-performing approach that has become even more robust with the introduction of the automated Soxhlet extraction instruments, for example, the Soxtec systems from Foss.

Among many published applications of the Soxhlet extraction prior to TLC analysis, isolation of 6-gingerol from an Ayurvedic formulation originating from Orissa, India, and dedicated to curing stomach problems has been reported [4]. Another application of Soxhlet extraction was in the area of Indian, Sri Lankan, and South African folk medicine [5]. This chapter describes Soxhlet extraction with methanol followed by the silica gel HPTLC analysis of shanziside methyl ester and barlerin from the aerial parts of *Barleria prionitis* L. In the countries on the Indian Ocean rim, *B. prionitis* L. is very popular for its multiple bioactivities, including antiarthritic, anti-inflammatory, hepatoprotective, and immunorestorative. Yet another example was given by Patil et al. [6] in their investigation of curcuminoids contained in turmeric (*Curcuma longa* L.), a plant highly valued in Indian and Far Southeast Asian cuisine. Curcumin is the major constituent of curcuminoids, and it is reported to be a natural antioxidant with inhibition effects for Parkinson's disease, cytotoxicity, and cancer. In all of the described cases, hot methanol was used as an extractant, due to the wide spectrum of compound classes that can be isolated from the plant material with its use (even those that considerably differ in their polarities and structures).

7.4   SOLID-PHASE EXTRACTION

Solid-phase extraction (SPE) has been used prior to the TLC separations for many chemical and life science applications, including pharmaceutical, biomedical, and biological analyses. SPE cartridges are packed with a sorbent, and the process can be operated either manually or automatically. Samples are passed through a conditioned cartridge, and then a series of solvents is passed through to elute various analytes in different fractions prior to TLC. Impurities (which may be the compounds outside the scope of a given study and not the impurities in a direct sense of the word) are eluted in separate fractions and discarded or remain on the column after elution of the analytes. SPE is also carried out in disks, but cartridges have been mostly used for TLC applications. In many cases, cleanup by both liquid–liquid extraction (LLE) and SPE is carried out.

The most widely applied SPE sorbent reported for sample preparation prior to TLC analyses is octyldecyl silane (C_{18}) chemically bonded silica gel, but a number of other sorbents have also been used. For example, octyl silyl (C_8) chemically bonded silica gel was used in the TLC profiling of impurities in 3,4-methylenedioxymeth-amphetamine synthesis from piperonal [7]; aminopropyl (NH_2) bonded silica for fractionation of lipid classes [8] and recovery of lipids from Gaucher and Krabbe's disease patient tissue [9]; $AgNO_3$-modified silica gel for determination of docosa-hexaenoic acid in bovine milk [10]; SBD-1 (styrene divinyl benzene) for screening of organophosphorus pesticides in water [11]; and Florisil for determination of two oxycholesterols in raw and cooked meat [12].

The following are more extensively discussed practical cases chosen to illustrate the large scope of the use of SPE prior to TLC analysis. Namur et al. [13] described a novel method of preconcentrating tizoxanide (a metabolite of nitazoxanide, a broad-spectrum antiparasitic agent) from human plasma with use of a cation-exchange SPE cartridge followed by HPTLC on silica gel 60 F_{254} plates developed with toluene–ethyl acetate–acetic acid (6:13.6:0.4) mobile phase and scanning at 313 nm for the internal standard metronidazole and 410 nm for tinoxanide. Ali et al. [14] introduced a validated C_{18} SPE–TLC method of preconcentrating and quantifying certain anti-depressant drugs from human plasma. The same authors introduced an SPE–TLC method [15] for characterization and quantification of drug–receptor interactions; the contraceptive drugs were first isolated with the use of C_{18} cartridges from the blood plasma. Several modern extraction methods preceding the TLC analysis of phospholipids contained in biological samples were reviewed, and SPE was consid-ered a leading method [16]. The SPE technique can also be used in order to precon-centrate biologically active compounds from herbal material; a nice illustration of this application was provided by Apers et al. [17], who introduced a validated TLC method for quantifying aescin (a mixture of saponins employed for the treatment of peripheral vascular disorders) in herbal medicinal products that included preconcen-tration of saponins with the use of C_{18} type SPE cartridges.

Another important area of application of the SPE technique is environmental analysis, for example, the analysis of water and wastewater. There is a broad spec-trum of water pollutants, among which human and veterinary drugs as well as pes-ticides are predominant. A comparison was made of different SPE materials used for sample extraction prior to the TLC analysis (cyano [CN] bonded F_{254} plates with 0.5 M oxalic acid–methanol (81:19) mobile phase) of veterinary drugs contained in water samples. Eight veterinary drugs of diverse structure were chosen as the test compounds, and over 10 different types of the SPE cartridge sorbents were studied; a general statement was made that the silica gel matrices covalently bonded with the C_8 and C_{18} ligands and the synthetic organic polymers proved to be the best performing sorbents and were very well suited for the intended purpose [18]. Two other papers by the same research group reported the use of SPE in the TLC analysis of sulfonamides (silica gel 60 F_{254} plates) [19] and antibiotics (CN F_{254} s plates) [20] contained both in spiked water and in wastewater samples.

Another vital analytical problem is the monitoring of pesticide contamination of lake, river, and sea waters. Tuzimski [21] compared selected SPE sorbents with respect to their extraction performance when preconcentrating pesticides of different polarity and molecular structure. At the second step, the extracted pesticides were analyzed by means of a unique silica gel HPTLC–diode array detector (DAD) densitometer system. The general conclusion regarding applicability of the different SPE sorbents was similar to that found in the studies presented above, that is, C_{18} sorbent performs best with the widest spectrum of the investigated pesticides.

7.5 LLE AND LIQUID–LIQUID MICROEXTRACTION (LLME)

LLE is a term that is used in the literature both for initial extraction of an original liquid sample with an insoluble solvent and for cleanup of a liquid extract of a solid sample by partitioning with an insoluble liquid (probably best called liquid–liquid partitioning). The precondition of a liquid sample applies to medicinal and pharmacokinetic analysis with body fluids (blood, serum, plasma, bile, saliva, urine, etc.) as the analyzed samples. All body fluids contain water as the predominant liquid moiety; hence, liquid extractants should be water-immiscible organic solvents. LLE can also be applied in environmental analysis, for example, for isolation of different pollutants contained in water and wastewater and in food analysis, very often in combination with additional sample preparation techniques. For example, multistep sequential liquid–liquid chromatography can be used to fractionate extracts derived from medicinal plants prior to the TLC analysis of the resulting less complex mixtures. The simplest way to perform LLE is with use of a separatory funnel, although other practical solutions are also possible.

An application of LLE to a pharmacokinetic study with prospective usability for pharmacotherapy adherence studies was described by Mennickent et al. [22]. The authors developed a validated method for silica gel HPTLC quantification of propranolol (a common antihypertension drug) in human serum. The procedure consisted of separating serum from human blood by means of centrifugation, followed by LLE of propranolol from the serum with n-heptane–isoamyl alcohol (95.5:1.5). The separated organic layer was evaporated to dryness under mild conditions (in a gentle stream of dry nitrogen at 40°C), and the dry residue was dissolved in methanol for the HPTLC quantification by densitometry at 290 nm. Part of the sample preparation procedure was addition of an internal standard, verapamil, to the serum prior to analysis; use of an internal standard is common in TLC-densitometric analysis of bodily fluids as is performed in clinical and forensic chemistry laboratories, but usually not for other types of samples.

Faraji et al. [23] applied LLME to isolate six phenols from a clean (blank) well water sample purposely spiked with the phenol standards. 1-undecanol was selected as an extraction liquid. The entity composed of the phenol-containing water and 1-undecane was stirred for certain period of time in a closed vial, which was then placed on ice in a beaker. The solidified organic phase with the extracted phenols was easily separated from the liquid water layer and then melted, and, finally, 1-undecanol was removed by evaporation. The dry residue was dissolved in methanol and analyzed by means of HPTLC on reversed-phase (RP) $C_{18} F_{254}$ plates. The developed analytical method was further employed for the analysis of phenolic contaminants in environmental water samples.

Ostry et al. [24] published an application of LLE to food analysis by HPTLC. In view of an increasing concern of the European Commission and the European Food Safety Authority (EFSA) with the spread of mycotoxins in the environment and the impact thereof on human health [25], the authors developed a validated method for quantification of the major *Alternaria alternata* mycotoxins, which are the recognized plant pathogens ubiquitous in the environment. The investigated pathogens, which originated from Czech winter wheat and grape samples, were cultivated on rice as a known substrate for *Alternaria* production. Rice culture of *A. alternata* was first blended with methanol and centrifuged, then the supernatant was mixed with a 5% aqueous solution of ammonium sulfate. In the LLE step, mycotoxins of interest were extracted from the water-based solution with dichloromethane. The dichloromethane extract was evaporated to dryness, and the extracted mycotoxins were reconstituted in methanol for the HPTLC analysis using silica gel 60 plates precoated with 10% oxalic acid in methanol.

Applications of the sequential LLE technique can be found in the literature to fractionate plant extracts preliminarily obtained through maceration with a solvent at room temperature, or with a traditional liquid extraction at the solvent's boiling point under reflux. Ghosh and Katiyar [26] used four different solvents of increasing polarity (*n*-hexane, chloroform, ethyl acetate, and *n*-butanol) to fractionate the traditionally obtained methanol extract of walnut (*Juglans regia* L.) bark, a popular medicine in the traditional system of India. Their targets were the selected plant phenolics (juglone, quercetin, myricetin, rutin, caffeic acid, and gallic acid), which at the end were quantified by means of HPTLC using C_{18} plates developed with methanol–water–formic acid–acetic acid (48.8:46.4:2.4:2.4) mobile phase. Extraction yields with use of each individual solvent were chromatographically evaluated, and recommendations for the sequential extraction of other herbal raw materials with use of the same procedure were made.

An analogous sequential LLE approach was proposed by Kaur et al. [27]. This time, *Crataegus oxyacantha* L., another Indian medicinal plant from the Rosaceae family, that is useful in cardiovascular disorders and known for its pronounced antioxidative properties was investigated. Five flavonoids (vitexin, vitexin-2″-*O*-rhamnoside, hyperoside, quercetin, and apigenin) were extracted from the plant leaves using three solvents of increasing polarity (*n*-hexane, ethyl acetate, and *n*-butanol). Again, the extraction performance of each individual extraction step was evaluated by means of C_{18} HPTLC.

7.6 SUPERCRITICAL FLUID EXTRACTION

Supercritical fluid extraction (SFE) is the process of separating a component of interest from a solid or liquid matrix with use of a supercritical fluid. The concept of SFE has been developed for two different practical applications, that is, analytical (sample preparation) and technological (preparative: isolation of higher quantities of compounds). The most often employed supercritical fluid is CO_2 used above its critical temperature and pressure, which are 31°C and 74 bar, respectively. The frequent use of CO_2 as the extracting supercritical fluid is due to its convenient, that is, low critical parameters, which make it relatively easy to handle in a supercritical state. Other advantages of CO_2 are that it is safe (nontoxic and nonhazardous), inexpensive, and readily available (SFE is a *green* technology). Its disadvantage is, however, that it is nonpolar (NP) and, hence, has a limited extraction power. Therefore, CO_2 can be used as an extractant without any modifier to extract low or medium polarity target compounds only. Slight enhancement of its extracting performance can be obtained in two different ways, that is, by modification of CO_2 with polar solvents (e.g., methanol, ethanol, or acetone) and/or by employing pressures considerably higher than the critical CO_2 pressure. This brief description is enough to grasp that SFE is not a universal extracting technique and its applications cover selected practical tasks only. In spite of that, one sporadically encounters examples of SFE employed as a sample preparation method preceding TLC analysis. These usually are applications to the best suited analytical targets, that is, to samples of natural origin containing low-polarity essential oils and/or lipids.

An interesting application of SFE in combination with TLC is given in the paper by Yee et al. [28]. The goal of this research project was to test if SFE can be used in a technological process of lowering fat contents in Cheddar and Parmesan cheeses without negatively affecting their flavor. For this purpose, both cheese types were extracted by means of SFE with CO_2 under different pressures, and then the extracted lipids were analyzed by means of TLC. Silica gel plates were developed with the polar mobile phase chloroform–methanol–water (65:25:4) or NP petroleum ether–ethyl ether–glacial acetic acid (85:15:2); separated zones were detected with iodine vapor, and plates were scanned with a Biorad Model GS-700 imaging densitometer. The obtained results showed that the only fatty compounds extracted from the cheese samples were the NP lipids (i.e., saturated triglycerides and free fatty acids), whereas phospholipids were retained in the cheese matrix. Quantitative reduction of the NP lipid fraction under the optimized working parameters was up to 50%, so that the technological application of SFE to the production of the low-fat varieties of the Cheddar and Parmesan cheese seems justified. Additionally, by means of gas chromatography–mass spectrometry (GC–MS) it was confirmed that SFE did not remove flavor components from the cheese matrices, which was an additional bonus of the approach.

An analytical application of SFE–TLC was investigation by Pereira et al. [29] of the plant known as *Tabernaemontana catharinensis*, which is widespread in South America and used in folk medicine as a potent infusion antidote for snake bites. The authors employed SFE for the fractionated extraction of the plant alkaloids with use of CO_2 containing several polar cosolvents (methanol, ethanol, 2-propanol, water, and mixtures thereof) in order to enhance and fine tune the extraction power of the fluid. The obtained extract fractions underwent further fractionation using TLC, and then they were tested for their antioxidant and antimycobacterial activities. Silica gel 60 F_{254} plates with chloroform–methanol (9:1) mobile phase and Dragendorff alkaloid detection reagent were used.

Hamrapurkar et al. [30] compared the efficiency of the extraction of phyllantin, a hepatoprotective lignan present in *Phyllanthus amarus*, with use of methanol Soxhlet extraction or SFE with use of CO_2 plus methanol as a cosolvent. It was concluded that the elaborated SFE-based sample preparation method was more efficient than the Soxhlet extraction. The optimized conditions for SFE included 150 bar CO_2 pressure, 45°C temperature, and 30 min extraction time; the type of instrument used was not elaborated. HPTLC-densitometry analysis of the extract was carried out on silica gel 60 F_{254} plates with hexane–toluene–ethyl acetate (2:2:1) mobile phase.

Plander et al. [31] compared several extraction techniques (including SFE with use of pure CO_2 and with addition of ethanol, Soxhlet extraction, and hydrodistillation [see Section 7.14]) to isolate essential oil from botanical material (*Satureja hortensis* L.) for further analysis of the major antioxidants contained therein (basically polyphenolic acids and flavonoids). A comparison of the obtained extracts was performed with the use of TLC-densitometry, and no measurable superiority was attributed to any extraction technique employed. That is no surprise since the authors targeted polar compounds, for which SFE is not the extraction method of choice.

The Spe-ed 2 is a commercial instrument (Figure 7.1) that has been used to prepare samples for TLC analysis; an example is given in Section 7.18.7. Features include oven capacity of two vessels, temperature to 240°C, pressure up to 10,000 psi (680 bar), and pump flow rates up to 400 L min^{-1}.

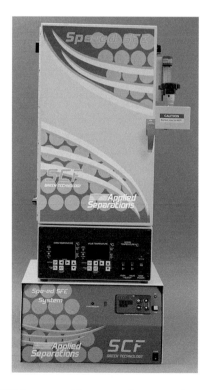

FIGURE 7.1 Spe-ed SFE-2 system. (Courtesy of Dan Lewis, Applied Separations, Inc., Allentown, PA.)

7.7   PRESSURIZED LIQUID EXTRACTION

The simplest way to execute pressurized liquid extraction (PLE; also referred to as accelerated solvent extraction [ASE]) is with use of a fully automated commercial instrument. In PLE, solvent is pumped into an extraction vessel containing the sample and is heated (e.g., 60°C–200°C) and pressurized (e.g., 3.5–20.0 MPa).

The number of developed PLE–TLC analytical methods is considerably lower than those employing other and more cost-friendly extraction techniques, yet certain important features emerge from the information already available from the literature. First, PLE can considerably reduce both the extraction time and consumption of the extractants. Second, the extraction yields obtained with PLE can be considerably higher than those from any other available extraction technique. As a consequence, PLE is particularly suitable for the extraction of the compounds embedded in the so-called *difficult* matrices (e.g., in biological cells of plants or in animal tissues). However, a combination of elevated temperature and pressure can result in the extractant's boiling temperature being considerably higher than its boiling point under an atmospheric pressure, which in turn can decompose thermally less stable compounds of interest. Therefore, the working parameters of a PLE system should be optimized by means of a chemometric approach in order to establish an efficient PLE–TLC methodology.

PLE–TLC has been applied to the analysis of the antimicrobial activity of certain extracts. Jaime et al. [32] investigated the antioxidant activity of carotenoids contained in *Haematococcus pluvialis* unicellular microalgae. The authors used an ASE 200 commercial instrument and showed using TLC (silica gel layer, petroleum ether–acetone [75:25] mobile phase) and high-performance liquid chromatography (HPLC) with a DAD that extraction with ethanol was much better than with lower polarity *n*-hexane and that extraction temperature had a positive effect on extraction yield. A photograph of the ASE 200 instrument is shown in Figure 7.2 and a schematic diagram of the PLE process with this instrument in Figure 7.3.

Ha et al. [33] studied the extraction of the *Calophyllum inophyllum* fruit peels (a waste product from the production of oils from this fruit) for further TLC

FIGURE 7.2   Dionex ASE 200 PLE extraction instrument. (Courtesy of Pixie Kempel, Thermo Fisher Scientific, Sunnyvale, CA.)

ASE schematic

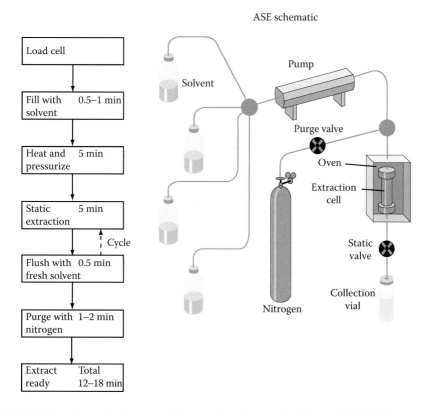

Load cell	
Fill with solvent	0.5–1 min
Heat and pressurize	5 min
Static extraction	5 min
Flush with fresh solvent	0.5 min
Purge with nitrogen	1–2 min
Extract ready	Total 12–18 min

FIGURE 7.3 Schematic diagram of the operational steps of the Dionex ASE 200 instrument for PLE. (Courtesy of Pixie Kempel, Thermo Fisher Scientific, Sunnyvale, CA.)

investigation (silica gel 60 F_{254} aluminum sheets with three different mobile phases and zone detection by UV light and ferric chloride spray reagent) of the antimicrobial activity of the crude extract; it was demonstrated that PLE with an ASE 100 apparatus (1500 psi, 120°C, 15 min) in many ways outperformed automated Soxhlet extraction; however, the authors pointed out the necessity of optimizing the PLE working parameters in order to avoid thermal decomposition of the extraction targets and, thus, ensure the highest possible extraction yields.

The response surface method (RSM) was used to optimize duration, extraction temperature, and the quantity of a dispersant (hydromatrix) required for an efficient PLE of lipids from oleaginous yeast (*Rhodotorula glutinis*). TLC was used to compare optimized PLE to Soxhlet and modified Bligh and Dyer conventional extraction methods, and it showed higher extraction recovery, reduction of analysis time by 10, and solvent quantity reduction by 70% [34].

The RSM method with central composite design (CCD) was employed in order to optimize the time and solvent composition for the PLE of the phenolics from six different Thymus species and compares results with those obtained using Soxhlet extraction [35]. Due to its sensitivity to elevated temperatures, rosmarinic acid that was abundantly present in all thyme species was used as a test compound, and the extraction conditions were adjusted to maximize the rosmarinic acid extraction yield. An ASE 200 used for PLE (Figure 7.3) with methanol–water in this study after herbal material was pre-extracted by *n*-hexane to at least partially remove lipids and chlorophyll. TLC analysis was carried out on silica gel 60 F_{254} plates with ethyl acetate–formic acid–acetic acid–water (100:11:11:13) mobile phase and 10% methanolic sulfuric acid detection reagent.

7.8 ULTRASONIC-ASSISTED EXTRACTION

Natural products and their derivatives constitute ca. 50% of the raw materials used in modern therapy. Currently, new curative materials of natural origin are being extensively searched for, and traditional botanical drugs are being scrutinized with the aid of modern analytical techniques including TLC. A dynamically growing demand for natural, mostly botanical, products gave rise to the search for novel and effective extraction procedures that might accelerate sample preparation, and ultrasonic-assisted extraction (UAE) is one option offering enhanced extraction efficiency due to the propagation of ultrasound pressure waves through the extraction solvent and the resulting cavitation phenomenon. Extraction efficiency of UAE is usually attributed to cavitation, mechanical, and thermal effects, which result in disruption of the cell walls, particle size reduction, and enhanced mass transfer across cell membranes. For nonindustrial, mostly analytical, purposes, UAE is predominantly carried out in an indirect manner, that is, vessels containing the material to be extracted and an extractant are immersed in an ultrasonic bath.

A review was presented [36] of various applications of UAE both for industrial and for application purposes, pointing out its rapidly growing popularity for the extraction of active chemical compounds from natural products in pharmaceutical and chemical industries with reduced processing time and solvent consumption. The review presented an exhaustive overview of different aspects of UAE of various natural products, including the mechanism, recommendations for optimizing operating conditions for maximizing extraction yields, different applications, and possible intensification of UAE by coupling with traditional solvent extraction and SFE. The most successful areas of application include extraction of phenolics, antioxidants, essential oils, hemicelluloses from natural products, natural dyes and pigments, saponins, and, particularly, medicines of botanical origin. In TLC laboratories, samples for analysis are increasingly being prepared with the aid of UAE. The following three papers reported on this type of sample preparation of botanical material preceding TLC analysis.

Kac et al. [37] reported the use of UAE to rapidly (in 10 min) isolate xanthohumol from hop pellets using methanol. An interest in xanthohumol is due to its cancer chemopreventive, antioxidant, and radical-scavenging activity, on top of its positive effect on the specific bitter taste of beer. The contents of xanthohumol in hop were determined by HPTLC using silica gel 60 plates, toluene–dioxane–acetic acid (77:20:3) mobile phase, and scanning at 368 nm. Quantification of hop samples of diverse geographical origin was carried out.

UAE was employed to isolate the alkaloid, trigonelline, and the major free amino acid, 4-hydroxyleucine, contained in fenugreek seeds, which are known for their pungent aromatic properties and for their firm position in the Ayurvedic and Unani medicine systems (e.g., for the treatment of epilepsy, paralysis, or chronic cough) [38]. A 20 min UAE with methanol proved very efficient, enabling establishment of a validated TLC method to rapidly quantify trigonelline and 4-hydroxyleucine in the fenugreek seeds based on silica gel F_{254} plates, n-butanol–methanol–acetic acid–water (4:1:5:11) mobile phase, and scanning at 266 nm and at 395 nm after detection with ninhydrin spray reagent, respectively.

A UAE was developed as part of a method to rapidly evaluate the phenolic contents and antioxidant and antimicrobial activity of *Bauhinia purpurea* leaves, which are used in traditional medicine systems of Southeast Asia to cure such conditions as body pain, rheumatism, stomach tumors, diarrhea, or skin diseases. The authors compared conventional maceration (3 days), Soxhlet extraction (48 h), and UAE (1 h) with use of 99.5% ethanol and demonstrated advantages of the latter technique in terms of both rapidity and the extraction yields. Then the obtained extracts were extensively evaluated with aid of silica gel HPTLC-densitometry in terms of the total phenolic yields and radical-scavenging and antimicrobial activity [39].

Of course UAE is not limited to research problems in the area of phytochemistry or pharmacognosy alone, but it has also been applied in many other nonbotanical types of chemical, life science, and environmental TLC analysis. For example, UAE of histamine was applied to cheese samples with the use of several solvents [40]. The extraction results were compared with those obtained using reflux extraction and microwave-assisted extraction (MAE; see Section 7.9), and it was demonstrated that UAE with methanol provided the highest extraction yields of histamine contained in cheese. Histamine was isolated from the obtained extract by two-dimensional (2D) preparative layer chromatography, and then it was quantified using chronopotentiometry. 2D TLC was carried out on 0.25 mm thick silica gel G layers developed with ethyl acetate–methanol–ammonia (4:3:2) in both directions (at right angles), and the histamine was recovered by scraping the layer around its spot and elution with the supporting electrolyte.

Surfactants contained in the wastewater from a hospital laundry were isolated by UAE with use of chloroform, and then a TLC assay of the surfactants was developed to assess the contents thereof in the *real life* samples [41]. UAE was found to be faster and more efficient when compared to traditional shake-flask extraction, and it was also economically and ecologically favorable because much less solvent was used. The TLC method comprised silica gel 60 F_{254} plates, 2-propanol mobile phase, and quantification using a CAMAG videodensitometer with VideoStore and VideoScan software.

7.9 MICROWAVE-ASSISTED EXTRACTION

Standard operations that employ microwaves for the purpose of heating are nowadays increasingly popular in chemical laboratories throughout the world. One such operation is running chemical syntheses in specially designed microwave reactors. Another microwave-assisted operation is MAE, which has been in use since 1985 for heating in the extraction of compounds. Contrary to the Soxhlet extraction, which is carried out in an open system and hence the temperature of the boiling extractant strongly depends on atmospheric pressure, MAE is carried out in closed vessels. Thus, the extracting medium can obtain temperatures considerably higher than those available with use of a Soxhlet apparatus, and, in principle, these temperatures are controlled by the resistance of a sealed ampule. The main advantages of MAE are considerably higher extraction yields as compared with traditional extraction under the reflux and Soxhlet extraction and a considerable shortening of the extraction time.

MAE has been used for various analytical purposes, for example, the extraction of botanical material and environmental samples as reported in the following three examples. Genetic algorithm-based optimization of selected parameters (temperature, solvent amount, and duration of the extraction process) of MAE prior to the TLC analysis of the obtained extract of soil samples spiked with atrazine and simazine herbicides was described [42]. A CEM Corp. Mars X MAE system was used, and TLC was carried out on silica gel 60 F_{254} layers using hexane–chloroform–acetone (60:25:15) mobile phase and quantification at 254 nm with a CAMAG videodensitometer.

A validated method of analysis of the soil samples contaminated with α-cypermethrin, a popular synthetic insecticide, included extraction of the soil extracted with use of MAE, UAE, and the shake-flask method [43]. A comparison of the extraction yields of the three techniques concluded that MAE was best. MAE was carried out in a Mars 5 (Figure 7.4) with XP-1500 Plus vessels; the solvent was hexane–dichloromethane (1:1), microwave power was 1600 W, and temperature was raised to 75°C in 5 min (holding time 5 min) and then to 100°C in 5 min. Silica gel 60 F_{254} HPTLC plates were developed with hexane–toluene (1:1) and scanned at 220 nm for quantification.

Another chemometric approach (CCD) was used to optimize four parameters (duration and temperature of extraction, mixed extractant ratio, and extractant volume) for the extraction of a pharmaceutically active compound, withaferin A, from *Withania somnifera* Dunal (a medicinal herb recognized by the Ayurvedic curative system) [44]. The chosen time was 150 s, temperature was 68°C, and volume of methanol–water (25:75) extractant was 17 mL. For the purpose of the chemometric CCD approach, silica gel HPTLC quantification of the targeted compound

FIGURE 7.4 MARS 5 MAE instrument with open door allowing view of vessels inside. (Courtesy of Elaine Hasty, CEM Corporation, Matthews, NC.)

was used to evaluate the extraction performance with ethyl acetate–toluene–formic acid–2-propanol (7:2:0.5:0.5) mobile phase and scanning at 220 nm.

7.10   IMMUNOAFFINITY EXTRACTION AND CLEANUP

Immunoaffinity (IA) methods are based on the principle of molecular recognition via very selective antigen–antibody interactions. Preparation of the tailor-made IA sorbents is regarded as a challenging and sophisticated synthetic task, and such materials are very sensitive to inappropriate handling (e.g., they are vulnerable to even minor pH changes of the employed solvent, which can easily deform or even destroy the antigen or antibody peptide ligands attached to an organic matrix). The use of IA sorbents for sample preparation prior to TLC has been mostly for determination of toxins. The following are selected examples: the steroid animal drug trenbolone and its metabolite in bovine urine to detect illegal use in Holland [45]; fumonisin B1 in corn with methanol–water (80:20) extraction followed by IA column cleanup and TLC on C_{18} plates with acetonitrile–water mobile phase and fluorescence densitometry after postchromatographic derivatization with a fluorogenic reagent [46]; aflatoxins B1, B2, G1, and G2 in foods regulated within the European Community using methanol extraction, IA column cleanup, and densitometric quantification [47]; deoxynivalenol (vomatoxin) in wheat flour and malt by extraction with water and polyethylene glycol, IA cleanup on Vicam Dontest columns, and HPTLC-fluorescence densitometry at 366/420 excitation/emission wavelengths [48]; and ochratoxin A (OTA) in green coffee using extraction with methanol-aqueous $NaHCO_3$, IA cleanup on columns containing OTA antibodies, and NP or RP TLC-densitometry [49].

7.11   DIALYSIS

Dialysis is a cleanup method in which the analyte passes from one solution to another through a membrane while impurities do not transfer. The process can be set up to remove either low or high molecular weight interferences prior to TLC analyte determinations. As examples, dialysis was used to extract the mycotoxin patulin from apple juice using ethyl acetate in a diphasic system [50], and to remove low molecular weight constituents from bark extracts prior to carbohydrate and phenol analysis [51].

7.12   EXTRELUT COLUMN LIQUID EXTRACTION

Extrelut columns (Merck KGaA) can be used to perform LLE in place of a separatory funnel. The tube is packed with diatomaceous earth, and an aqueous sample is poured in. When a suitable organic solvent is then passed through, the analyte is eluted while water and polar compounds in the sample are retained on the column. Applications are illustrated by the use of Extrelut extraction prior to TLC for the determination of nicotine and its metabolite cotinine in urine samples of children of smoking parents [52] and of caffeine in coffee [53].

7.13   QuEChERS (QUICK, EASY, CHEAP, EFFICIENT, RUGGED, AND SAFE)

The QuEChERS sample preparation technique was introduced at the very beginning of the third millennium [54], so it is among the youngest techniques mentioned in this review. QuEChERS was originally designed for multiresidue analysis of pesticides in fruits and vegetables, but it was rather quickly applied to other sample types as well, such as soil, water, animal tissue, and sewage sludge. The QuEChERS approach combines salting-out LLE (SALLE) in the first step, followed by dispersive SPE (dSPE) in the second step.

Originally, the chromatographic quantification techniques following the QuEChERS extraction were highly instrumental, like column HPLC or ultra-performance column liquid chromatography (UPLC) with DAD or mass spectrometry (MS) detection. However, a recent publication [55] demonstrated the fully satisfactory potential of HPTLC-densitometry for quantification of the QuEChERS extract contents. This study was devoted to investigating residues of the anthranilic diamide insecticide chlorantraniliprole on cauliflower, which is an important crop in North India that is seriously endangered by various insects and, therefore, protected by insecticides. However, accumulation of insecticide residues in the crop can ultimately endanger the human population and for this reason should undergo strict control, necessitating the development of new analytical methods. The extraction solvent was ethyl acetate; NaCl was added to aid phase separation; and dSPE cleanup was carried out using PSA (primary secondary amine) sorbent, $MgSO_4$, and graphitized carbon black. The results of HPTLC-densitometry analysis of the QuEChERS extract of the cauliflower samples for the contents of chlorantraniliprole residues were compared with those of HPLC–DAD. That comparison univocally confirmed applicability of HPTLC for quantification of the QuEChERS extracts. Like most modern HPTLC quantitative analyses, automated instruments were used for various steps, that is, a CAMAG Linomat 5 for application of sample and standard bands onto silica gel 60 F_{254} aluminum plates, ethyl acetate mobile development in a CAMAG ADC2 developing chamber, and quantification by densitometry with a CAMAG TLC Scanner 3 using winCATS software.

Similar confirmation was provided in the use of a QuEChERS method with tetrahydrofuran or acetone and NaCl salt for a simple and quick extraction of mycotoxins (aflatoxins and OTA) from the dried figs, followed by NP and RP TLC-densitometry analysis of the obtained extract using a Tidas TLC 2010 fiber optics scanner [56]. Thus, a robust technique was developed for the rapid control and screening of the dried fruits, which are internationally traded on a large scale.

7.14   HYDRODISTILLATION

Hydrodistillation (or steam distillation) is a well-established sample preparation technique for the extraction of essential oils from plant material. On a laboratory scale, most often it is carried out in the Clevenger-type (recommended by the European Pharmacopoeia) or the Deryng-type (described in the Polish

Pharmacopoeia) apparatus. These two types of apparatuses are glassware that slightly differ in construction but work based on the same principle. Each apparatus consists essentially of three parts: a round-bottomed flask in which is placed the material containing volatile oil and water, a separator in which oil is automatically separated from the distillate in a graduated tube, and a convenient condenser. This type of equipment works for volatile oils that are lighter than water. Although GC–MS or headspace-GC–MS (which omits the hydrodistillation step) is often the analytical technique of choice for the essential oil analysis, reports in the literature [57,58] have described the successful fractionation and fingerprinting of essential oils isolated by hydrodistillation by means of the low-temperature TLC carried out in a refrigerator at ca. 10°C.

7.15   PRE-TLC DERIVATIZATION

Pre-TLC derivatization has been used in some analyses when a derivative of the analyte is better separated than the underivatized compound, or to provide detection in place of the usual post-TLC application of reagents by spraying or dipping. Derivatives can be formed in solution and then applied to the layer, or the reaction can be carried out *in situ* at the origin by overspotting the reagent after applying the underivatized sample, often with the use of an automated application instrument. Three examples of this procedure are given below.

Twenty-one amino acids were separated and detected as butylthiocarbamyl (BTC) derivatives formed by prechromatographic reaction at 40°C for 30 min with butyl isothiocyanate in a test tube or *in situ* by overspotting. The BTC derivatives were separated on silica gel 60 F_{254} aluminum plates by development in a horizontal DS chamber with ethanol–methanol–chloroform (1:1:2), which was chosen as best after testing 15 binary and ternary mobile phases. Detection at picomole levels was achieved with iodine-azide detection reagent, and separation of the difficult leucine–isoleucine pair was achieved as well as other amino acid mixtures [59].

A validated HPTLC-densitometry method was described for the simultaneous determination of oleanolic and ursolic acids in some common spices plant extract involving treatment of extract samples directly on a silica gel 60 F_{254} plate by 1% iodine solution in chloroform; this prechromatographic derivatization was necessary for separation of the acids because of their similar structures. The mobile phase was toluene–petroleum ether–ethyl acetate–acetonitrile (5:5:1:0.3), detection of the acids was by use of 10% sulfuric acid in ethanol spray reagent, and quantification was by scanning at 400 nm with a computer-controlled Desaga CD 60 densitometer [60].

Acrylamide was determined in ground coffee after prechromatographic *in situ* derivatization with dansulfinic acid. After acetonitrile extraction in an ASE 200 system (Figure 7.3), the extract was cleaned up by carbon SPE, applied to a silica gel 60 HPTLC plate, oversprayed with dansulfinic acid, and heated at 120°C for 1 h to form the fluorescent derivative dansylpropanamide. The chromatographic separation with ethyl acetate-*tert*. butyl methyl ether (8:2) was followed by densitometric quantification at 254 nm/>400 nm. The method was validated for commercial coffee [61].

7.16   HIGH THROUGHPUT PLANAR SPE

Olleig and Schwack introduced this new method for sample preparation prior to the determination of the residues of seven pesticides (acetamiprid, azoxystrobin, chlorpyrifos, fenarimol, mepanipyrim, pencolazole, and primicarb) in organic black and green teas. Teas were extracted with acetonitrile without addition of water followed by dSPE cleanup with $MgSO_4$ and PSA and C18 sorbents. After application of acetonitrile-dSPE extracts onto a silica gel 60 F_{254} plate, planar solid phase cleanup was performed by development in an ADC2 with twin trough chamber using acetonitrile–water (19:1) for a migration distance of 85 mm followed by acetone–water (7:1) for 31 mm in the opposite direction with 33% relative humidity control ($MgCl_2$). Target pesticide zones were eluted with acetonitrile–10 mM ammonium formate buffer (1:1) at a flow rate of 0.2 mL min^{-1} using the CAMAG TLC–MS interface into autosampler vials for detection and quantification by HPLC/electrospray ionization (ESI)–MS. Mean recoveries for all of the studied pesticides were 72%–111% at 0.01 and 0.1 mg/kg spiking levels with RSD of 0.7%–4.7% [62].

7.17   DIRECT TLC WITHOUT SAMPLE PREPARATION

An advantage of TLC is that a plate is used only once, so every sample is separated on fresh stationary phase without carryover or cross-contamination; therefore, less purified samples with strongly sorbed impurities can often be applied compared to HPLC where injected sample impurities can build up and destroy column performance for analysis of later samples. In some cases, samples can be applied and directly analyzed by TLC with no prior extraction or cleanup steps. Four examples involving quantitative determinations after separation on silica gel HPTLC plates with a preadsorbent zone are cholesterol in human saliva and blood serum (visible mode scanning after detection with cupric acetate-phosphoric acid reagent) [63]; caffeine in regular and diet colas and noncola beverages (fluorescence quenching mode scanning) [64]; quinine in tonic waters (fluorescence mode scanning) [65]; and sorbic, benzoic, and dehydroacetic acid preservatives in sodas and iced teas (fluorescence quenching mode, C_{18} and silica gel preadsorbent plates were used) [66]. These preadsorbent plates have a diatomaceous earth strip below the analytical layer to which samples can be applied quickly and diffusely as spots by hand, after which mobile phase development forms an ideal band-shaped initial zone at the interface of the layers; the preadsorbent must have no sorption capacity for the analyte but may retain some sample impurities to provide cleanup.

Thin layer radiochromatography (planar radiochromatography) is a widely used method for qualitative and quantitative analysis of radiolabeled compounds [67]. Its most numerous and important application is radiochemical purity measurement of nuclear medicines, usually carried out on instant TLC binderless glass microfiber chromatography paper sheets impregnated with silica gel (ITLC–SG). The samples for these analyses are simply the medicine prepared for patient administration that is directly applied to the sheet, followed by mobile phase development and detection of radioactivity with a specialized instrument such as an *in situ* scanner or phosphor imager.

7.18   ADDITIONAL APPLICATIONS

This review is the first published comprehensive coverage of sample preparation techniques since Chapter 1 of the third edition of the *Handbook of Chromatography* [68]. That book also contains details of the theory, methods, and instruments for all steps of TLC as well as chapters on applications to 19 different types of analytes. Additional recent applications of the methods described above are cited in the sample preparation sections of biennial reviews of planar chromatography published in 2012 [69] and 2014 [70]. This section contains selected additional applications organized by analyte(s) published after coverage in the second review [70] ended. Complete details of sample collection, sample preparation, TLC/HPTLC procedures, and results and discussion of the studies can be found in the cited references. In all but the last entry, silica gel 60 layers with or without F_{254} indicator were used for the analyses described below in this section.

7.18.1   ESCULENTOSIDE P PHYTOLACCASAPONIN ACARICIDE

Sample: Different organs of *Phytolacca americana* L.

Sample preparation: Extraction with solvents of increasing polarity followed by extract cleanup on a silica gel column eluted with a stepwise solvent gradient of increasing polarity

Mobile phase: Petroleum ether–ethyl acetate (1:5) and ethyl acetate–methanol (9:1 and 1:1)

Detection: 254 nm UV light or iodine vapor

Reference: [71].

7.18.2   GALLIC ACID

Sample: *Terminalia nigrovenulosa* bark

Sample preparation: Methanol extraction, fractionation of the crude extracts with different solvents, silica gel column chromatography with two different mobile phases, and Sephadex LH-20 size exclusion chromatography column chromatography.

Mobile phase: Chloroform–ethyl acetate–formic acid (70:30:15)

Detection: 254 nm UV light

Reference: [72].

7.18.3   GARCINOL AND ISOGARCINOL

Sample: *Garcinia indica* fruits

Sample preparation: UAE at 40°C for 45 min with five different solvents

Mobile phase: *n*-pentane–ethyl acetate–formic acid (7:3:0.5)

Detection/quantification: Densitometric scanning at 327 nm

Reference: [73].

7.18.4 TALLOW

Sample: Adulterated cow ghee
Sample preparation: Saponification with 20% methanolic KOH under reflux
    followed by LLE to obtain saponified and unsaponifiable fractions
Mobile phase: *n*-hexane–diethyl ether–glacial acetic acid (4:6:0.2 and
    6.5:3.5:0.2)
Detection: 10% methanolic sulfuric acid spray reagent
Quantification: Reflectance-absorbance mode densitometric scanning
Reference: [74].

7.18.5 MAGNOLOL AND HONOKIOL

Sample: Herbal medicine *Magnoliae officinalis* Cortex
Sample preparation: UAE with methanol for 30 min with 260 W power and
    33 kHz frequency
Mobile phase: Toluene–methanol (20:2)
Detection: Dipping in 2,2-diphenyl-1-picrylhydrazyl radical (DPPH*) reagent
Quantification: Densitometric scanning at 550 nm
Reference: [75].

7.18.6 BETULINIC AND OLEANIC ACIDS

Sample: *Achyranthes aspera* herb
Sample preparation: UAE with 95% aqueous methanol for 15 min at 60 kHz
Mobile phase: Benzene–ethyl acetate–formic acid (67.9:22.7:9.4)
Detection: Densitometric scanning at 210 nm
Reference: [76].

7.18.7 ALKALOIDS, FLAVONOIDS, AND TERPENOIDS

Sample: Mango leaves
Sample preparation: Low pressure solvent extraction with 99% ethanol by per-
    colation for 2 h and SFE using a Spe-ed Model 7017 unit with CO_2 flow rate
    of 8.3×10^{-5} kg/s for 480 min
Mobile phase: Hexane–chloroform (25:75)
Detection: Anisaldehyde, DPPH*, Dragendorff, and NP/PEG (2-aminoethyl-
    diphenylborinate in methanol) color forming reagents for different classes
    of compounds
Reference: [77].

7.18.8 AMITRIPTYLINE

Sample: Gastric lavage
Sample preparation: Lavage adjusted to pH 10.8 with 1 M NaOH and then
    LLE with ethyl acetate–*n*-heptane (1:1)

Mobile phase: Methanol–28% ammonia (98.5:1.5)
Detection/quantification: Densitometric scanning at 210 nm
Reference: [78].

7.18.9   BETA-SITOSTEROL

Sample: Leaves, stems, and roots of blue and white *Clitoria ternatea* L.
Sample preparation: Soxhlet extraction with dichloromethane–methanol (1:1)
   for 24 h
Mobile phase: *n*-Hexane–acetone (8:2)
Detection: 5% methanolic sulfuric acid reagent
Quantification: Densitometric scanning at 414 nm
Reference: [79].

7.18.10   BERGENIN AND GALLIC ACID

Sample: Rhizome of *Bergenia ciliate*, *stracheyi*, and *ligulata*
Sample preparation: Coarse powder samples of rhizome of the three species
   were fractionated successively in a Soxhlet apparatus with hexane, chloro-
   form, acetone, and ethanol (2 × 25 mL)
Mobile phase: Toluene–ethyl acetate–formic acid (3.5:5.5:1.0)
Detection/quantification: Densitometric scanning at 314 nm
Reference: [80].

7.18.11   ESSENTIAL OILS

Sample: Dill
Sample preparation: UAE with *n*-hexane–diethyl ether (1:1) was most effec-
   tive in a comparative study against maceration and MAE using a home-
   made apparatus
Mobile phase: Petroleum ether–dichloromethane (30:70)
Detection: Vanillin spray reagent
Quantification: Densitometric scanning at 600 nm
Reference: [81].

7.18.12   FLUMEQUINE ANTIBIOTIC

Sample: Milk
Sample preparation: Addition of (1:1) acetonitrile to milk proved best after
   testing of 17 extraction procedures based on precipitation of proteins
Mobile phase: Methanol–dichloromethane–2-propanol–25% aqueous ammo-
   nia solution (3:3:5:2)
Detection: Direct bioautography (DB)
Quantification: Digitization of bioautograms by a scanner and measurement of
   inhibition zones of flumequine with a planimeter
Reference: [82].

7.18.13   ENT-PIMARA-8(14),15 DIENE

Sample: Engineered *Aspergillus nidulans*
Sample preparation: ASE in a Dionex Model 200 (Figure 7.2) with ethyl acetate at 90°C with 12.07 MPa pressure
Mobile phase: Ethyl acetate–heptane–acetic acid (9:90:1)
Detection: 0.01% Rhodamine 6B GO spray reagent
Reference: [83].

7.18.14   AFLATOXIN M-1

Sample: Milk
Sample preparation: IA column VICAM (VF) eluted with methanol–water (80:20) and C_{18} SPE cartridge eluted with dichloromethane–acetone (95:5)
Mobile phase: Dichloromethane–acetone–2-propanol (85:10:5), diethyl ether–methanol–water (94:4.5:4), and dichloromethane–acetone–methanol (90:10:2)
Detection: Fluorescence under 366 nm UV light
Reference: [84].

7.18.15   CARBOFURAN PESTICIDE

Sample: Stomach contents and liver tissue in forensic cases
Sample preparation: Ethanol extraction, chloroform partitioning after adding saturated brine solution for salting out, and cleanup on a Florisil minicolumn using elution with 6%–15% diethyl ether in petroleum ether.
Mobile phase: n-Hexane–acetone–toluene–ethyl acetate (4:4:2:2) and n-hexane–acetone–toluene (10:2:2)
Detection: Exposure to bromine vapor and bromophenol blue–$AgNO_3$ (1:3) spray reagent
Reference: [85].

7.18.16   ENVIRONMENTAL POLLUTANTS

Sample: Surface and sewage water
Sample preparation: C18 SPE tubes preconditioned with 5 × 1 mL 100% methanol and 5 × 100 mL methanol–water (1:1), then the sample was applied, and analytes were eluted with 4 × 1 mL 100% methanol.
Layer: RP18 W F_{254} s
Mobile phase: Methanol–water (8:2)
Detection: Direct fluorescence and derivatization with phosphomolybdic acid
Quantification: Image analysis with digital processing
Reference: [86].

7.19   CONCLUSIONS

Traditional solvent extraction and extract cleanup by LLE and column chromatography continue to be the primary methods for TLC analysis. Of the newer techniques, use of SPE, PLE, UAE, and MAE has increased most rapidly, and it is to be expected that this trend will continue in order to achieve better speed, efficiency, and/or convenience in certain analyses. Simple dissolving of samples (after grinding, if necessary, e.g., for tablets) by stirring and/or UAE with filtering of undissolved non-active ingredients is the method of choice for pharmaceutical product analysis, which is probably the most active research area of TLC-densitometry at this time.

Majors [87] reviewed some sample preparation technologies that are used prior to HPLC and GC, including SALLE. Salting out to force analytes from a water-rich phase into an immiscible solvent phase has been used widely as a sample preparation method prior to TLC for many years. For example, the organophosphorus pesticide prophenphos was extracted from minced stomach, intestine, and liver tissue by extraction with diethyl ether after addition of 10 g ammonium sulfate per 50 g sample to improve extraction efficiency [88]. However, the SALLE method discussed by Majors involves addition of salt to water soluble extractants such as acetonitrile to help formation of two phases with aqueous samples and aid extraction of analytes into the organic solvent; as mentioned above, this extraction is the first step of the QuEChERS method, followed by dSPE cleanup. It has been noted [89] that acetonitrile with added $MgSO_4$ and $NaCl$ can be used to extract pesticides from water-rich fruits and vegetables followed by application of the organic phase directly to the layer for analysis, without dSPE cleanup, but to date SALLE with water soluble solvents has not been widely used prior to TLC. It is expected that research will proceed on developing new TLC applications of SALLE with water soluble solvents as a separate method using conventional procedures or with a two syringe miniaturized system described by Majors [87], or as the first step of many more deserved applications of the QuEChERS method, which has grown from its original use for pesticide residues in fruits and vegetables to many types of analytes and sample matrixes prior to column LC and LC/MS as tabulated by Majors [87]. A modified QuEChERS method developed for preparation of food samples prior to pesticide residue determination by GC–MS or HPLC–MS involving homogenization, extraction with nonacidified acetonitrile plus magnesium sulfate and sodium acetate, and PSA sorbent cleanup also seems to have good potential for successful use prior to TLC determinations [90].

Miniaturized planar chromatography using office peripherals and ultrathin layers has been proposed by Morlock and coworkers [91]. Very low sample volumes (pL to low nL) are applied in this system, so samples must have sufficient concentration and purity in order that weights spotted meet the LOD and LOQ of the analysis and chromatograms are not interfered with by coextractives. Any special requirements for sample preparation prior to application must be carefully considered as practical applications of miniaturized planar chromatography are developed.

REFERENCES

1. Olennikov, D.N. and Partilkhayev, V.V., Isolation and densitometric HPTLC analysis of rutin, narcissin, nikotiflorin, and isoquercitrin in *Caragana spinosa* shoots, *J. Planar Chromatogr. – Mod. TLC*, 25, 30–35, 2012.
2. Sajewicz, M., Staszek, D., Waksmundzka-Hajnos, M., and Kowalska, T., Comparison of TLC and HPLC fingerprints of phenolic acids and flavonoids fractions derived from selected sage (Salvia) species, *J. Liq. Chromatogr. Relat. Technol.*, 35, 1388–1403, 2012.
3. Staszek, D., Orłowska, M., Waksmundzka-Hajnos, M., Sajewicz, M., and Kowalska, T., Marker fingerprints originating from TLC and HPLC for selected plants from the Lamiaceae family, *J. Liq. Chromatogr. Relat. Technol.*, 36, 2463–2475, 2013.
4. Rout, K.K. and Mishra, S.K., Efficient and sensitive method for quantitative analysis of 6-gingerol in marketed Ayurvedic formulation, *J. Planar Chromatogr. – Mod. TLC*, 22, 127–131, 2009.
5. Ghule, B.V., Palve, P.P., Rathi, L.G., and Yeole, P.G., Validated HPTLC method for simultaneous determination of Shanzhiside methyl ester and barlerin in *Barleria prionitis*, *J. Planar Chromatogr. – Mod. TLC*, 25, 426–432, 2012.
6. Patil, M.B., Taralkar, S.V., Sakpal, V.S., Shewale, S.P., and Sakpal, R.S., Extraction, isolation, and evaluation of anti-inflammatory activity of curcuminoids from *Curcuma longa, Int. J. Chem. Sci. Appl.*, 2, 172–174, 2011.
7. Kochana, J., Wilamowska, J., and Parczewski, A., TLC profiling of impurities of 1-(3,4-methylenedioxyohenyl)-2-nitropropene, an intermediate in MDMA synthesis, *J. Liq. Chromatogr. Relat. Technol.*, 27, 2297–2314, 2004.
8. Flurkey, W.H., Use of solid phase extraction in the biochemistry laboratory to separate different lipids, *Biochem. Mol. Biol. Edu.*, 33, 357–360, 2005.
9. Bodennec, J., Pelled, D., and Futerman, A.H., Aminopropyl solid phase extraction and 2D TLC of neutral glycosphingolipids and neutral lysoglycosphingolipids, *J. Lipid Res.*, 44, 218–226, 2003.
10. Kozutsumi, D., Kawashima, A., Adachi, M., Takami, M., Takemoto, N., and Yonekubo, A., Determination of docosahexaenoic acid in milk using affinity solid phase purification with argentous ions and modified thin layer chromatography, *Int. Dairy J.*, 13, 937–943, 2003.
11. Hamada, M. and Wintersteiger, R., Fluorescence screening of organophosphorus pesticides in water by an enzyme inhibition procedure on TLC plates, *J. Planar Chromatogr. – Mod. TLC*, 16, 4–10, 2003.
12. Janoszka, B., Warzecha, L., Dobosz, C., and Bodzek, D., Determination of 7-ketocholesterol and 7-hydroxycholesterol in meat samples by TLC with densitometric detection, *J. Planar Chromatogr. – Mod. TLC*, 16, 186–191, 2003.
13. Namur, S., Cariño, L., and Gonzalez-de la Parra, M., Development and validation of a high-performance thin-layer chromatographic method, with densitometry, for quantitative analysis of tizoxanide (a metabolite of nitazoxanide) in human plasma, *J. Planar Chromatogr. – Mod. TLC*, 20, 331–334, 2007.
14. Ali, I., Hussain, A., Saleem, K., and Aboul-Enein, H.Y., Separation and identification of antidepressant drugs in human plasma by SPE-TLC method, *J. Appl. Biopharm. Pharmacokinet.*, 1, 12–17, 2013.
15. Ali, I., Hussain, I., Saleem, K., and Aboul-Enein, H.Y., Development of efficient SPE-TLC method and evaluation of biological interactions of contraceptives with progesterone receptors, *Arab. J. Chem.*, 5, 235–240, 2012.
16. Dyńska-Kukulska, K. and Ciesielski, W., Methods of extraction and thin-layer chromatography determination of phospholipids in biological samples, *Rev. Anal. Chem.*, 31, 43–56, 2012.

17. Apers, S., Naessens, T., Pieters, L., and Vlietinck, A., Densitometric thin-layer chromatographic determination of aescin in a herbal medicinal product containing *Aesculus* and *Vitis* dry extracts, *J. Chromatogr. A*, 1112, 165–170, 2006.
18. Mutavdžić, D., Babić, S., Ašperger, D., Horvat, A.J.M., and Kaštelan-Macan, M., Comparison of different solid-phase extraction materials for sample preparation in the analysis of veterinary drugs, *J. Planar Chromatogr. – Mod. TLC*, 19, 454–462, 2006.
19. Babić, S., Ašperger, D., Mutavdžić, D., Horvat, A.J.M., and Kaštelan-Macan, M., Determination of sulfonamides and trimethoprim in spiked water samples by solid-phase extraction and thin-layer chromatography, *J. Planar Chromatogr. – Mod. TLC*, 18, 423–426, 2005.
20. Ašperger, D., Mutavdžić, D., Babić, S., Horvat, A.J.M., and Kaštelan-Macan, M., Solid-phase extraction and TLC quantification of enrofloxacin, oxytetracycline, and trimethoprim in wastewater, *J. Planar Chromatogr. – Mod. TLC*, 19, 129–134, 2006.
21. Tuzimski, T., Application of SPE-HPLC-DAD and TLC-HPTLC-DAD to the analysis of pesticides in lake water, *J. Planar Chromatogr. – Mod. TLC*, 22, 235–240, 2009.
22. Mennickent, S., Mario Vega, M., Mario Vega, H., De Diego, M., and Fierro, R., Quantitative determination of propranolol in human serum by high-performance thin-layer chromatography, *J. Planar Chromatogr. – Mod. TLC*, 25, 54–59, 2012.
23. Faraji, H., Saber-Tehrani, M., Mirzale, A., and Waqif-Husain, S., Application of liquid-liquid microextraction - high-performance thin-layer chromatography for preconcentration and determination of phenolic compounds in aqueous samples, *J. Planar Chromatogr. – Mod. TLC*, 24, 214–217, 2011.
24. Ostry, V., Skarkova, J., and Ruprich, J., Densitometric high-performance thin-layer chromatography method for toxigenity testing of *Alternaria alternata* strains isolated from foodstuffs, *J. Planar Chromatogr. – Mod. TLC*, 25, 388–393, 2012.
25. European Food Safety Authority (EFSA) Panel on contamination in the food chain (CONTAM), *EFSA J.*, 9, 1–97, 2011.
26. Ghosh, P. and Katiyar, A., Densitometric HPTLC analysis of juglone, quercetin, myricetin, rutin, caffeic acid, and gallic acid in *Juglans regia* L., *J. Planar Chromatogr. – Mod. TLC*, 25, 420–425, 2012.
27. Kaur, P., Chaudhary, A., Katiyar, A., Singh, B., Gopichand, and Singh, R.D., Rapid validated RP-HPTLC method for the quantification of major bioactive constituents of *Crataegus oxyacantha* L., *J. Planar Chromatogr. – Mod. TLC*, 25, 415–419, 2012.
28. Yee, J.L., Khalil, H., and Jimenez-Flores, R., Flavor partition and fat reduction in cheese by supercritical fluid extraction: process variables, *Le Lait*, 87, 269–285, 2007.
29. Pereira, C.G., Leal, P.F., Sato, D.N., and Meireles, A.A., Antioxidant and antimycobacterial activities of *Tabernaemontana catharinensis* extracts obtained by supercritical CO_2 + cosolvent, *J. Med. Food*, 8, 533–538, 2005.
30. Hamrapurkar, P., Pawar, S., and Phale, M., Quantitative HPTLC analysis of phyllanthin in *Phallanthus amarus*, *J. Planar Chromatogr. – Mod. TLC*, 23, 112–115, 2010.
31. Plander, S., Gontaru, L., Blazics, B., Veres, K., Kery, A., Kareth, S., and Simandi, B., Major antioxidant constituents form *Satureja hortensis* L. extracts obtained with different solvents, *Eur. J. Lip. Sci. Technol.*, 114, 772–779, 2012.
32. Jaime, L., Rodriguez-Meizoso, I., Cifuentes, A., Santoyo, S., Suarez, S., Ibañez, E., and Señorans, F.J, Pressurized liquids as an alternative process to antioxidant carotenoids' extraction from *Haematococcus pluvialis* microalgae, *Food Sci. Technol.*, 43, 105–112, 2010.
33. Ha, M.H., Nguyen, V.T., Nguyen, K.Q.C., Cheah, E.L.C., and Heng, P.W.S., Antimicrobial activity of *Calophyllum inophyllum* crude extracts obtained by pressurized liquid extraction, *Asian. J. Trad. Med.*, 4, 141–146, 2009.
34. Cescut, J., Severac, E., Molina-Jouve, C., and Uribellarea, J.-L., Optimizing pressurized liquid extraction of microbial lipids using the response surface method, *J. Chromatogr. A*, 1218, 373–379, 2011.

35. Orłowska, M., Stanimirova, I., Staszek, D., Sajewicz, M., Kowalska, T., and Waksmundzka-Hajnos, M., Optimization of extraction based on the thin-layer chromatographic fingerprints of common thyme, *J. AOAC Int.*, 97, 1274–1281, 2014.

36. Shirsath, S.R., Sonawane, S.H., and Gogate, P.R., Intensification of extraction of natural products using ultrasonic irradiations – A review of current status, *Chem. Eng. Process.: Process Intens.*, 53, 10–23, 2012.

37. Kac, J., Mlinarič, A., and Umek, A., HPTLC determination of xanthohumol in hops (*Humulus lupulus* L.) and hop products, *J. Planar Chromatogr. – Mod. TLC*, 19, 58–61, 2006.

38. Gopu, C.L., Gilda, S.S., Paradkar, A.R., and Mahadik, K.R., Development and validation of a densitometric TLC method for analysis of trigonelline and 4-hydroxyisoleucine in fenugreek seeds, *Acta Chromatogr.*, 20, 709–719, 2008.

39. Annegovda, H.V., Mordi, M.N., Ramanathan, S., Hamdan, M.R., and Mansor, M.N., Effect of extraction techniques on phenolic content, antioxidant and antimicrobial activity of *Bauhinia purpurea*: HPTLC determination of antioxidants, *Food Anal. Methods*, 5, 226–233, 2012.

40. Švarc-Gajić, J. and Stojanović, Z., Deterrmination of histamine in cheese by chronopotentiometry on a thin film mercury electrode, *Food Chem.*, 124, 1172–1176, 2011.

41. Rezić, I. and Bokić, L., Ultrasonic extraction and TLC determination of surfactants in laundry wastewaters, *Tenside Surf. Deterg.*, 42, 274–279, 2005.

42. Babić, S., Horvat, A.J.M., Mutavdžić, D., Čavić, D., and Kaštelan-Macan, M., Sample preparation for TLC-genetic algorithm-based optimization of microwave-assisted extraction, *J. Planar Chromatogr. – Mod. TLC*, 20, 95–99, 2007.

43. Acikkol, M., Semen, S., Turkmen, Z., and Mercan, S., Determination of α-cypermethrin from soil by using HPTLC, *J. Planar Chromatogr. – Mod. TLC*, 25, 48–53, 2012.

44. Mirzajani, F., Ghassempour, A., Jalali-Heravi, M., and Mirjalili, M.H., Optimisation of a microwave-assisted method for extracting withaferin A from *Withania somnifera* Dunal. using central composite design, *Phytochem. Anal.*, 21, 544–549, 2010.

45. van Ginkel, L.A., van Blitterswijk, H., Zoontjes, D., and van den Bosch, R.W.S., Assay of trenbolone and its rnetabolite 17 alpha-trenbolone in bovine urine based on immunoaffinity chromatographic cleanup and off-line high performance liquid chromatography-thin layer chromatography, *J. Chromatogr.*, 445, 385–392, 1988.

46. Preis, R.A. and Vargas, E.A., A method for determining fumonisin B1 in corn using immunoaffinity column cleanup and thin layer chromatography-densitometry, *Food Addit. Contam.*, 17, 463–468, 2000.

47. Stroka, J., van Otterdijk, R., and Anklam, E., Immunoaffinity column cleanup prior to thin layer chromatography for the determination of aflatoxins in various food matrices, *J. Chromatogr. A*, 904, 251–256, 2000.

48. Ostry, V. and Skarkova, J., Development of an HPTLC method for the determination of deoxynivalenol in cereal products, *J. Planar Chromatogr. – Mod. TLC*, 13, 443–446, 2000.

49. Santos, E.A. and Vargas, E.A., Immunoaffinity column cleanup and thin layer chromatography for determination of ochratoxin A in green coffee, *Food Addit. Contam.*, 19, 447–458, 2002.

50. Prieta, J., Moreno, M.A., Blanco, J.L., Suarez, G., and Dominguez, L., Determination of patulin by diphasic dialysis extraction and thin layer chromatography, *J. Food Prot.*, 55, 1001–1002, 1992.

51. Churms, S.C. and Stephen, A.M., Chromatographic separation and examination of carbohydrate and phenolic components of the non-tannin fraction of black wattle (*Acacia mearnsii*) bark extract, *J. Chromatogr.*, 550, 519–537, 1991.

52. Diab, A.M., Abdul-Kawy, A., Abdel-Rahman, M., and Abou-Amer, A., Nicotine and cotinine in urine of passively smoking children, *Bull. Nat. Res. Cent. (Egypt)*, 22, 43–50, 1997.

53. Sommer, K. and Venke, S.C., An experiment on the isolation of natural products. *Naturwissenschaften im Unterricht Chemie*, 15, 18–21, 2004.
54. Anastassiades, M., Lehotay, S.J., Stajnbaher, D., and Schenck, F.J., Fast and easy multiresidue method employing acetonitrile extraction/partitioning and "dispersive solid phase extraction" for the determination of pesticides residues in produce, *J. AOAC Int.*, 86, 412–431, 2003.
55. Kar, A., Mandal, K., and Singh, B., Environmental fate of chlorantraniliprole residues on cauliflower using QuEChERS technique, *Environ. Monit. Assess.*, 185, 1255–1263, 2013.
56. Broszat, M., Welle, C., Wojnowski, M., Ernst, H., and Spangenberg, B., A versatile method for quantification of aflatoxins and ochratoxin A in dried figs, *J. Planar Chromatogr. – Mod. TLC*, 23, 193–197, 2010.
57. Sajewicz, M., Wojtal, Ł., Staszek, D., Hajnos, M., Waksmundzka-Hajnos, M., and Kowalska, T., Low temperature planar chromatography–densitometry and gas chromatography of essential oils from different sage (*Salvia*) species, *J. Liq. Chromatogr. Relat. Technol.*, 33, 936–947, 2010.
58. Sajewicz, M., Wojtal, Ł., Hajnos, M., Waksmundzka-Hajnos, M., and Kowalska, T., Low-temperature TLC-MS of essential oils from five different sage (*Salvia*) species, *J. Planar Chromatogr. – Mod. TLC*, 23, 270–276, 2010.
59. Kazmierczak, D., Ciesielski, W., and Zakrzewski, R., Detection and separation of amino acids as butylthiocarbamyl derivatives by thin layer chromatography with iodide-azide detection system, *J. Liq. Chromatogr. Relat Technol.*, 28, 2261–2271, 2005.
60. Nowak, R., Wojciak-Kosior, M., Sowa, I., Sokolowska-Krzaczek, A., Pietrzak, W., Szczodra, A., and Kocjan, R., HPTLC-densitometry determination of triterpenic acids in *Origanum vulgare*, *Rosmarinus officinalis*, and *Syzygium aromaticum*, *Acta Pol. Pharm.*, 70, 413–418, 2013.
61. Alpmann, A. and Morlock, G., Rapid and cost effective determination of acrylamide in coffee by planar chromatography after derivatization with dansulfinic acid, *J. AOAC Int.*, 92, 725–729, 2009.
62. Schwack, W. and Oellig, C., Solid phase extraction as cleanup for pesticide residue analysis using planar chromatographic developing techniques, *CAMAG Bibliography Service (CBS)*, 110, 12–15, 2013.
63. Touchstone, J.C., Hansen, G.J., Zelop, C.M., and Sherma, J., Quantitation of cholesterol in biological fluids by TLC with densitometry. In *Advances in Thin Layer Chromatography-Clinical & Environmental Applications*, Touchstone, J.C. (ed.), John Wiley & Sons, Inc., New York, NY, 1982, pp. 219–228.
64. Sherma, J. and Miller, R.L. Jr., Quantification of caffeine in beverages by densitometry on preadsaorbent HPTLC plates, *Am. Lab.*, 16(2), 126–127, 1984.
65. Sherma, J. and Targan, D.A., Determination of quinine in tonic water by quantitative thin layer chromatography on high performance preadsorbent plates, *Acta Chromatogr.*, 5, 7–11, 1995.
66. Khan, S.H., Murawski, M.P., and Sherma, J., Quantitative high performance thin layer chromatographic determination of organic acid preservatives in beverages, *J. Liq. Chromatogr.*, 17, 855–865, 1994.
67. Sherma, J. and DeGrandchamp, D., Review of advances in planar radiochromatography, *J. Liq. Chromatogr. Relat. Technol.*, 38, 381–389, 2015.
68. Sherma, J. and Fried, B. (eds.), *Handbook of Thin Layer Chromatography*, 3rd ed., Marcel Dekker, Inc., New York, NY, 2003.
69. Sherma, J., Biennial review of planar chromatography: 2009–2011, *J. AOAC Int.*, 95, 992–1009, 2012.
70. Sherma, J., Biennial review of planar chromatography: 2011–2013, *Cent. Eur. J. Chem.*, 12, 427–452, 2014.

71. Ding, L.J., Ding, W., Zhang, Y.Q., and Luo, J.X., Bioguided fractionation and isolation of esculentoside P from *Phytolacca americana* L., *Ind. Crop. Prod.*, 44, 534–541, 2013.

72. Nguyen, D.M.C., Seo, D.J., Nguyen, V.N., Kim, K.Y., Park, R.D., and Jung, W.J., Nematicidal activity of gallic acid purified from terminalia nigrovenulosa bark against the root-knot nematode meloidogyne incognita, *Nematology*, 15, 507–518, 2013.

73. Bharate, J.B., Vishwakarma, R.A., Bharate, S.B., Thite, T.B., Kushwaha, M., and Gupta, A.P., Quantification and validation of two isomeric anticancer compounds, garcinol and isogarcinol, in ultrasound assisted extracts of *Garcinia indica* fruits using high performance thin layer chromatography, *J. Planar Chromatogr. – Mod. TLC*, 26, 480–485, 2013.

74. De, S., Nariya, P., and Jirankalgikar, N., Development of a novel high performance thin layer chromatographic-densitometric method for the determination of tallow adulteration in cow ghee, *J. Planar Chromatogr. – Mod. TLC*, 26, 486–490, 2013.

75. Gu, L., Zheng, S., Wu, T., Chou, G., and Wang, Z., High performance thin layer chromatographic-bioautographic method for the simultaneous determination of magnonol and honokiol in *Magnoliae officinalis* Cortex, *J. Planar Chromatogr. – Mod. TLC*, 27, 5–10, 2014.

76. Pai, S.R., Upadhya, V., Hegde, H.V., Joshi, R.K., and Kholkute, S.D., New report of triterpenoid betulinic acid along with oleanic acid from *Achyranthes aspera* by reversed phase ultra flow liquid chromatographic analysis and confirmation using high performance thin layer chromatographic and Fourier transform infrared spectroscopic techniques, *J. Planar Chromatogr. – Mod. TLC*, 27, 36–41, 2014.

77. Prado, I.M., Prado, G.H.C., Prado, J.M., and Meireles, M.A.A., Supercritical CO_2 and low pressure solvent extraction of mango (*Mangifera indica*) leaves: global yield, extraction kinetics, chemical composition and cost of manufacturing, *Food Bioprod. Process.*, 91, 656–664, 2013.

78. Turkmen, Z., Mercan, S., Bavunoglu, I, and Cenzig, S., Development and validation of a densitometric high performance thin layer chromatographic method for quantitative analysis of amitriptyline in gastric lavage, *J. Planar Chromatogr. – Mod. TLC*, 26, 496–501, 2013.

79. Rout, K.K., Swain, S.S., and Chand, P.K., Quantification of beta-sitosterol in hairy root cultures and natural plant parts of butterfly pea (*Clitoria ternatea* L.), *J. Planar Chromatogr. – Mod. TLC*, 27, 42–46, 2014.

80. Srivastava, N., Verma, S., Pragyadeep, S., Srivastava, S., and Rawat, A.K.S., Evaluation of successive fractions for optimum quantification of bergenin and gallic acid in three industrially important *Bergania* species by high performance thin layer chromatography, *J. Planar Chromatogr. – Mod. TLC*, 27, 69–71, 2014.

81. Stan, M., Lung, I., Opris, O., and Soran, M.-L., High performance thin layer chromatographic quantification of some essential oils from *Anethum graveolens* extracts, *J. Planar Chromatogr. – Mod. TLC*, 27, 33–37, 2014.

82. Choma, I.M., Grzelak, E.M., and Majer-Dziedzic, B., Comparison of deproteinization methods used before TLC-DB and HPLC analysis of flumequine residues in milk, *Med. Chem.*, 8, 95–101, 2012.

83. Bromann, K., Viljanen, K., Moreira, V.M., Yli-Kauhaluona, J., Ruohonen, L., and Nakari-Setala, T., Isolation and purification of ent-pimara-8(14),15 from engineered *Aspergillus nidulans* by accelerated solvent extraction combined with HPLC, *Anal. Methods*, 6, 1227–1234, 2014.

84. Mulunda, M. and Mike, D., Occurrence of aflatoxin M-1 from rural subsistence and commercial farms from selected areas of South Africa, *Food Control*, 39, 93–96, 2014.

85. Tennakoon, D.A.S.S., Karunarathna, W.D.V., and Udugampala, U.S.S., Carbofuran concentrations in blood, bile and tissues in fatal cases of homicide and suicide, *Forens. Sci. Int.*, 227, 106–110, 2013.

86. Zarzycki, P.K., Slaczka, M.M., Wlodarczyk, Z., and Baran, M.J., Micro-TLC for fast screening of environmental samples derived from surface and sewage water, *Chromatographia*, 76, 1249–1259, 2013.
87. Majors, R.E., Application of existing sample preparation technologies to new areas, *LCGC North Am.*, 31, 914, 916, 918, 920, 922, 924, 2013.
88. Chavan, V.R. and Mali, B.D., High performance thin layer chromatographic detection of profenfos in biological material, *J. Planar Chromatogr. – Mod. TLC*, 27, 66–68, 2014.
89. Spangenberg, B., Poole, C.F., and Weins, C., *Quantitative Thin Layer Chromatography: A Practical Survey*, Springer, New York, NY, p. 105, 2011.
90. Krol, W.J., Eitzer, B.D., Arsenault, T., Mattina, M.J.I., and White, J.C., Significant improvements in pesticide residue analysis in food using the QuEChERS method, *LCGC North Am.*, 32, 116, 118–122, 124, 125, 2014.
91. Morlock, G.E., Oellig, C., Bezuidenhout, W., Brett, M.J., and Schwack, W., Miniaturized planar chromatography using office peripherals, *Anal. Chem.*, 82, 2940–2946, 2010.

8 Modeling of HPLC Methods Using QbD Principles in HPLC

Imre Molnár, Hans-Jürgen Rieger, and Robert Kormány

CONTENTS

8.1   INTRODUCTION

The increasing demand for quality by design (QbD) in analytical science is a logical consequence of the often chaotic method development practices in high-performance liquid chromatography (HPLC) that result from a trial and error approach, during which stumbling over new or disappearing peaks in the ultra high-performance liquid chromatography (UHPLC) validation process is a common pitfall. To ensure a higher standard of method quality, in 2002, the International Conference on Harmonization (ICH) and Food and Drug Administration (FDA) started demanding solid and scientific work using Design of Experiments (DoE).

The Molnár-Institute has been promoting this type of approach for almost 30 years by contributing to the development of DryLab® software through cooperation with LC resources, under the leadership of Lloyd R. Snyder. Using DryLab 4, the systematic and accurate preparation of experiments was initiated, achieving useful and reproducible results. Better peak tracking in DryLab ensured safe and precise data entry before the model was built.

Designed by UHPLC experts, DryLab 4 offers chromatographers unprecedented insight into how a substance can best be separated and efficiently support the success of their chromatographic work.

DryLab is the world standard for chromatography modeling in both method development and training applications. The following time schedule shows the long and well-documented development history of DryLab—from the very beginning in 1986 to the essential UHPLC method development tool we have today.

8.2   HISTORY OF DRYLAB DEVELOPMENT

Modeling HPLC started with Csaba Horváth at Yale University in 1975 by systematic measurements of retention phenomena in reversed-phase chromatography (RPC) [1–3].

In 1986, Lloyd Snyder, John Dolan, Tom Jupille, and Imre Molnár started a modeling tool by first learning how to optimize and model isocratic HPLC, and then creating software with several extensions called DryLab 1–5 [4, 5, 6]. It was first programmed by John Dolan in Basic language, under Microsoft *DOS*, for modeling capacity factors, calculating flow rate changes, and critical resolution values as well. Learning the influence of column dimensions led to a first column optimization part (DryLab 1). The software was then extended to RP-%B optimization (DryLab 2) and to normal phase HPLC (DryLab 3), followed by a module for ion pair-RPC (DryLab 4). The first calculations of retention in gradient modeling created DryLab 5. In the very first versions, chromatograms were plotted with stars*.

In 1987, DryLab I (I for isocratic) was born as a combination of DryLab 1,2,3, and 4 which were programmed all in isocratic modeling; this means, the peak widths were increasing with the retention time. With the addition of new graphics in DOS, where chromatograms could be plotted, one could create chromatograms similar to well-known ones from real instruments. This step of visualization of chromatographic science was a unique property of DryLab already as early as in 1988 and it is still a revolutionary approach today.

The next extension was a combination of DryLab 1,2,3 and DryLab 5 to DryLab G—*G* for gradient modeling.

In 1989, DryLab I/plus, DryLab G/plus came out and were the first versions of DryLab programmed in *C*, which included a number of new features, such as peak name options, zoom and scale of chromatograms, resolution maps for partial peak sets, ASCII files for data input and storage, and the ability to import data system files.

- 1992 DryLab I/mp: Isocratic multiparameter version was born with a wholly graphical interface for the Windows® 1.0 operating system, including functions for the mouse control of the program.
- 1998 DryLab version 2.0 was launched, a first version with two-dimensional (2D) modeling capabilities, incorporating simultaneous modeling of two separation parameters, for example, gradient time (tG) versus temperature (*T*) or %B versus *T*. It was also capable of modeling six to seven additional method parameters as in all other versions included new features, such as automated peak matching for maximum eight peaks at that time.
- In year 2000 "DryLab 2000", version 3.0 was launched.
- In 2002 DryLab 2000 plus v. 3.1 was adjusted to Windows® 3.1 and NT for network applications. It was also the first version to be released in C++. It could do 2D modeling for any combination of variables called *anything versus anything*.
- In 2005 DryLab was acquired by the Molnár-Institute and further developed in Berlin. The first addition of a new module—called *PeakMatch* v.1.0—was the first peak tracking software introduced for DryLab for easier alignment of peaks in four to six different chromatograms, and was running on Windows XP.
- In 2006 DryLab and PeakMatch v. 2.0 came out and allowed the automated generation of experiments with Agilent 1100.
- 2007 DryLab 2000 plus v. 3.5 included DryLab v. 3.9 and PeakMatch v. 3.5, running three tG-*T*-models running on Windows Vista.
- 2009 DryLab v. 3.9—introduction of the revolutionary *3D Cube* at the HPLC meeting in Dresden: a 3D resolution map allowing and simultaneously optimizing three critical parameters and calculating further seven factors, which was truly a leapfrog step in the development toward more flexible methods. It was also the first modeling tool, which was programmed in C# (C-sharp) with new user-friendly windows management and the amazing 3D Cube, compatible from Windows® XP through Windows 8 [7, 8].
- 2010 Followed the introduction of DryLab 2010—a combination of the DryLab Core and PeakMatch.
- 2012 DryLab 4.0 was launched with a completely new windows management and an automated data acquisition procedure for the Shimadzu HPLC-line in Europe.
- 2013 DryLab 4.1—Introduction of the robustness module, which could calculate six factors at three levels 3^6 = 729 runs in 20 s.
- 2014 Introduction of the knowledge management protocol, a complete documentation of a method development process including all input data, the peak tracking, the model validation, the robustness study all together in a pdf document.

8.3 KNOWLEDGE MANAGEMENT DOCUMENT IS MAINLY USED FOR POST-APPROVAL CHANGES

The actual DryLab version is 4.2, in which the robustness module was extended also to step gradients with 4 points: this means 6 + 2 + 2 additional factors at 3 levels = 3^10 ca. 60,000 runs in ca. 3 min.

DryLab is trying to answer the following questions: What are our most important goals in separation science?

1. Get the *best and fastest separation*
2. Find the *most robust conditions for routine work* that is, get critical resolution $(R_{s, crit})$ maximized
3. Select the *best column* for the application, based on multifactorial studies for the creation of *maximum robustness and flexibility* by working inside of the design space

The most efficient DoE is using 12 experiments and predicts more than 10^6 precise model chromatograms with a precision in retention times of better than 99% compared to the corresponding real runs [9, 10, 11, 12].

8.4 ISOCRATIC MODELS

This type of chromatography was used in the 1970–1980s in quality control, where only a few peaks were analyzed. The peak width in isocratic work is linearly proportional to the retention time. This means, that peaks, which are strongly retarded, become very wide and therefore very flat and cannot be recognized anymore, so that the analysis is *de facto* incomplete. In gradient elution the late retarded peaks are just as visible as any other peak, so in the meantime, gradient elution dominates in the field.

8.4.1 CASE STUDY 1: pH MODEL

This example shows the dependence of the critical resolution (y-axis) on the pH in the range of 2–5 (Figure 8.1). In RPC, this is the most common pH range, simply because here the silanol groups are protonated and therefore fairly homogeneous. This means, band spreading due to heterogeneous silanols is at a minimum; the peaks are sharp, instead to be broader and/or having tailing. The model is based on only 3 runs at pH values of 2.9, 3.5, and 4.1. The y-axis is showing the critical resolution values. Baseline resolution is above $R_{s, crit} > 1.5$.

We can see, that there is only a small pH interval from 2.9 to 3.1 (shown by the arrow), where we have a high critical resolution, $R_{s, crit}$ with baseline separation 1.5. At all other pH values we would have problems with more or less peak overlaps, where $R_{s, crit}$ is approaching the value 0, leading even to missing peaks. In this way, one can reduce a possible pH range *screening* from 2.0 to 5.0 to the range of 2.9–3.1.

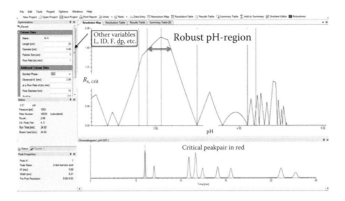

FIGURE 8.1 The case study is showing a model for the separation of nine organic acids. The critical resolution map (top right) is exhibiting the dependence of the critical resolution from the pH in the first place. However, other variables (top left) indicate other possibilities to change f.e. the column length, ID, particle size and flow rate and see, how the chromatogram, ($R_{s,\,crit}$), retention times, and peak elution orders are changing.

Outside of the range 2.9–3.1 the screening does not make sense at all and would only keep the instrument and the lab worker busy without satisfactory results.

8.5 GRADIENT MODELS

8.5.1 CASE STUDY 2: GRADIENT TIME (τG) MODEL AND DEVELOPMENT OF STEP GRADIENTS

In this study the influence of the gradient steepness (or tG) is shown. The critical band spacing is depending on the gradient time also strongly as we can see this on the resolution map [12]. There are three peak overlaps ($R_{s,\,crit} = 0$) shown. The best separation is possible not only at ca. tG: 100 min, but also in the region of tG = 20 min with the same quality. This is another proof that *screening* without a scientific basis is a waste of time (Figure 8.2).

Step gradients are a common way of trying to improve a separation. However, due to misinterpretation of the timescale of the gradient program, retention times in the resulting chromatograms are often very different from the expected results, even for experienced chromatographers (Figure 8.3).

8.6 SIMULTANEOUS CHANGE OF TWO FACTORS—2D MODELS

8.6.1 CASE STUDY 3

This example was generated in the development for the separation of 15 compounds in a pharmaceutical company. One worked on it for quite a long period of several months, without any success. Using DryLab 2D design, the individual peak movements could be understood and the final method allowed a robust control of the product as shown in Figure 8.4.

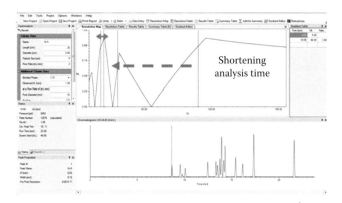

FIGURE 8.2 Reduction of *screening* to the meaningful values of gradient time. Based on only two experimental gradient runs the selection of the best gradient time can be rapidly achieved. The time for ca. 100 potential, but unnecessary screening runs can be saved.

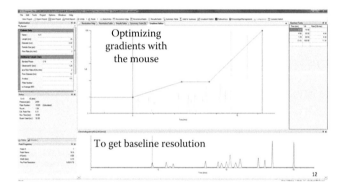

FIGURE 8.3 The timescale of the chromatogram and the timescale of the gradient program are different. In the top chromatogram the gradient line shows the composition of the *eluent* in the detector cell, while the *sample* composition is registered in the lower chromatogram. This important detail enables development of a robust gradient method in a short time. *Screening experiments* with step gradients are *trial and error* and they cannot be justified with *statistics* in analytical Quality by Design (AQbD).

A presentation of a great number of scientific papers on DryLab modeling is compiled in Molnár [13].

8.7 WHICH MODEL IS THE MOST EFFICIENT?

In statistical DoE there are a great number of the so-called Designs which are rather confusing for a chromatographer. In HPLC we reduced these many different designs to a reasonable set of experiments and recommend the following steps:

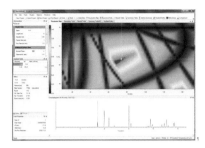

FIGURE 8.4   2D Resolution map, gradient time, tG (*x*-axis) versus temperature (*y*-axis) showing the best separation at the red area. The map is based on only 4 runs and it corresponds to ca. 10,000 model experiments and the optimum is found in seconds instead of running screening experiments in weeks.

- Start with the tG-*T*-model (4 runs only)
- Continue to form a tG-*T*-pH-Cube (12 runs) for polar compounds or form a tG-*T*-tC-Cube (12 runs) for neutral compounds
- Test the robustness of the separation for variabilities using 6 factors at 3 levels => $3^6 = 729$ runs in 30 s

The most efficient design is certainly the Cube. It needs only 12 runs and produces more than 10^6 different chromatograms with varying selectivities [13–15].

8.8   WHICH MODELING TOOLS TO SELECT?

Three different types of modeling tools are available.

1. Tools for drug design, where molecular structures play an important role, are the following tools available: Pallas (Budapest, Hungary), ChromSword (S. Galushko, Darmstadt, Germany), and ACD-Lab (Toronto, Canada)
2. Tools for separation understanding, mainly in QC:
   DryLab 4 (Molnár-Institute) based on gradient elution, works with complex unknown mixtures, helping to visualize peak movements:
   1. Eluent (pH-, ternary-, temperature-, tG, etc.) influences are all measured
   2. Influences of column dimensions (L, ID), dp, flow rate, in gradient elution are calculated
   3. Instrument influences (Vd, Vext.col.) are considered
   4. 3D resolution maps based on only 12 runs to find the best separation out of >1 million choices with visualized design space are increasing flexibility with Out of Specification (OoS)
   5. Robustness metrics allows safe industrial QC work
3. Tools based on statistics: Fusion (S-Matrix, USA) is fixed to an instrument (Waters) and automates the creation of runs with different possible experimental designs

8.9   MULTIFACTORIAL MODELING: THE 3D RESOLUTION CUBE

8.9.1   CASE STUDY 4

In 2009 there was a shortage of acetonitrile in the market and everyone feared what to do with all the QC work, if acetonitrile as the organic mobile phase was not available anymore. We thought, we would try out a different concept of changing the organic eluent and replace the acetonitrile amount in the eluent by HPLC-grade methanol. It turned out, it was not just a success to save acetonitrile (and money) but also to improve critical resolution and at the same time develop a design space according to QbD requirements. At the same time the pH influence on selectivity was also investigated between 2.4 and 3.6 as shown in Figures 8.5 and 8.6.

The first ternary Cube (the *original Cube*) was calculated out of 3 tG-*T*-sheets with three different eluent B mobile phases: AcN, (50 : 50) mix, and MeOH. Then DryLab calculated another 97 sheets in between the measured ones and so the Cube could be filled out and allow to visualize over 10^6 different selectivities. The best one with the highest robustness (= highest critical resolution) could be found by one mouse click. The method operable design region (MODR) is visualized as one or more irregular geometric bodies shown in Figure 8.6 on the very right.

8.10   HOW TO FIND THE BEST COLUMN?

8.10.1   CASE STUDY 5

There are several ways how to select a column. The most universal columns are certainly the C18 or octadecylsilica reversed-phase materials. However, we have ca. 500 different types of them and other RP-column variants on the market; some are as

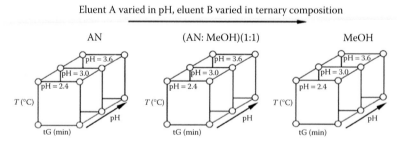

Eluent A varied in pH, eluent B varied in ternary composition

Multifactorial optimization strategy of 4 measured critical HPLC method parameters: Gradient time (tG), temperature (*T*), pH, and ternary composition (B1:B2), based on $12 \times 3 = 36$ experiments.

FIGURE 8.5   Multifactorial design of experiments to study the influence of four measured and seven calculated factors on chromatographic selectivity for higher method robustness. The measured factors were tG, *T*, pH, ternary composition tC (AcN: MeOH ratio). Calculated influences are by the column length, inner diameter, particle size, flow rate, dwell volume, extra column volume, starting and final %B. (From Molnár, I. et al., *J. Chromatogr. A*, 1217, 3193–3200, 2010.)

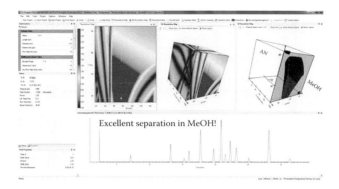

Excellent separation in MeOH!

FIGURE 8.6 The ternary Cube from the original paper shows in the further developed version DryLab 4 with a new windows management, where baseline resolution is occuring (red body in the right top figure). It turned out, that baseline separation was not possible in acetonitrile, as the back side of the cube remained white, as long as the front side of the Cube (eluent B is MeOH) was largely red and was offering excellent baseline separation of all peaks involved, as shown in the corresponding chromatogram (below).

old as 30+ years and some are very new. As the stationary phase geometry, the reaction of the chemical modifications is all developing further year by year. Snyder and Dolan tried to classify this great variety on columns in their hydrophobicity subtraction database, which is also included in DryLab 4. This database has the advantage that all the many columns are comparable based on a scientific experimental comparison [16,17]. So it is easy to find a reference column, a replacement column, or find columns with very different selectivities. The comparison is based on an isocratic experiment at a fixed temperature and fixed eluent.

Snyder started to use another approach to look at column selectivity by using gradient elution, variable temperature, and pH or ternary composition of the eluent B [18]. In this way, column selectivity might be optimized better than in isocratic conditions. In the following we are reporting about an old Pharmacopoeia method, which caused a lot of trouble in routine industrial applications, producing permanent OoS results, such as missing peaks, false order of elution, and critical resolution out of compliance [19]. The very long analysis time took over 50 min, so one hesitated to make a new time-consuming validation. The plan was to reestablish the understanding of peak movements and reduce the analysis time below 10 min. One assumed that the pH would be the main reason for the confusion, so a pH Cube was created with 12 runs.

After data generation for a pH Cube and data import, peak tracking was carried out to align peaks in a table so that each peak was put in a horizontal line. Based on the data in the table (retention times and peak area pairs) the Cube could be calculated.

In the finished Cube the best working point for each column could be selected. The robustness of the method could be challenged using the robustness module of DryLab. The best column was found by comparisons of the best $R_{s,\,crit}$ values for all columns (Table 8.1). The technique was used to select the best column for each different sample.

TABLE 8.1

Comparison of 25 Different Column Chemistries and the Accuracy between Predicted and Real Retention Times, Compared Over the Cube Evaluation Method Showing Excellent Predictability of the Average Retention Time with a Precision of Typically MT 99%

Columns	Properties of the Columns				Predicted Parameters					Average of Retention Time	
	Silica Type	Pore Size (Å)	Surface Area (m²/g)	Surface Coverage (μmol/m²)	pH	T (°C)	tG (min)	$R_{s,\,crit}$	Critical Imp. Peak Pair	Difference[a] (min)	% Error[b]
1. Acquity BEH C18	Hybrid	130	185	3.0	2.1	13.5	8.1	2.54	G–H	0.007	0.23
2. Acquity BEH Shield RP 18	Hybrid	130	185	3.3	2.0	38.3	9.8	2.16	B–G	−0.017	−0.79
3. Acquity BEH C8	Hybrid	130	185	3.3	2.5	33.0	9.8	2.27	D–F	−0.018	−0.85
4. Acquity BEH Phenyl	Hybrid	130	185	3.0	2.0	29.3	9.8	2.32	G–B	−0.001	0.41
5. Acquity CSH C18	Hybrid	130	185	2.3	3.0	13.5	9.8	3.13	D–F	0.017	0.88
6. Acquity CSH Phenyl-Hexyl	Hybrid	130	185	2.1	2.1	13.5	2.9	1.92	D–F	0.005	0.60
7. Acquity CSH Fluoro-Phenyl	Hybrid	130	185	2.4	3.0	13.5	2.7	1.22	D–F	−0.002	−0.55
8. Triart C18	Hybrid	110	370	1.5	3.0	13.5	7.4	2.49	D–F	0.011	0.57
9. Acquity HSS C18	Fully porous	100	230	3.2	2.1	24.0	9.8	2.50	G–H	−0.038	−1.95
10. Acquity HSS C18 SB	Fully porous	100	230	1.8	2.0	30.0	9.8	2.04	D–F	−0.014	−0.37
11. Acquity HSS T3	Fully porous	100	230	1.7	2.0	31.5	9.6	2.16	G–H	−0.023	−0.94
12. Acquity HSS PFP	Fully porous	100	230	3.2	2.0	19.5	9.8	1.58	D–F	−0.005	−0.27
13. Acquity HSS CN	Fully porous	100	230	2.0	3.0	13.5	7.9	1.95	D–F	0.000	−0.15
14. Hypersil GOLD C18	Fully porous	175	220		3.0	41.3	9.8	2.72	D–F	−0.003	−0.10
15. Hypersil GOLD C8	Fully porous	175	220		2.7	42.0	9.8	2.55	D–F	−0.009	−0.24
16. Hypersil GOLD CN	Fully porous	175	220		2.9	27.8	9.0	1.67	G–B	0.002	0.56
17. Zorbax SB-C18	Fully porous	80	180	1.8	2.2	29.3	9.8	2.13	G–H	−0.016	−0.36

(Continued)

TABLE 8.1 (Continued)

Comparison of 25 Different Column Chemistries and the Accuracy between Predicted and Real Retention Times, Compared Over the Cube Evaluation Method Showing Excellent Predictability of the Average Retention Time with a Precision of Typically MT 99%

	Properties of the Columns				Predicted Parameters					Average of Retention Time	
Columns	Silica Type	Pore Size (Å)	Surface Area (m²/g)	Surface Coverage (μmol/m²)	pH	T (°C)	tG (min)	$R_{s, crit}$	Critical Imp. Peak Pair	Difference[a] (min)	% Error[b]
18. Zorbax SB-C8	Fully porous	80	180	1.6	2.8	13.5	6.1	2.03	D–F	0.013	1.09
19. Zorbax SB-Phenyl	Fully porous	80	180	2.1	2.0	13.5	8.9	1.52	D–F	−0.022	−2.42
20. Kinetex XB-C18	Core shell	100	200	1.8	2.2	13.5	9.8	2.24	D–F	0.013	0.81
21. Aeris XB-C18	Core shell	100	200	1.7	3.0	15.0	9.8	2.50	G–H	0.001	0.35
22. Kinetex C18	Core shell	100	200	3.0	2.5	20.3	9.8	2.38	D–F	−0.011	−0.54
23. Kinetex C8	Core shell	100	200	3.5	2.4	13.5	9.8	2.52	D–F	0.002	0.14
24. Kinetex Phenyl-Hexyl	Core shell	100	200	3.4	2.2	33.8	9.8	2.22	D–F	−0.010	−0.28
25. Kinetex PFP	Core shell	100	200	3.2	2.4	16.5	9.8	2.44	G–H	0.019	1.67

Source: Kormány, R. et al., *J. Pharm. Biomed. Anal.*, 80, 79–88, 2013. With permission.

Note: The table shows furthermore 3 different critical peak pairs (G–H, D–F, B–G), meaning, that it is important not to fix only one critical peak pair, leading often to noncompliance in inspections (23). The best results are achieved with column number 1, 5, 14, 21, and 23. These columns are the *alternative* columns for the method.

[a] Difference (min): Predicted Retention Time − Experimental Retention Time.

[b] % Error: ([Predicted Retention Time − Experimental Retention Time] / Experimental Retention Time) × 100 Columns: 50 × 2.1 mm with sub-2 μm particles (porous shell 0.23 μm on core shell types).

We are trying to demonstrate this on the following industrial example:

- To have comparable experimental conditions, all columns were having the same geometry: 50 mm long, 2.1 mm ID, 1.7–1.8 μm.
- To find the best gradient slopes scouting gradients were carried out, first with MeOH, followed by AcN. DoE was based on two gradient time values (tG1 and tG2) with a factor 3 difference, that is, tG2 = 3 × tG1 (3 and 9 min from 0%B to 100%B) at two different temperatures T_1 and T_2 (T_1: 30°C, T_2: 70°C).
- Eluent A was varied with three different pH1, pH2, and pH3 distances of 0.5–0.6 pH units. The design is carried out as follows:
  - Running six experiments first at the low temperature T_1 (1, 2, 5, 6, 9, 10), followed by running six experiments at the high temperature T_2 (3, 4, 7, 8, 11, 12).
  - The next step is to import the 12 experimental data in AIA-format (*.cdf) and go to peak tracking.
  - In peak tracking, the reduction of peak areas in mV × second might help to have a better visual control on how peaks are moving, as they are used as labels over the peaks (Figures 8.7 and 8.8).

This observation did not change, if the pH was changed in the investigated region, so in the study the pH could be excluded as the reason for lack of robustness. It turned out that the reasons were rather small changes in the temperature or in the gradient slope. DryLab could show the MODR as the space, where robust routine work could be performed as a red irregular body, in which the working point might be moving [20–22]. However, the robustness of the working point in a multifactorial space might be changing. Also different columns might have different best working point parameters, but differently robust methods (Figure 8.9).

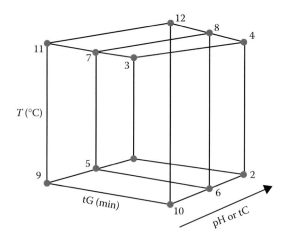

FIGURE 8.7 A revolutionary design of experiments (DoE) for the simultaneous optimization of gradient time (tG), temperature (T), and ternary composition (tC) of the eluent B or the pH of the eluent A. (From Kormány, R. et al., *J. Pharm. Biomed. Anal.,* 80, 79–88, 2013. With permission.)

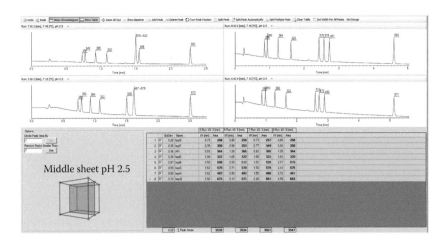

FIGURE 8.8 Peak tracking: the goal here is to align individual peaks in a horizontal line and to separate double or triple peaks, do turn over peaks, if necessary. It is important to consider peak areas not as quantitation tools, but more as peak-identification tools, like spectral data. In the middle group of three compounds the largest peak is in the first run (lower left), the last one in the middle group; in the second run (lower right) it is in the middle position; in the third run (top left) it coelutes with another peak; and in the fourth run (top right) it is the first peak. So this substance is moving around quite a bit and if we do not understand these movements, we will always have OoS results in the QC work. (From Kormány, R. et al., *J. Pharm. Biomed. Anal.*, 80, 79–88, 2013. With permission.)

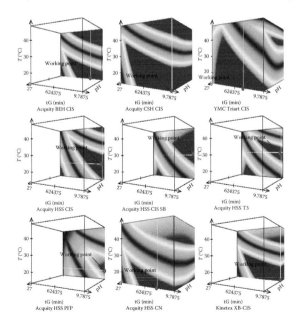

FIGURE 8.9 Selection of the best working point for different column chemistries. As it can be seen, the best working point with the highest critical resolution is different from column to column. Blue regions mean sample overlap; red areas are showing where baseline separation can be achieved. (From Kormány, R. et al., *J. Pharm. Biomed. Anal.*, 80, 79–88, 2013. With permission.)

8.11   ROBUSTNESS MODELING

8.11.1   Case Study 6

There are a number of papers discussing robustness issues [23–26]. In 2011 the break-through Robustness Tool was started, which is based on modeling of 3^n (n = number of factors) experiments using DryLab 4. In the first version the following six factors were evaluated in three levels (-1, 0, $+1$): Gradient time tG, temperature T, pH, as measured factors and flow rate, starting %B, and end %B as calculated factors. In the meantime this model was extended to step gradients with three steps, including the dwell volume also as an important factor in method transfer.

The advantage of this technology is to model every one of the $3^6 = 729$ experiments, collect them in an Excel-like table, sort them, and additionally by clicking on any of them with the mouse one can see each of them as a chromatogram. Furthermore, the so-called % success rate (100% failure rate) is calculated with the analytical target profile, which is the critical resolution $R_{s, crit}$—typically as baseline resolution with the value of 1.5. In this way we can calculate in advance, how a method will perform in routine analysis and how much *OoS* results will it produce (Figure 8.10).

It is possible to see also which factors are influencing the results in the strongest way, so one can act to reduce that influence. This means, if a method produces OoS data, one can see, if the instrument is under control or not, because the accuracy in gradient mixing, temperature, flow rate, etc., are included. Therefore, it is easy to find in a case of OoS the reason for declining instrument performance and it is easy to correct it (Figure 8.11). Generally, modeled experiments are highly precise and allow in this way great amount of time saving and faster development of new drugs [27,28].

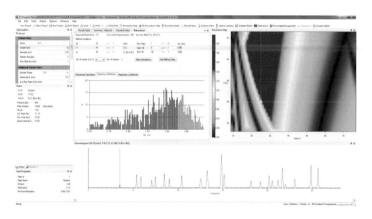

FIGURE 8.10   The fact to have a baseline resolution $R_{s, crit} > 1.5$ does not mean that we have a robust method. If we work at the edge of failure (EoF) (top right) (tG: 25 min, T: 45°C), that is, at the border (edge) of the baseline-separation (red) region, we might have only ca. 25% runs with $R_{s, crit} > 1.5$ (blue lines at top left) as long ca. 75% runs are $R_{s, crit} < 1.5$ (red lines top left) (OoS results).

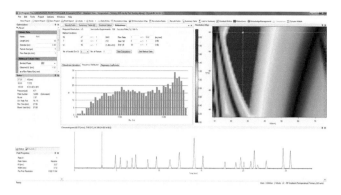

FIGURE 8.11 Moving away from the EoF to the middle of the red area (tG: 30 min, *T*: 60°C) and increase the accuracy of the pump from 0.1 to 0.02 mL/min tolerance limit, we have no OoS results, but 100% success rate (blue lines in top left) in routine QC operations.

The investigations were further extended as it is shown in Tables 8.1 and 8.2 for 25 commercially available columns.

We can see in the table, 8 columns with $R_{s,\,crit}$ > 2.5, at different combinations of pH, temperature, and gradient time tG. These are the best columns, delivering the most robust methods. Twenty columns are at the working point better than baseline resolution $R_{s,\,crit}$ > 1.5 and only one column shows a weak performance with $R_{s,\,crit}$ = 1.22 (Acquity CSH Fluoro-Phenyl). The critical peak pairs at the four best columns are D–F and G–H. The average retention time error is mostly under 1%.

However, these results do not equal to a 100% performance in routine QC operation, where multifactorial changes might reduce the success rate. Therefore we were looking at those columns, where we can assume—based on modeling multifactorial changes of six factors—that the failure rate is 0% (or success rate is 100%), which is shown in Table 8.2. The five equivalent columns with 0% failure rate are Acquity BEH C18, Acquity CSH C18, Hypersil Gold C18, Kinetex C18, and Kinetex C8.

8.12 MODELING PROTEIN SEPARATIONS

As presented in this chapter, HPLC or UHPLC modeling using DryLab 4 provides several advantages on various fields of pharmaceutical analysis including assays, impurity profiling, protein analysis, or even charge heterogeneity analysis of mAbs in IEX chromatography [29,30]. The time spent for method development can drastically be shortened and with the resolution maps in our hands, further adjustment and transfer of the methods is easier and more flexible.

In 2013, the Molnár-Institute launched the DryLab HPLC-Knowledge Management Document, which offers comprehensive method documentation for better knowledge sharing. It automatically collects all relevant method data directly from DryLab 4 and offers a platform for comments and the justification of method criteria. DryLab Knowledge Management is a new reporting tool for documenting and archiving an HPLC method. It encourages a QbD approach to method development and ensures that the method conforms to these standards by providing a comprehensive method

TABLE 8.2
With Each Column Robustness, Studies Were Carried Out and a Failure Rate Was Established with the DryLab Robustness Module

| Best Columns for the Sample | Predicted Parameters | | | | | Robustness | | | | Average of Retention Time | |
| Columns | PH | T (°C) | tG (min) | $R_{s,\,crit}$ | Critical Peak Pair | Failed Experiments | | Failure Rate (%) | | Difference[c] (min) | % Error[d] |
						a	b	a	b		
Acquity BEH C18	2.1	13.5	8.1	2.54	ImpG–ImpH	0	0	0.00	0.00	0.007	0.23
Acquity BEH Shield RP 18	2.0	38.3	9.8	2.16	ImpB–ImpG	0	228	0.00	31.28	−0.017	−0.79
Acquity BEN C8	2.5	33.0	9.8	2.27	ImpD–ImpF	0	0	0.00	0.00	−0.018	−0.85
Acquity BEH Phenyl	2.0	29.3	9.8	2.32	ImpG–ImpB	0	0	0.00	0.00	−0.001	0.41
Acquity CSH C18	3.0	13.5	9.8	3.13	ImpD–ImpF	0	0	0.00	0.00	0.017	0.88
Acquity CSH Phenyl-Hexyl	2.1	13.5	2.9	1.92	ImpD–ImpF	0	81	0.00	11.11	0.005	0.60
Acquity CSH fluoro-Phenyl	3.0	13.5	2.7	1.22	ImpD–ImpF	729	729	100.00	100.00	−0.002	−0.55
Triart C18	3.0	13.5	7.4	2.49	ImpD–ImpF	0	0	0.00	0.00	0.011	0.57
Acquity HSS C18	2.1	24.0	9.8	2.50	ImpG–ImpH	0	0	0.00	0.00	−0.038	−1.95
Acquity HSS C18 SB	2.0	30.0	9.8	2.04	ImpD–ImpF	0	0	0.00	0.00	−0.014	−0.37
Acquity HSS T3	2.0	31.5	9.8	2.16	ImpG–ImpH	0	149	0.00	20.47	−0.023	−0.94
Acquity HSS PFP	2.0	19.5	9.8	1.58	ImpD–ImpF	489	570	67.17	78.30	−0.005	−0.27
Acquity HSS CN	3.0	13.5	7.9	1.95	ImpD–ImpF	0	39	0.00	5.36	0.000	−0.15
Hypersil GOLD C18	3.0	41.3	9.8	2.72	ImpD–ImpF	0	0	0.00	0.00	−0.003	−0.10
Hypersil GOLD C8	2.7	42.0	9.8	2.55	ImpD–ImpF	0	0	0.00	0.00	−0.009	−0.24
Hypersil GOLD CN	2.9	27.8	9.0	1.67	ImpG–ImpB	567	646	77.78	88.61	0.002	0.56
Zorbax SB-C18	2.2	29.3	9.8	2.13	ImpG–ImpH	62	211	8.50	28.94	−0.016	−0.36
Zorbax SB-C8	2.8	13.5	6.1	2.03	ImpD–ImpF	0	243	0.00	33.33	0.013	1.09

(Continued)

TABLE 8.2 (Continued)
With Each Column Robustness, Studies Were Carried Out and a Failure Rate Was Established with the DryLab Robustness Module

Best Columns for the Sample	Predicted Parameters					Robustness				Average of Retention Time	
						Failed Experiments		Failure Rate (%)			
Columns	PH	T (°C)	tG (min)	$R_{s,\,crit}$	Critical Peak Pair	a	b	a	b	Difference[c] (min)	% Error[d]
Zorbax SB-Phenyl	2.0	13.5	8.9	1.52	ImpD–ImpF	299	345	41.02	47.33	−0.022	−2.42
Kinetex XB-C18	2.2	13.5	9.8	2.24	ImpD–ImpF	0	0	0.00	0.00	0.013	0.81
Aeris XB-C18	3.0	15.0	9.8	2.50	ImpG–ImpH	0	0	0.00	0.00	0.001	0.35
Kinetex C18	2.5	20.3	9.8	2.38	ImpD–ImpF	0	0	0.00	0.00	−0.011	−0.54
Kinetex C8	2.4	13.5	9.8	2.52	ImpD–ImpF	0	0	0.00	0.00	0.002	0.14
Kinetex Phenyl-Hexyl	2.2	33.8	9.8	2.22	ImpD–ImpF	0	84	0.00	11.52	−0.010	−0.28
Kinetex PFP	2.4	16.5	9.8	2.44	ImpG–ImpH	0	242	0.00	33.20	0.019	1.67

Source: For another sample other columns might be the best ones.

Note: The results from 729 virtual experiments of 6 factors at 3 levels are shown in the table under column *Robustness*. The columns a and b under *Robustness* have different tolerance limits, in column a smaller, in column b twice as large. As visible, the tighter the tolerance limits can be maintained, the more robust the method is.

[a] Robust parameter ± tolerances: tG ± 0.1 min, T ± 1°C, pH ± 0.1, Flow Rate ± 0.005 mL/min, Start%B ± 0.5, End%B ± 0.5.

[b] Robust parameter ± tolerances: tG ± 0.2 min, T ± 2°C, pH ± 0.2, Flow Rate ± 0.010 mL/min, Start%B ± 1.0, End%B ± 1.0.

[c] Difference (min): Predicted Retention Time − Experimental Retention Time.

[d] % Error: ([Predicted Retention Time − Experimental Retention Time/Experimental Retention Time]) × 100.

report, including a platform for the step-by-step justification of method choices. By implementing DryLab Knowledge Management, one can achieve excellent good manufacturing practice (GMP) documentation of a method, but one can also more easily and effectively collaborate between departments, and support analytical method transfer during development and manufacturing. DryLab Knowledge Management Document provides an analytical method summary to be signed and dated by the author and supervisor, making it GMP compliant and to be the perfect and safe documentation for inspections.

8.13 SUMMARY

HPLC method modeling is becoming a powerful tool to be used in the communication about method quality in HPLC between different labs, different companies, and between companies and regulatory agencies. The understanding of simple rules of peak movements will facilitate the development of new drugs, which are badly needed for smaller patient populations.

The new features of HPLC modeling software, such as 3D resolution map, the modeled robustness testing, a practicable method transfer, or a method knowledge management offer a closed loop of all information about the birth and practical use of a method, and it further suggests the use of such software solutions in regulated laboratories to make analyst's life easier—especially in the pharmaceutical industry.

ACKNOWLEDGMENTS

The authors would like to thank Jenö Fekete (TU-Budapest) and Szabolcs Fekete (University of Geneva) for valuable contributions to this chapter.

REFERENCES

1. Cs. Horváth, W. Melander, I. Molnár, Solvophobic interactions in liquid chromatography with nonpolar stationary phases (solvophobic theory of reversed phase chromatography, part I.), *J. Chromatogr.*, 125 (1976) 129–156.
2. Cs. Horváth, W. Melander, I. Molnár, Liquid chromatography of ionogenic substances with nonpolar stationary phases (Part II.), *Anal. Chem.*, 49 (1977) 142–154.
3. Cs. Horváth, W. Melander, I. Molnár, Liquid chromatography of ionogenic substances with nonpolar stationary phases (Part III.), *Anal. Chem.*, 49 (1977) 2295–2305.
4. L.R. Snyder, High performance liquid chromatography. In *Advances and Perspectives*, Cs. Horváth (ed.), Academic Press, New York, 1980, vol. 1, pp. 330.
5. J.W. Dolan, L.R. Snyder, M.A. Quarry, Computer simulation as a means of developing an optimized reversed-phase gradient-elution separation, *Chromatographia*, 24 (1987) 261–276.
6. L.R. Snyder, J.L. Glajch (eds.), *Computer-Assisted Method Development for High Performance Liquid Chromatography*, Elsevier, Amsterdam, 1990; *J. Chromatogr.*, 485 (1989) 1–640.
7. I. Molnár, H.J. Rieger, K.E. Monks, Aspects of the "design space" in high pressure liquid chromatography method development, *J. Chromatogr. A*, 1217 (2010) 3193–3200.

8. R. Kormány, J. Fekete, D. Guillarme, S. Fekete, Reliability of simulated robustness testing in fast liquid chromatography, using state-of-the-art column technology, instrumentation and modelling software, *J. Pharm. Biomed. Anal.*, 89 (2014) 67–75.

9. ICH Q8 (R2) – *Guidance for Industry, Pharmaceutical Development*, 2009. http://www.ich.org/about/history.html (accessed on 24 May, 2016).

10. F. Erni, Presentation at the Scientific Workshop *Computerized Design of Robust Separations in HPLC and CE*, July 31, 2008, Molnár-Institute, Berlin, Germany.

11. K.E. Monks, H.J. Rieger, I. Molnár, Expanding the term "Design Space" in high performance liquid chromatography (I), *J. Pharm. Biomed. Anal.*, 56 (2011) 874–879.

12. J.W. Dolan, Dwell volume revisited, *LC GC N. Am.*, 24 (2006) 458–466.

13. I. Molnár, Computerized design of separation strategies by reversed-phase liquid chromatography: Development of DryLab software, *J. Chromatogr. A*, 965 (2002) 175–194.

14. I. Molnár. K.E. Monks, From Csaba Horváth to quality by design: Visualizing design space in selectivity exploration of HPLC separations, *Chromatographia*, 73 (2011) S5–S14.

15. K. Monks, I. Molnár, H.J. Rieger, B. Bogáti, E. Szabó, Quality by design: Multidimensional exploration of the design space in high performance liquid chromatography method development for better robustness before validation, *J. Chromatogr. A*, 1232 (2012) 218–230.

16. L.R. Snyder, J.W. Dolan, P.W. Carr, The hydrophobic-subtraction model of reversed-phase column selectivity, *J. Chromatogr. A*, 1060 (2004) 77–116.

17. C. Chamseddin, I. Molnár, T. Jira, Intergroup cross-comparison for the evaluation of data-interchangeability from various chromatographic tests, *J. Chromatogr. A*, 1297 (2013) 146–156.

18. J.W. Dolan, L.R. Snyder, T. Blanc, L.Van Heukelem, Selectivity differences for C18 reversed-phase columns as a function of temperature and gradient steepness. I. Optimizing selectivity and resolution, *J. Chromatogr. A*, 897 (2000) 37–50.

19. R. Kormány, I. Molnár, H.J. Rieger, Exploring better column selectivity choices in ultra-high performance liquid chromatography using Quality by Design principles, *J. Pharm. Biomed. Anal.*, 80 (2013) 79–88.

20. A.H. Schmidt, I. Molnár, Using an innovative Quality-by-Design approach for development of a stability indicating UHPLC method for ebastine in the API and pharmaceutical formulations, *J. Pharm. Biomed. Anal.*, 78–79 (2013) 65–74.

21. S. Fekete, J. Fekete, I. Molnár, K. Ganzler, Rapid high performance liquid chromatography method development with high prediction accuracy, using 5 cm long narrow bore columns packed with sub-2 μm particles and Design Space computer modeling, *J. Chromatogr. A*, 1216 (2009) 7816–7823.

22. R. Kormány, I. Molnár, J. Fekete, Quality by design in pharmaceutical analysis using computer simulation with UHPLC, *LC GC N. Am.*, 32 (2014) 354–363.

23. N.S. Wilson, M.D. Nelson, J.W. Dolan, L.R. Snyder, R.G. Wolcott, P.W. Carr, Column selectivity in reversed-phase liquid chromatography I. A general quantitative relationship. *J. Chromatogr. A*, 961 (2002) 171–193.

24. B. Dejaegher, Y. Vander Heyden, Ruggedness and robustness testing, *J. Chromatogr. A*, 1158 (2007) 138–157.

25. Y. Vander Heyden, A. Nijhuis, J. Smeyers-Verbeke, B.G.M. Vandeginste, D.L. Massart, Guidance for robustness/ruggedness tests in method validation, *J. Pharm. Biomed. Anal.*, 24 (2001) 723–753.

26. J.J. Hou, W.Y. Wu, J. Da, S. Yao, H.L. Long, Z. Yang, L.Y. Cai, M. Yang, X. Liu, B.H. J., D.A. Guo, Ruggedness and robustness of conversion factors in method of simultaneous determination of multi-components with single reference standard, *J. Chromatogr. A*, 1218 (2011) 5618–5627.

27. R. Kormány, I. Molnár, J. Fekete, D. Guillarme, S. Fekete, Robust UHPLC separation method development for multi-API product containing amlodipine and bisoprolol: The impact of column selection, *Chromatographia*, 77 (2014) 1119–1127.

28. R. Kormány, J. Fekete, D. Guillarme, S. Fekete, Reliability of computer-assisted method transfer between several column dimensions packed with 1.3–5 μm core–shell particles and between various instruments, *J. Pharm. Biomed. Anal.*, 94 (2014) 188–195.

29. S. Fekete, S. Rudaz, J. Fekete, D. Guillarme, Analysis of recombinant monoclonal antibodies by RPLC: Toward a generic method development approach, *J. Pharm. Biomed. Anal.*, 70 (2012) 158–168.

30. S. Fekete, A. Beck, J. Fekete, D. Guillarme, Method development for the separation of monoclonal antibody charge variants in cation exchange chromatography, part I: Salt gradient approach, *J. Pharm. Biomed. Anal.*, 102 (2015) 33–44.

Index

Note: Page numbers followed by f and t refer to figures and tables, respectively.